Energy: Principles Problems Alternatives

ENERGY

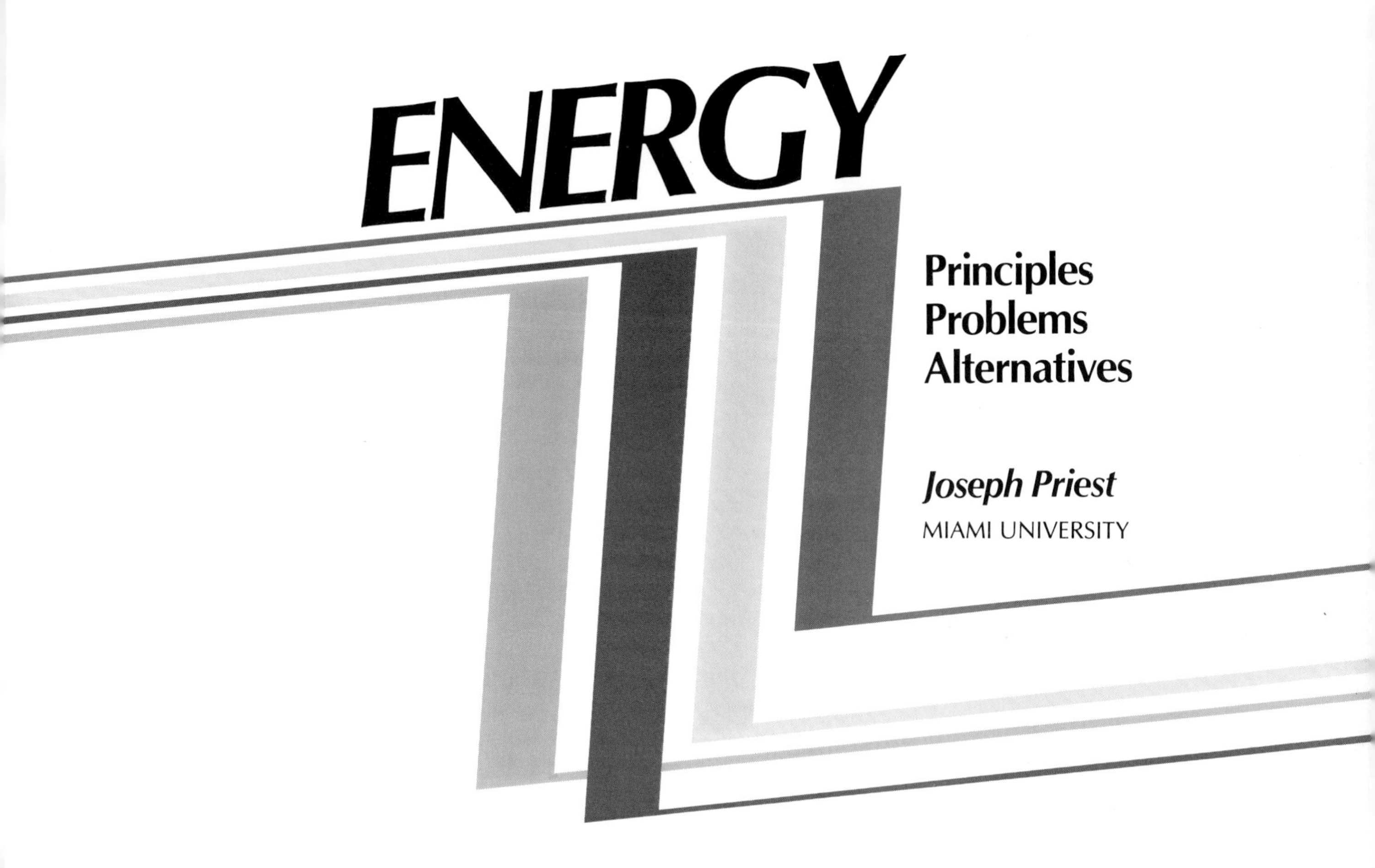

Principles Problems Alternatives

Joseph Priest
MIAMI UNIVERSITY

THIRD EDITION

ADDISON-WESLEY PUBLISHING COMPANY

Reading, Massachusetts • Menlo Park, California
London • Amsterdam
Don Mills, Ontario • Sydney

SPONSORING EDITOR:
Bob Rogers
PRODUCTION MANAGER:
Herbert Nolan
PRODUCTION EDITOR:
William J. Yskamp
TEXT AND COVER DESIGNER:
Patricia O'Hare Williams
ART COORDINATOR:
Robert C. Forget
MANUFACTURING SUPERVISOR:
Hugh J. Crawford

Library of Congress Cataloging in Publication Data

Priest, Joseph M.
Energy: Principles, Problems, Alternatives.

Bibliography: p.
Includes index.

1. Power resources. 2. Power (Mechanics) I. Title.
TJ163.2.P74 1984 621 83-15810
ISBN 0-201-06004-3

ABCDEFGHIJ-HA-8987654

To Mary Jean, Mary Jo, John, Catherine, and Daniel
for their love and inspiration

To the Instructor

Energy: Principles, Problems, Alternatives is the third edition of a book published in 1975. The earlier editions served as a text in a variety of introductory courses. The experience of instructors in these courses figured heavily in the development of this third edition. Almost sixty instructors provided input to the revision. The author has capitalized on the competency of physicists for:

1. Teaching physical principles as they relate to the societal uses of energy;
2. Examining the consequences of working with finite energy resources;
3. Examining the effects, extent and control of by-products that are usually produced in energy conversion processes; and
4. Evaluating the technology of energy converters and assessing the trade-offs involved in their societal deployment.

The book is organized as follows:

Chapter 1	Energy and Society
Chapter 2	Physical Basis for Energy
Chapter 3	Fossil Fuels
Chapter 4	Electric Energy
Chapter 5	Electric Power Plants and the Environment
Chapter 6	Automobiles and the Environment
Chapter 7	Thermodynamic Principles
Chapter 8	Atmospheric Problems
Chapter 9	Nuclear Power
Chapter 10	Solar Energy
Chapter 11	Other Energy Systems
Chapter 12	Nuclear Breeder and Nuclear Fusion Reactors
Chapter 13	Energy Conservation

This chapter sequence differs somewhat from the earlier editions. Chapter 2, "Physical Basis for Energy," was Chapter 3 in the first two editions. This move introduces energy concepts and energy units before delving into energy uses and technology. The lengthy chapter entitled "Environmental Effects of Burning Fossil Fuels" in the first two editions forms the basis of Chapter 5, "Electric Power Plants and the Environment," and Chapter 6, "Automobiles and the Environment." Two chapters entitled "Thermodynamic Principles and the Steam Turbine" and "Disposal of Rejected Heat from Steam Turbines" in the earlier editions were combined to make Chapter 7, "Thermodynamic Principles." "Nuclear Physics Principles" and "Nuclear-fueled Electric Power Plants" were combined to make Chapter 9, "Nuclear Power." The chapter on solar energy has been expanded significantly. A concentrated effort was made to refine the physics, update the energy data, and add new topics. About 25 multiple-choice questions with answers were added to each of the end-of-chapter sections.

Some familiarity with elementary algebra and numerical manipulation is required, but no college physics prerequisite is presumed. An effort has been made to integrate the physics with the energy issues. Generally, the text is used in a course that meets three hours a week for 15 weeks. To help the student understand the textual material and to prepare for examinations there is included at the end of each chapter:

1. Selected "References" that provide an accurate source of information and a guide to the types of journals and books students should peruse themselves;
2. A "Review" section that includes questions highlighting the main topics and ideas in the chapter and about 25 multiple-choice questions with answers; and
3. A "Questions and Problems" section that emphasizes short written answers and practical numerical calculations.

Oxford, Ohio
January 1984

J.P.

To the Student

We live in a society totally dependent on massive uses of energy, but choosing energy sources is problematic. We have an abundance of coal but burning coal produces sulfur oxides and carbon dioxide. Sulfur oxides lead to the formation of acid precipitation. Carbon dioxide accumulates in the atmosphere and creates a greenhouse effect that may raise the average temperature of the earth. Nuclear power provides a significant portion of our electricity and there is great potential for an increasing role. However, accidents in nuclear power plants have persuaded much of the public that we should not pursue the nuclear option. Solar energy technology is exciting, but it will not do everything we would like and we are only beginning to understand and accept what solar energy can do for us. There is great hope for nuclear fusion energy, but commercial use is more than a quarter of a century away.

Other energy technologies include geothermal energy, trash energy, hydroelectricity, magnetohydrodynamics, wind energy, biomass energy, hydrogen energy, and tidal energy. Energy conservation will play a significant role no matter what energy route we choose. All aspects of this wide variety of energy technologies involve important decisions. Often these decisions require our input. Citizens are being asked to cast a vote for or against the use of nuclear power in their states. It is the author's hope that studying this text will help you make meaningful energy decisions whether in the public poll booth or, for example, in deciding on a residential solar heating system.

Physical principles are central to the teaching philosophy in this text. A knowledge of physical principles helps you understand energy technologies and their inherent benefits, risks, and problems, and opens up a whole new world of adventure into physics. The principles of physics you glean from studying this text will not change but much of the energy information will. Federal and state energy regulations will change. Emphasis on particular energy technologies will change. Changes in energy policy follow switches in presidential administration. We can keep abreast only by perusing current literature. Selected references are included at the end of each chapter. These references are not intended to be the last word, for you should look to recent editions of the journals referenced for current energy information.

When you begin a chapter examine the illustrations and captions to get a general feel for the content. After you have studied the chapter take advantage of the review material, especially the multiple-choice questions for which answers are given. Look to the questions and problems for more in-depth study.

Like it or not, numbers are a part of our life. We make change at the grocery. We fill out income tax returns. We are conscious of the speed registered by a speedometer. We evaluate our electricity bills and pay them. The examples are endless. Many of these transactions are algebraic although we may not be consciously aware that they are. Algebra facilitates reasoning. In discussions of physical principles and energy issues, we always arrive at a point where we must ask "how much?" and "why?" Answering requires numbers and reasons; therefore numbers and elementary algebra are a part of this book. You will be dealing with physical quantities like kilowatt-hours and British thermal units, but the manipulations are not at all unlike those you do on an everyday basis. It is my wish that you would adopt this attitude and seek out the analogies with everyday situations. For a small investment, the rewards are big.

Hand-held calculators and personal computers have freed us from much of the drudgery and fear of numerical manipulations. With impressive speed, they perform numerical calculations involving numbers of any size. These fascinating instruments are making an enormous impact on our lives. Take advantage of them! They will help you evaluate physics and energy problems as well as save you money at the grocery. An appendix is included to help you feel comfortable with the wide range of numbers encountered in the book.

Societal problems with energy are very serious and require careful consideration by all of us. Our life-styles may very well change from coping with these problems. Physics can also be very serious business. But physics can also be fascinating. I hope that your comprehension of energy problems and physics grows as a result of studying this book. Above all, I hope that physics is made more meaningful to you.

J.P.

Acknowledgments

The continued interest of students in physics and energy prompted the third edition of this text. I appreciate that interest very much. Responses from instructors who took the time to fill out a rather detailed questionnaire were enormously useful. The honest and comprehensive reviews of the manuscript by Professors David Uhrich, Kenneth Mantei, William Kemper, and Roger Mills were especially helpful. I am grateful to them. To the physics department, to the library and audiovisual staffs at Miami University who cooperated in numerous ways, to Ula Cooper who never balked at a last-minute typing job, thank you very much.

Dear Mary, what more can I say? I haven't fixed the dryer, the porch, the washer . . . (and I probably never will).

Contents

1

Energy and Society

2

Physical Basis for Energy

3

Fossil Fuels

Electric Energy

5

Electric Power Plants and the Environment

6

Automobiles and the Environment

7

Thermodynamic Principles

8

Atmospheric Problems

9

Nuclear Power

10

Solar Energy

11

Other Energy Systems

12

Nuclear Breeder and Nuclear Fusion Reactors

13

Energy Conservation

1

Energy and Society

1.1 ENERGY AND SOCIETY

Every society, from the most primitive to the most highly industrialized, incorporates an assortment of motions. Animals move, waters flow, plants grow, arrows fly, automobiles roll, and so on. Human beings must move to gather food and to perform those basic functions for existing. Modern societies employ automobiles, airplanes, trucks, and trains for movement of people and commodities. They use engines for machines producing necessities as well as luxuries, and electricity for motors and lights. There is an aspect of energy and energy conversion to all these examples. To run, animals convert energy from food. Plants grow through the grace of solar energy. Arrows fly by converting the potential energy of a stretched bow. Water flowing downward from the top of a dam

This array of solar collectors that converts sunlight to electricity typifies the exciting new energy technology that is finding its way into our society. (Photograph courtesy of Arizona Public Service Company.)

converts gravitational potential energy into energy of motion. Combustion engines in motor vehicles convert chemical energy in gasoline and oxygen into energy of motion.

Solar energy streams to us continually and without it and the food-producing plants it supports, we could not survive. For sundry other functions we rely on the fossil fuels (coal, petroleum, and natural gas) and uranium. None of these sources stream to us like the rays of the sun. They are buried in the earth's crust and, whatever their abundances, they are limited and will eventually be depleted. In fact, as we begin to see the end of some energy resources, we scurry to develop alternatives. The development of alternatives is exciting but fraught with decisions involving the best judgment of us all. Whatever energy technology emerges—whether it be a refurbishing of an old one like coal or a new one like nuclear fusion—there will be unwanted by-products to deal with and social, political, scientific, and economic factors to consider. Hopefully, this study of energy and society will encourage you to get involved in energy planning and development, and help you to arrive at acceptable alternatives.

1.2 OVERVIEW OF ENERGY USE IN THE UNITED STATES

Since the founding of our country, the energy supply has been dominated by wood, coal, petroleum, and natural gas. However, the mix of these fuels has changed rather systematically with time as illustrated in Fig. 1.1. As the use of wood declined in the latter half of the nineteenth century, the use of coal increased accordingly. Coal use peaked around 1910 after which there occurred a corresponding rise in the use of petroleum and natural gas. Will this cyclic trend in energy use continue? It appears that the use of petroleum and natural gas peaked around 1980 and that the use of nuclear energy is making a move upwards. Perhaps the use of coal is making a new resurgence. Other prospects like solar energy and nuclear fusion energy are promising but have yet to make a mark on the energy-use graph. The fate of any type of energy is not clear-cut because each has its limits and peculiarities. As we delve into our study we will address the salient features of existing and future energy sources.

Many contemporary concerns for societal uses of energy originated in the 1950s and progressively

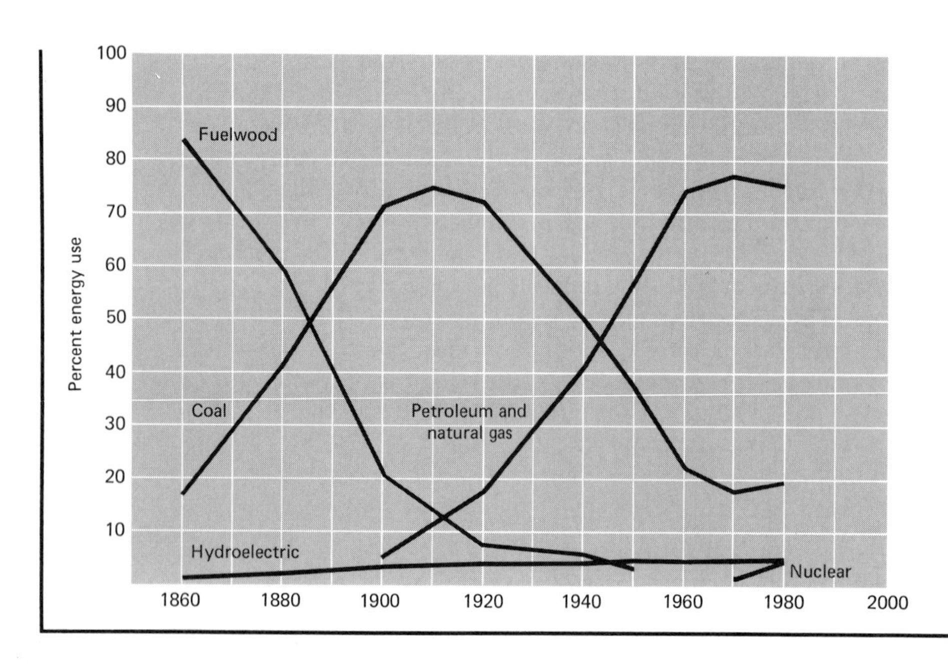

FIGURE 1.1 Historical dependence of the United States on various sources of energy. The use of a given fuel type is presented as a percentage of the total energy consumption. Fossil fuels (coal, petroleum, and natural gas) have provided more than 90% of the energy since the turn of the century. However, the mix has changed significantly. In 1900 coal provided 71.3% of the total energy; petroleum and natural gas combined contributed 5%. In 1981 these figures changed to 22% for coal, 27% for natural gas, and 43% for petroleum. The data are from the Department of Energy and are tabulated in Table A.1.

magnified. Let us focus on a three-decade period beginning with 1950 by examining the energy-consumption graph in Fig. 1.2. Because we have yet to specify measuring units, energy is expressed in terms of the heat energy liberated in burning a 42-gallon barrel of crude oil. In 1950 the total energy consumed in the United States was equivalent to burning 5.9 billion barrels of crude oil. To obtain a feeling for the magnitude of this much energy, consider that a barrel stands about three feet high. Stacked end-to-end 5.9 billion barrels would extend 3.4 million miles, or about 1000 times the coast-to-coast distance of the United States. The energy content is equivalent to the energy liberated from burning 0.27 trillion gallons of gasoline, enough for about 1800 gallons for every person in the United States in 1950. Even for 1950, the annual energy use was impressive. The growth rate of energy consumption and production in the 1950s and 1960s is unprecedented in the United States. Economic growth moved in lockstep with energy growth. The great bulk of the growth was provided by petroleum and natural gas. Natural gas usage nearly quadrupled and petroleum usage more than doubled between 1950 and 1970. To meet the petroleum demand, the United States relied increasingly on imports. Although the growth of energy use waned between 1970 and 1980, the fossil fuels continued to supply more than 90% of the total energy. Hydroelectric power more than doubled in the three-decade period but its percentage contribution to the total energy usage was essentially constant at 4%. Nuclear power made little impact until 1970 but by 1980 it contributed 3.5% to the total effort.

Energy supplies are used in four main sectors and converted to several forms—electric, heat, light, and mechanical.

1. Electric power plants producing electricity for lights, powering electric motors, producing heat, etc.

2. Residential and commercial facilities for space heating, heating water, etc.

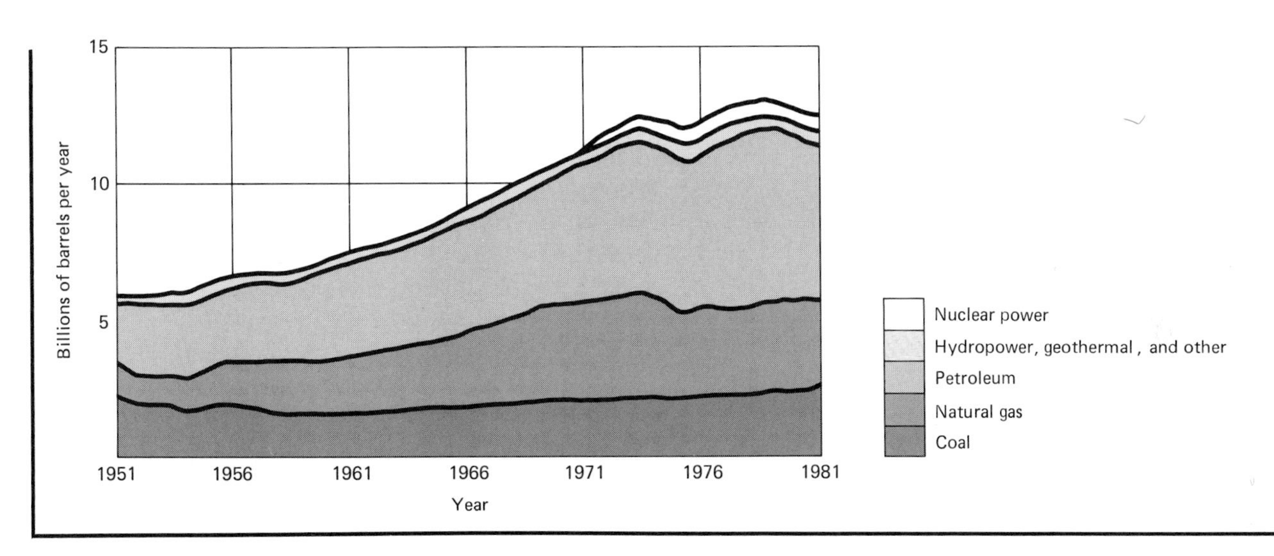

FIGURE 1.2 Annual consumption of energy in the United States for a three-decade period beginning in 1950. The energy units are in terms of the energy equivalent of burning crude oil. Note the rather constant annual consumption for coal and the rapid growth of the use of petroleum and natural gas in the 1950s and 1960s. Nuclear energy began making its mark around 1970. The data for this graph are from the Department of Energy and are presented in Table A.1.

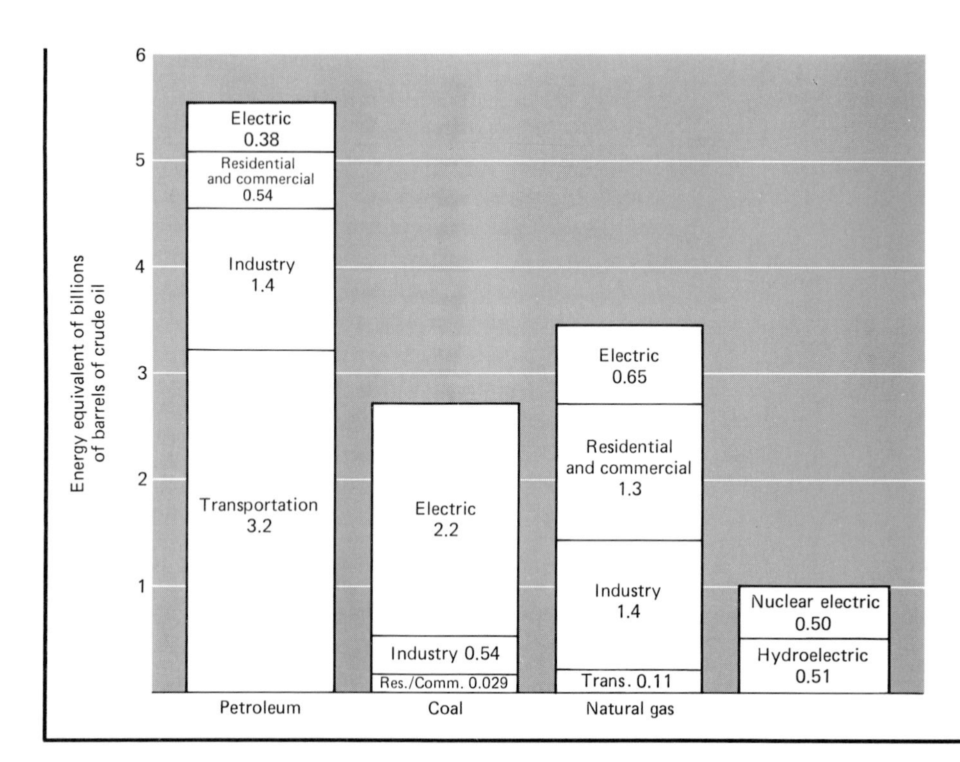

FIGURE 1.3 Contribution of the major sources of energy to the transportation, industrial, residential and commercial, and electric utilities sectors for the year 1981. The number in each sector-block represents the annual use of the designated type of energy in terms of the energy equivalent of burning crude oil. For example, petroleum contributed the energy equivalent of burning 5.52 billion barrels of oil of which 3.2 billion barrels were used in transportation. The data for the graph are from the Department of Energy and are presented in Table A.2.

3. Industries for making steam for manufacturing processes, heating materials, etc.
4. Transportation systems for moving people and commodities.

Figure 1.3 presents a sector breakdown for each significant energy type for the year 1981. The numerical amounts refer to inputs before conversion to the useful energy form. When you read that 2.2 energy units of coal were channeled to electric power plants, it means that the coal burned annually produced energy equivalent to burning 2.2 billion barrels of crude oil. Because of energy losses in producing and delivering electric energy to consumers, the electric energy delivered would be about 30% of the input energy.

Note in Fig. 1.3 that the fuel used for transportation is almost exclusively petroleum. Although some electric energy is fed to the transportation sector to run electric locomotives and to recharge batteries for electric automobiles, the contribution is less than 0.2%. Electric-powered vehicles may someday make an impact on energy use but we can expect petroleum to dominate the transportation sector for several decades. All fossil fuels contribute to the production of electricity, but coal is "king." Whether or not nuclear energy recovers its growth rate of the 1970s remains to be seen. Whatever happens, the fossil fuels will continue to play important roles in the production of electricity. The industrial sector draws from all three types of fossil fuels with coal the least important in terms of sheer amounts.

In Fig. 1.4 we sort the energy consumption according to three consumer sectors—transportation, industrial, and residential and commercial. This delineation is referred to as end-use sectors because it includes electric energy from electric power plants as well as the energy losses involved in the generating process. For example, if three units of electric energy were used in the industrial sector for lighting, then seven additional units would be credited to this sector because a power plant needs ten units of energy to deliver three units of electric energy to a consumer.

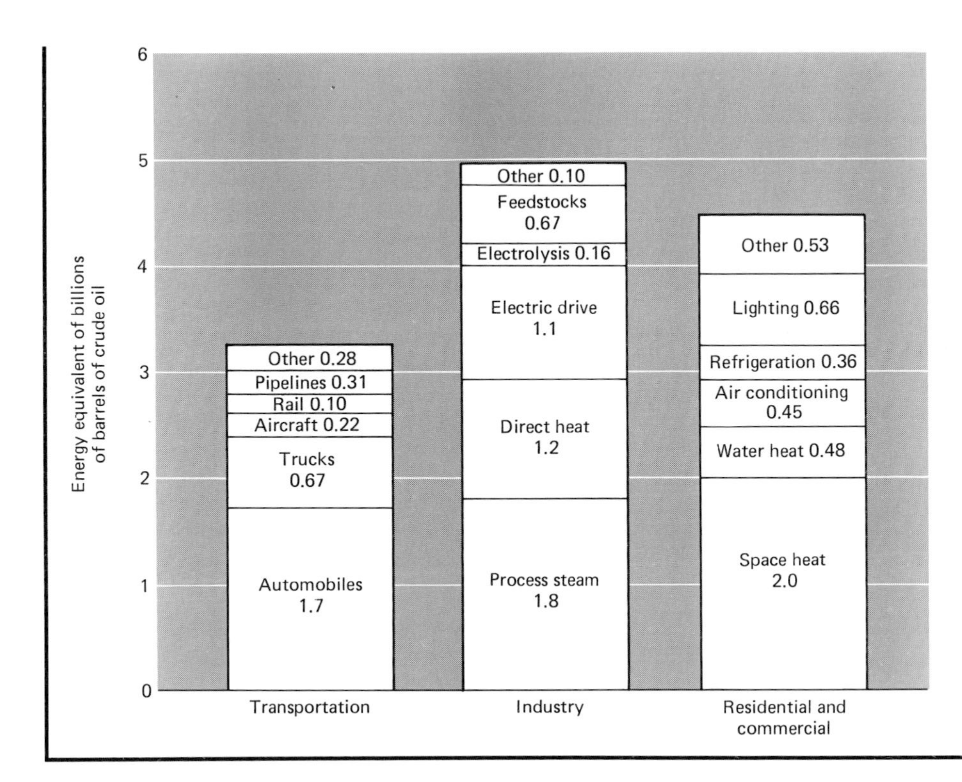

FIGURE 1.4 Breakdown of the energy end use sectors according to major user type. This breakdown includes energy for electricity generation and distribution. The percentage contribution of each type was taken from *Energy in America's Future*, Johns Hopkins University Press, Baltimore (1979). These data are recorded in Table A.3.

Over half the energy used for transportation is used by automobiles. This fact reflects our great dependence on private transportation. The feedstock category in the industrial sector may seem mysterious and insignificant. However, it represents petroleum that is used in the manufacture of such things as synthetic fibers and plastics. We only have to look around us to see the impact of these materials. It is often argued that the best use of petroleum is for feedstocks in the chemical industry rather than burning. The alternatives for producing heat are more evident than the alternatives for feedstocks. Energy conservation is an important element of any energy program. A perusal of how energy is used in the residential and commercial sector indicates clearly in which areas we as citizens have a personal opportunity to be involved in the efficient use of energy.

If we divide the total 1981 energy consumption by the 228 million American consumers that year we find an annual per person energy use equivalent to 56 barrels of crude oil. No country in the world has a larger per capita use of energy. With 5% of the world's population in 1979 the United States accounted for about 29% of all energy consumed. A decade earlier the figures were 6% for population and 35% for energy. It is generally argued that this high level of energy consumption produces a correspondingly high standard of living. To correlate standard of living with energy use in a quantitative way is not an easy task. In an exhaustive study we would surely consider such things as environmental quality, mental and physical health, and education. A numerical way that does not incorporate these intangible factors is to associate standard of living with the per capita gross national product.* A comparison of per capita energy use and per capita gross national product for some selected nations

* The gross national product (GNP) is defined as the total market value of all the goods and services produced by a nation during some specified period. Normally, the time period is a year.

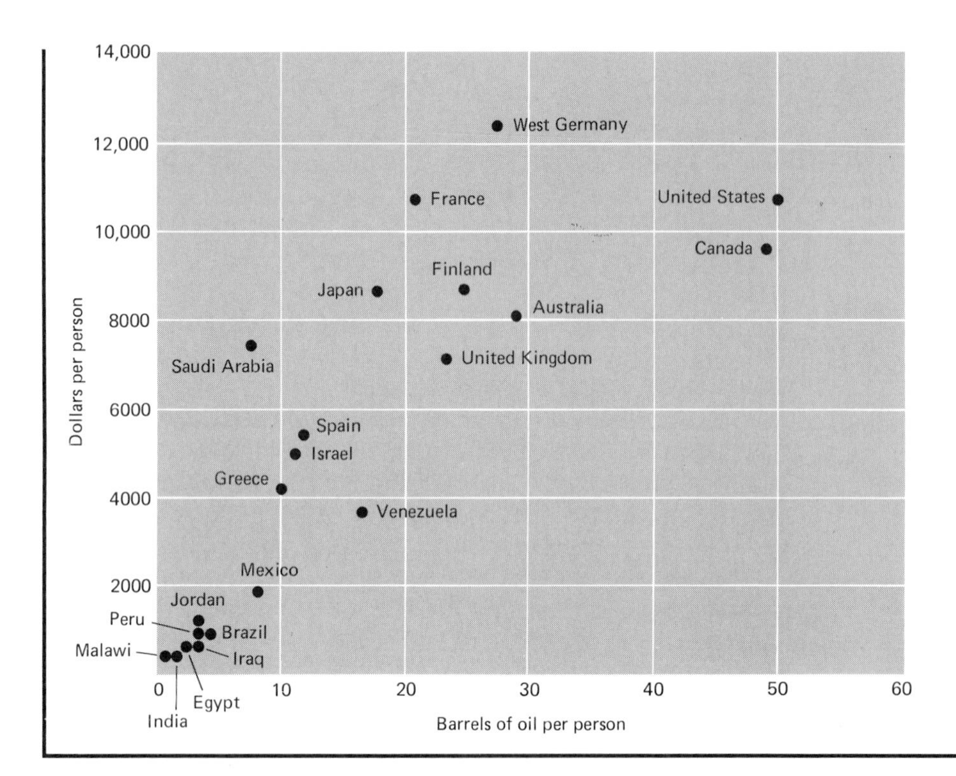

FIGURE 1.5 Comparison of energy use per capita and gross national product (GNP) per capita for some selected countries in the world for the year 1979. The GNP scale is in terms of 1979 United States dollars. The energy scale is in terms of the energy derivable from burning crude oil. Note that the United States, West Germany, Canada, and France have about the same GNP per capita but that there is about a factor of three difference between the lowest (France) and highest (United States) per capita use of energy. The GNP data are from *World Statistics in Brief,* United Nations Statistical Pocketbook, 6th Edition, 1981. The energy data are from the *1980 Yearbook of World Energy Statistics,* United Nations, New York, 1982. Both sets of data are presented in Table A.4.

is shown in Fig. 1.5. The correlation is strong. As we might expect, the industrialized nations dominate the upper portion of the plot. The less technologically oriented nations are eager to improve their standard of living and they recognize energy to be a necessary ingredient. This is the motivation for the Aswan Dam in Egypt and the development of nuclear energy in Argentina, South Korea, Pakistan, and Iraq, for example. For similar reasons, Mexico welcomed the discovery of significant petroleum deposits in the Yucatan peninsula. This natural desire for energy puts added strains on the world's available resources and adds to the world pollution problems by increasing the growth of high technology.

1.3 A CRISIS IN ENERGY SUPPLY

Few students escape college without facing all sorts of crises. A crisis develops when you perform poorly on an examination. While there is an element of surprise in the result, the cause probably stemmed from developments taking place over a period of time. Doing poorly on an examination may be the result of studying the wrong things. Off and on for more than a decade we have seen newspaper articles, magazine articles, and television programs with a title like "The Energy Crisis." Indeed, the monetary outlay for imported petroleum is of crisis proportion and the disruption of a petroleum supply creates a crisis. Although these crises seem to catch us by surprise, the stage for them was set in the mid-1950s. Until about 1955 the United States produced all the energy it consumed. Thereafter, consumption exceeded production and the nation met demands with foreign sources. The gap betwen production and consumption widened noticeably about 1970 (Fig. 1.6). This gap necessitated an increased reliance on foreign sources for petroleum. If for some reason a foreign source curtails its production or elects to escalate the price, a crisis results. This is the situation that evolved in 1973 when the Arab sources curtailed their oil shipments and effected a worldwide escalation of petroleum prices. Because the Arab countries control 56% of the world's known oil reserves, this cost aspect of the crisis is not likely to change even though they may choose to contribute to the world's petroleum markets. It is important to realize that the premise for this action was set in 1955 and the crisis atmosphere in the United States remains as long as reliance on foreign energy persists.

It is easy to look at production and consumption data and identify the crux of our energy dilemma. It is not as easy to say why the production–consumption gap developed and widened. Consumption increased for many reasons, mostly political and economic. Price structures and advertising encouraged energy usage. The development of the Interstate Highway System and the population shift to suburbia fostered travel by automobile. The widespread and expensive use of energy in air travel added significantly to energy consumption. Production lagged for many complicated reasons—political, economic, and technological. Through favorable tax structures, exploration for and production of foreign oil was encouraged. The development of offshore oil wells in the continental shelf was curtailed for one-and-a-half years following the massive oil spill in California's Santa Barbara Channel in 1969. The desire to protect the environment had its impact. The National Environment Policy Act (NEPA) of 1969 demands detailed assessments on environmental impact before energy sources may be developed. These demands led to delays in the development of the Alaskan pipeline and in the construction of nuclear power plants. In order to comply with the Clean Air Act of 1970, many industries shifted from burning coal to burning less-polluting natural gas and oil. A complicated picture with challenging problems emerges. Solving the problems involves environmental, economic, and social trade-offs. Hopefully, these trade-offs will be beneficial. Apart from these rather short-term problems, which can be solved by better planning and

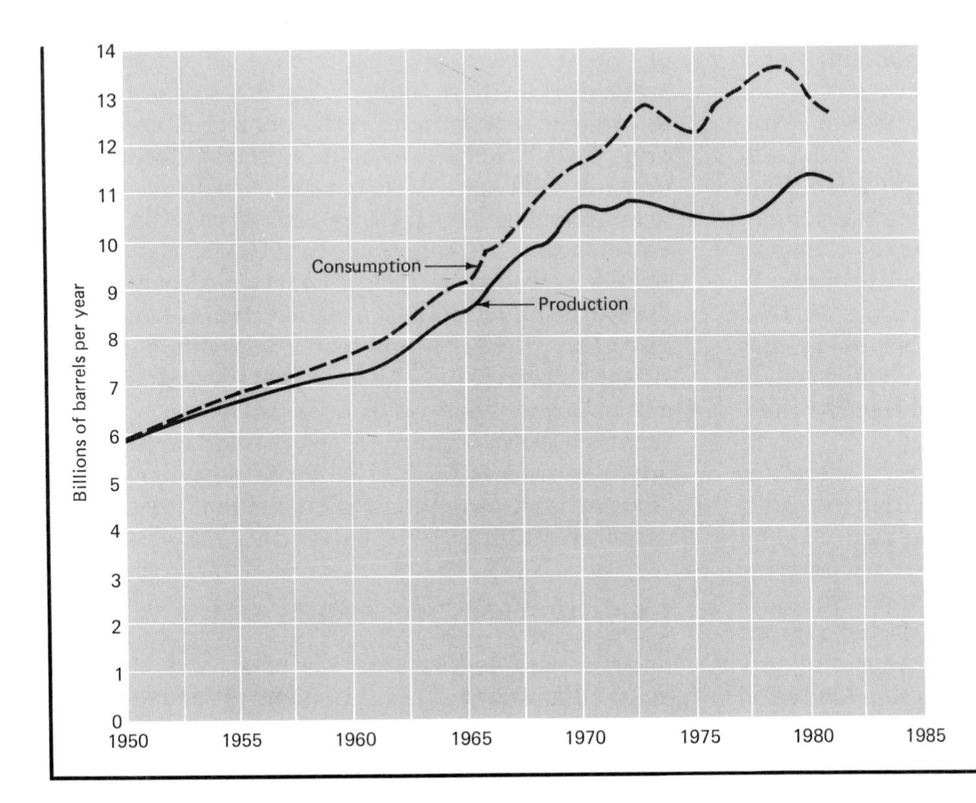

FIGURE 1.6 Yearly production and consumption of energy of all forms in the United States. The energy units are expressed in terms of the energy derivable from burning a barrel of crude oil. The difference between production and consumption is made up by imports of petroleum and natural gas. A substantial gap has existed since 1970. The data are from the Department of Energy and are presented in Table A.5.

conservation, there remains the fact that we are beginning to see the end of cheap energy resources. We are depleting these resources on a time scale which is staggeringly short. Although other energy resources exist, they present their own economic, political, environmental, and technical problems and they, too, involve inevitable trade-offs. Above all, their development takes time and support. It is the acceptable development of all our energy sources that constitutes our long-range solution to the energy problem and that is the essence of our study.

1.4 AN ENVIRONMENTAL CRISIS

Gasoline burned in automobiles and coal burned in electric power plants liberate heat that is converted to a more useful form of energy such as electricity and energy of motion. If only energy were garnered when fuels are burned, then society would have to cope only with the problems of depleting energy resources. But burning that produces unwanted by-products and mining that disrupts the land have created a spectrum of environmental crises. Automobiles produce gases that affect the human body and gases that in the presence of sunlight lead to the formation of photochemical smog, a condition that plagues nearly every major city. Power plants liberate gases that affect the human body and play roles in the formation of acid precipitation. Both automobiles and power plants produce gaseous carbon dioxide which neither has a direct biological effect nor plays any role in an environmental condition having a biological effect. But the accumulation of carbon dioxide in the atmosphere may prompt a warming of the earth that might lead to a shift in agricultural patterns and weather modifications.

Environmental problems have not befallen us suddenly. Isolated incidences of pollution caused by burning coal in this country have been recorded for several decades. The case most often cited occurred in the industrial city of Donora, Pennsylvania, in 1948 (see Fig. 1.7). A combination of stagnant air conditions and effluents from burning coal produced a polluted atmosphere in which over 40% of the population was affected to some extent. Problems caused by burning gasoline in cars were recognized in Los Angeles following World War II. The nation, and especially California, has not been completely negligent in trying to head off widespread problems. But with the growing use of energy and the tremendous growth in the number of automobiles, little headway was made nationwide until the Clear Air Act of 1970 was passed. Stringent maximum allowable levels for pollutants were invoked and emission limits for automobiles and industrial sources were set as a result of the act. Some advances toward bringing pollution under control have followed. But these gains have not come without opposition. Industries have insisted that the regulations are too severe and that technology for meeting the standards is not at hand. As energy resources become premium, the attraction for burning dirtier fuels blossoms and the pressures for relaxing the standards mount.

FIGURE 1.7 Donora, Pennsylvania, as it appeared on a polluted day in 1949. Air pollution was the cause of a major epidemic in this city in 1948. (Photograph courtesy of the Environmental Protection Agency.)

1.5 WHAT DOES THE FUTURE HOLD?

Every society needs energy resources, and modern industrial societies require them in abundance. There is no choice as to need. The choices for types of resources are limited and each is fraught with social, economic, political, and environmental trade-offs. Transportation systems thrive on petroleum but petroleum resources are dwindling and burning gasoline fouls the air. Electric power plants glean electricity from coal and uranium but burning coal produces environmental problems and using uranium presents some unique safety considerations. It is not a matter of turning away from these technologies because there are no "off-the-shelf" replacements. Rather, we must learn to cope with their problems while we reap the energy benefits. In the meantime, we develop new energy-conserving technologies, refurbish old methods and, above all, make energy conservation an integral part of our social ethics. It is an exciting prospect that we should look forward to. Many energy resources await development. Most are supplemental to conventional types but others have great potential.

The sun, which provides energy to grow our food, is a source of energy for warming (and cooling) our homes and factories and for heating water. Even an old technology like wood burning has merit for some. Solar energy and wood burning are not alternatives for every area but where they are economically viable, we should take advantage of them. The nuclear fusion process, which is responsible for the sun's energy, looms as a practical energy source for producing electricity or, perhaps, for disposing of trash. The geothermal energy that heats geysers and hot springs has been used as a clean energy source for domestic purposes and for generating electricity in isolated areas for over half a century. Geothermal energy is now being evaluated for large-scale use, and many scientists predict a bright future for it. The wind, which once propelled sailing vessels all over the world and still powers water pumps and electric generators for farms and remote areas, is being used to power giant electric generators providing supplementary electric power for conventional electric power plants. There is even some interest in developing oil tankers that would employ sails for propulsion. Heat energy from the oceans and energy from the ocean tides are intriguing. An environmentally acceptable method of trash and garbage disposal has boggled the minds of women and men for generations. The feasibility of burning trash and garbage to produce both electricity and heat for homes has been demonstrated. The development of these alternative energy sources may not be as glamorous as putting a man on the moon. However, it is no less difficult. Certainly their successful development will reward us to a degree that is hard to comprehend. The prospect of developing acceptable energy sources is what motivates this study of energy and its blessings and problems in our technological society. Many of the chapters that follow examine the energy sources mentioned above in detail with the aim of giving the reader a glimpse at both their potential problems and their potential successes.

REFERENCES

ENERGY AND SOCIETY

Energy and Society, Fred Cottrell, McGraw-Hill, New York, 1955. Reprinted in 1970 by Greenwood Press, Westport, Conn. An extremely interesting and prophetic book. Cottrell's discussions of the relations between energy, social change, and economic development apply to today's problems.

Man, Energy, Society, Earl Cook, W. H. Freeman, San Francisco, 1976. A scholarly and detailed treatment of the societal role of energy.

"Energy," Wolfgang Sassen, *Scientific American* **243,** 118 (September 1980).

HOW DOES AN ENERGY CRISIS DEVELOP?

"Energy: 1945–1980," *Wilson Quarterly,* Spring 1980. Adapted from *Energy Policy in Perspective: Today's Problems, Yesterday's Solutions,* Crawford Goodwin, Brookings Institution, Washington, D.C., 1981.

Energy in Transition: 1985–2010, Committee on Nuclear and Alternative Energy Systems, National Research Council, National Academy of Sciences, Washington, D.C., 1979. Published by W. H. Freeman, San Francisco.

Energy in America's Future, Sam H. Schurr, Project Director, Johns Hopkins University Press, Baltimore, 1979.

"Relaxed Energy Outlook Masks Continuing Uncertainties," Hans H. Landsberg, *Science* **218,** 973 (December 1982).

REVIEW

1. Plotting gross national product per person against per capita energy consumption for the nations in the world produces a fairly strong correlation. One finds
   a) many more countries at the bottom than at the top.
   b) many more countries at the top than at the bottom.
   c) about equal distribution of the countries.
   d) the industrial nations at the bottom.
   e) the underdeveloped countries at the top.
2. The GNP per capita in Saudi Arabia and the United Kingdom is roughly the same. However, the energy use per capita in the United Kingdom is nearly six times larger than per capita energy use in Saudi Arabia. This is because
   a) Saudi Arabia uses energy much more efficiently.
   b) Saudi Arabia is a large exporter of oil and is not highly industrialized.
   c) Saudi Arabia is highly industrialized.
   d) the United Kingdom is a major oil exporter.
   e) the United Kingdom's economy relies very little on oil.
3. Per capita energy use and per capita gross national product are
   a) generally unrelated.
   b) generally strongly correlated.
   c) the same for all countries.
   d) inversely related.
   e) never compared.
4. The major energy-consuming sectors of our society are usually categorized as industrial, transportation, and residential and commercial. Energy use in these three sectors is
   a) roughly the same.
   b) completely dominated by the industrial sector.
   c) completely dominated by the transportation sector.
   d) completely dominated by the residential and commercial sector.
   e) highly variable from year to year.
5. The meaning of an "energy gap" is
   a) a time gap between discovery and production of an energy reserve.
   b) a philosophical gap between energy producers and energy consumers.
   c) a technological gap between American and European manufacturers.
   d) an educational gap between the government and the public on energy issues.
   e) a gap between energy production and energy consumption in the United States.
6. An essential feature of the American oil crisis is that
   a) oil production has steadily declined since 1950.
   b) oil consumption has exceeded oil production for over a decade.
   c) we are committed to exporting most of our national oil.
   d) the federal government has forbidden the development of new oil fields.
   e) all of the above.
7. The United States with 5% of the world's population accounts for about ________ of the world's energy use.
   a) 30%
   b) 6%
   c) 50%
   d) 75%
   e) 100%

Answers

1. (a) 2. (b) 3. (b) 4. (a)
5. (e) 6. (b) 7. (a)

QUESTIONS AND PROBLEMS

1.1 ENERGY AND SOCIETY

1. Scan a daily newspaper for references to energy. Usually you will find energy-related articles in the editorial, business, and general news sections. Sometimes you will find references to energy in the comics and crossword puzzles.
2. The United States with 5% of the world's population in 1979 consumed about 29% of all the energy produced in the world. A decade ago the numbers were 6% and 35%. Forty years ago the United States had considerably fewer people and used substantially less energy. Recognizing the trend toward worldwide industrialization, speculate on whether the United States used more or less than 35% of the world's energy 40 years ago. Try to find a source of information to check your speculation.
3. How would you define energy for a concerned citizen?
4. To see how energy is used in your daily life, trace your "steps" for a few hours starting from the time you arise in the morning. Note how you use energy and how it was involved in the items you use.

5. Having thought about how energy is involved in your own life, consider the energy differences between your life and the lives of your parents and grandparents when they were your age. One factor to consider is the difference in transportation.
6. The development of a technological society can be characterized by the ways that its members have learned to increase the rate of using energy. Apply this idea to the development of transportation in the United States.
7. Choose a few countries of interest to you and think about how energy is used and how energy affects the life-style of the people. As a starter, consider Japan. Why is foreign petroleum so important to Japan's economy?
8. How has the availability of energy influenced suburban growth in the United States?

1.2 OVERVIEW OF ENERGY USE IN THE UNITED STATES

9. Using data from Fig. 1.2, show that the contribution of hydroelectric energy to the total energy input was about 4% in each of the years 1950, 1960, 1970, and 1980.
10. Using Fig. 1.4 as a source of information, determine the percentage reduction of the total input energy if there were a 10% reduction in energy used by automobiles.
11. Figure 1.4 delineates the disposition of energy in the three end-use sectors. Think about energy uses in a home and name at least five contributions to the "other" box in the residential and commercial sector.
12. To illustrate the dependence of the United States on imported energy, determine the annual imported energy every five years from 1950 to 1980 and present your results as a graph. The data may be taken from Fig. 1.6 or from Appendix A.
13. France, which ranks fairly high in both per capita energy use and per capita GNP (Fig. 1.5), has made a decision to derive much more energy from nuclear sources. What has prompted this decision by the French?
14. Determine from Fig. 1.6 the percentage of total energy used in the United States that is obtained from foreign sources.

1.3 A CRISIS IN ENERGY SUPPLY

15. The possibility of the energy dilemma seen in the 1970s was recognized at the time the United States began its program to put a man on the moon. If the American public had been given the choice of spending money to develop new energy sources or putting a man on the moon, what do you think the choice would have been?
16. For the particular part of the country in which you live, speculate on some possible alternative energy sources which could either supplement or replace the energy presently obtained from fossil fuel sources.
17. Look up the definition of crisis and judge whether the 1973 energy situation can be categorized as a crisis.
18. "Contrive," like "crisis," is a word with varied meanings. The oil interests have been accused of contriving the high gasoline prices in order to derive huge profits. They (the industries) say that the profits derived are needed to develop new energy sources. What is your interpretation of "contrive" when used this way? For further insight, read "How to Think about Oil Prices," by Carol J. Loomis, *Fortune* **89,** No. 4, 98 (April 1974).
19. The following quote appeared in an energy publication in 1924:

    In 1920 it was estimated that one half of the oil resources of the United States had been used. This estimate probably was too large, for new fields will be discovered; yet the country has been so thoroughly explored for evidences of oil that it is not probable that many important fields have been overlooked. We are consuming our petroleum with great rapidity, and its exhaustion seems only a few decades off, unless experiments with compressed air and water, forced into the oil-bearing rocks, shall result in the recovery of much oil now left in the ground.

    How do forecasts like this one influence public opinion about the current energy problems?

1.4 AN ENVIRONMENTAL CRISIS

20. As more and more oil is shipped across oceans, concerns for oil spills increase. To what extent are these concerns realized?
21. Pollution is a social cost for energy. What are some other social costs?

2

Physical Basis for Energy

2.1 MOTIVATION

Every country has its own monetary units. The United States uses dollars and cents, Mexico has pesos and centavos. Visitors to Mexico learn quickly to figure the equivalency of dollars and pesos. The mechanics of converting may be annoying but they do it and go about transacting their business, albeit inefficiently. Imagine how terribly inefficient it would be if several different monetary units were employed within a country. You might pay a merchant in dollars and receive change in rubles. In many ways this system prevails in energy transactions. Evaluating a room air conditioner is a good example. The air conditioner removes thermal energy, which we measure in British thermal units (Btu), from a room to maintain a temperature lower than the

Walking and relaxing in the warm spring sun are among the more enjoyable aspects of energy in our lives.

exterior of the building. Energy to operate the air conditioner is provided by an electric utility. We pay the electric utility for kilowatt-hours (kWh) of electric energy we use. We would like to know how much thermal energy is removed for a given investment of electric energy. Like figuring investment returns, the comparison is clearer if the measuring units are the same. Thus we must either convert the thermal energy from British thermal units to kilowatt-hours or convert the electric energy from kilowatt-hours to British thermal units. Then if a particular room air conditioner removes 1.3 energy units for every energy unit invested and if the energy unit were kilowatt-hours, we know that an investment of one kilowatt-hour of electric energy produces a return of 1.3 kilowatt-hours of thermal energy removed from the room.

Energy transactions employ measuring units that are not uniform. Therefore we must learn to cope in order to do business. Wanting to do energy business motivates this chapter. Our goal is two-fold. One is to learn more of the meaning of energy and power. The second is to become comfortable with the variety of units encountered in energy and power transactions.

2.2 RATES

Let us ease into the concepts of energy and power by reviewing and expanding on the idea of a rate. We deal with all manner of rates. Students fret over tuition rates, parents are concerned about telephone rates, scientists ponder the rate of accumulation of carbon dioxide in the atmosphere, and so on. All these rates involve a change in the amount of something during some time interval. The rate is determined numerically by dividing the change by the time interval. A student paying $1500 tuition sees his bank account change by $1500. If the tuition covers 15 weeks of schooling the weekly tuition rate is

$$\begin{aligned}\text{tuition rate} &= 1500\text{ dollars} \div 15\text{ weeks}\\ &= 100\text{ dollars/week.}\end{aligned}$$

The tuition rate is read as 100 dollars per week, per meaning "for each" as in 100 dollars for each week. As an equation, a rate is written in words as

$$\text{rate} = \text{change in the amount} \div \text{time interval}$$

Using symbols for brevity, the rate equation becomes

$$r = \frac{c}{t}. \tag{2.1}$$

The rate equation allows us to implement the physics idea of velocity and to describe the rate of heat loss through the windows of a house in winter. Where the rate is known, such as in the heat loss through a window, we can compute the total heat loss in some time interval by writing the rate equation as

$$c = rt. \tag{2.2}$$

For example, if heat is lost at a rate of 500 Btu per day, then the total heat loss in 30 days is

$$c = \frac{500\text{ Btu}}{\text{day}} \times 30\text{ days} = 15{,}000\text{ Btu.}$$

If some quantity of interest is used at a known rate, we can compute the depletion time by writing the rate equation as

$$t = \frac{c}{r}. \tag{2.3}$$

Perhaps a 100-gallon supply of fuel oil is burned at a rate of four gallons per day. Then the supply will be depleted in a time

$$t = \frac{100\text{ gallons}}{4\text{ gallons/day}} = 25\text{ days.}$$

Canceling numbers in an equation such as

$$\frac{25 \cdot 4}{4} = 25$$

is probably familiar to you. Canceling the units associated with the numbers may not be familiar but it is a very important bit of bookkeeping. When you are calculating a time, for example, the quantity you computed must have units of time, i.e., seconds, days, years, centuries. If you include the units with each numerical quantity and systematically keep track of them, then your answer will automatically have the appropriate units. If the units are wrong, the answer cannot be correct.

2.3 SPEED, VELOCITY, AND ACCELERATION

Speed, velocity, and acceleration are physics examples of rates. For speed, the change alludes to the distance traveled in some time interval so that

$$\text{speed} = \frac{\text{distance traveled}}{\text{time interval}},$$

$$s = \frac{d}{t}. \tag{2.4}$$

If a runner completes a 100-meter dash in ten seconds, her speed would be

$$s = \frac{100\,\text{meters}}{10\,\text{seconds}} = 10\frac{\text{meters}}{\text{second}}.$$

An automobile's speedometer continuously displays the automobile's speed. In most contemporary United States automobiles, the speed is denoted in both miles/hour and kilometers/hour. If the speedometer of a car reads 50 kilometers/hour and if this speed is maintained, then in a time interval of three hours the car will travel a distance

$$d = st \tag{2.5}$$

$$= 50\frac{\text{kilometers}}{\text{hour}} \times 3\,\text{hours}$$

$$= 150\,\text{kilometers}.$$

It is important to note that the concept of speed avoids all reference to direction. It is easy to imagine situations where the direction in which an object travels is just as important as the distance traveled. A pilot who wants to know how long it will take to fly between New York City and Chicago needs to know both the speed and direction of the plane. Velocity provides the means by which the direction of travel is included with the speed. A car traveling due north at 50 miles/hour has the same speed as one traveling due east at 50 miles/hour. Their velocities, however, are different because the directions are different. Although there is this distinction between speed and velocity, they are often used interchangeably. As a working rule it suffices to say that if no direction is involved, then you may use speed and velocity interchangeably. If a direction is implied, then velocity is the only appropriate term.

The dashman pushing off the starting blocks accelerates because his speed is changing.

Acceleration, like speed, finds everyday usage. Alvin Toffler in his book *Future Shock* talks about the "acceleration of change in our time." All drivers know that they must depress the accelerator pedal to increase the speed of their automobiles. All automobiles do not, however, respond in the same way to a depression of the accelerator. Some change speed much faster than others. The acceleration features of a car are often specified by the time taken to achieve a certain speed, say 60 mph, starting from a stop. In a comparison of two cars, the car achieving the designated speed in the *least* time has the *greater* acceleration. All these everyday observations and ideas are incorporated in a simple rate expression

$$\text{acceleration} = \frac{\text{change in speed}}{\substack{\text{time required for}\\\text{this change to occur}}}.$$

A car changing speed from zero to 60 kilometers per hour in five seconds has an acceleration of 12 kilometers per hour per second. The acceleration is written

$$\alpha = \frac{12\,\text{kilometers}}{(\text{hour}\cdot\text{second})}.$$

Maintaining this acceleration means its speed increases 12 kilometers/hour each second. If at some moment its speed were five kilometers/hour, then two seconds later the speed increases 24 kilometers/

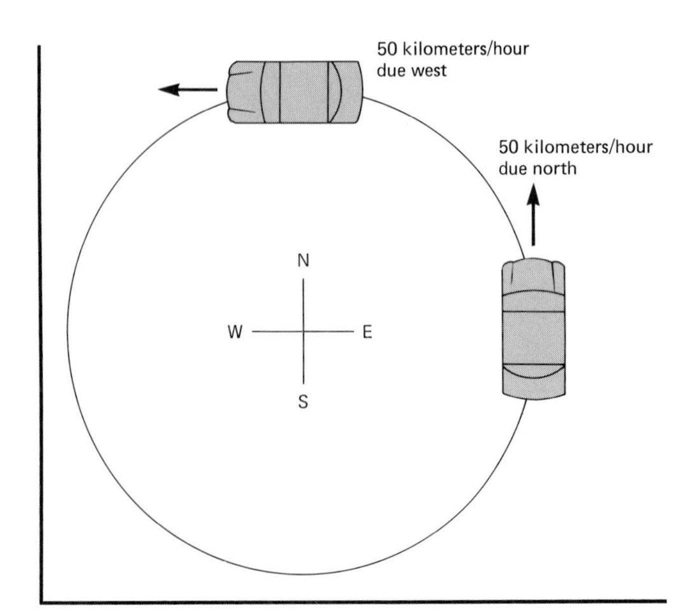

FIGURE 2.1 A car traveling at a constant speed of 50 kilometers per hour on a circular racetrack is accelerating because its direction is changing continually.

hour. It now has a speed of $5 + 24 = 29$ kilometers/hour.

We need not think of acceleration as being only a physics idea. You might read a newspaper article entitled, "The Nation's Coal Companies Accelerate Production." Production implies a rate in units like millions of tons per day and to accelerate production means to change the rate.

To be a little more rigorous, direction should be included with the speed, and acceleration should be defined in terms of a change in velocity rather than change in speed because an object can be accelerated by changing either (or both) the speed and direction. For example, a car traveling at a constant speed of 50 kilometers per hour on a circular race track is accelerating because its direction is continually changing (see Fig. 2.1).

2.4 NEWTON'S THREE LAWS OF MOTION

A father pushes a child in a swing to produce motion. A child pulls on her sled to move it up a hill. A father and his child pulling on opposite ends of a sled may produce no motion at all. Pushes and pulls are examples of forces. Combinations of forces like those exerted on a sled by a father and child may produce no motion. This balancing of forces with no ensuing motion is the desired result for a structure such as a house, a bridge, or a dam. If the combined forces on an object do not balance, then the object accelerates. Thus the child pulling harder and opposite to her father causes the sled to accelerate (change its speed) in her direction.

Newton's three laws of motion provide both a verbal and a quantitative base for discussions of forces acting on some object. Newton's statement of the first law reads "Every body continues in its state of rest, or of uniform motion in a straight line, unless it is compelled to change that state by forces impressed on it." Motion refers to velocity, and uniform means that the velocity does not change in number or direction as time progresses. If the velocity is constant, the acceleration is zero, and the velocity can be changed only by applying a force. Whenever an object is observed to be accelerating, Newton's first law suggests looking for the force responsible for the acceleration. Often the force producing the acceleration is obvious; other times it is not. A batted baseball is accelerated by the force of the bat on the ball. A ball dropped from the window of a building accelerates toward the ground. But what is the force that causes it to accelerate if nothing but air touches it? It turns out that the earth, though not in contact with the ball, pulls on the ball and accelerates it downward. This force is commonly called gravity.

In a fraction of a second a hockey puck struck by a stick changes speed from zero to upwards of a hundred miles per hour. If the stick were to contact the puck again with exactly the same force, then we should expect the same acceleration. Skilled hockey players believe strongly in this notion. But if a devilish referee were to slip a hockey puck with a hollow interior onto the ice, the same contact by the stick produces a larger acceleration. You may not have the chance to test this hypothesis with a hockey stick and hockey pucks but you might try kicking different objects on a smooth floor. The property of a system for resisting an acceleration is called inertia. For physical systems such as hockey pucks, the property is called inertial mass, or just mass for short.

Newton's second law of motion quantifies all these ideas by stating that the acceleration experienced by an object equals the resultant force acting on it divided by its mass.

$$\text{acceleration} = \frac{\text{resultant force}}{\text{mass}},$$

$$a = \frac{F}{m} \quad \text{or} \quad F = ma. \tag{2.6}$$

Resultant force means the result of all the forces acting on the mass. If a mass is pushed eastward with a force of 100 units and at the same time is pulled westward with a force of 30 units, the resultant force is 70 units eastward.

If two skaters are pushed along with the same constant force, we can use Newton's second law equation to write $F = ma$ for one skater having mass m and acceleration a and $F = MA$ for the other skater having mass M and acceleration A. Because the force F is the same, we can equate ma and MA to get

$$ma = MA$$

from which

$$M = \left(\frac{a}{A}\right) m.$$

If mass m is known, then measuring the accelerations a and A allows you to compute the value of mass M. In principle, this is how unknown masses are measured. A particular mass is assigned a mass of one kilogram. Then other masses are measured relative to this standard kilogram. In practice, masses are measured with devices like the pan balance shown in Fig. 2.2. Because mass has units of kilograms, and acceleration has units of meters per second per second, it follows from the equation $F = ma$ that force has units of kilogram meters per second per second. One kilogram meter per second per second is called a newton, symbol N.

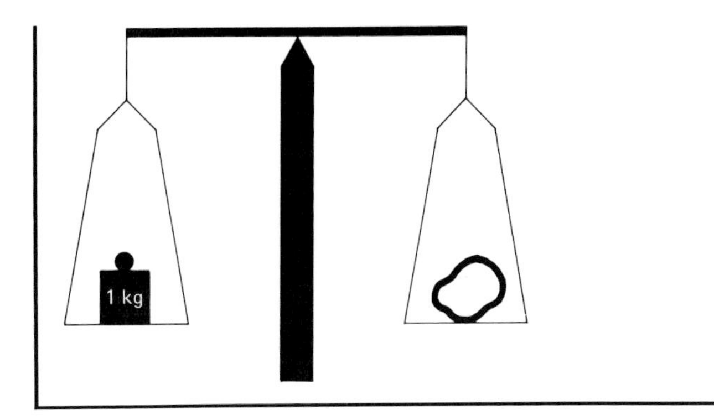

FIGURE 2.2 A simple balance used to determine the mass of an object. The pans on which the objects rest achieve the same level when the masses are equal.

Once the masses of objects are measured, then forces can be determined by measuring the acceleration produced when a force of interest acts on an object of known mass. For example, if a kilogram mass experiences an acceleration of one meter per second per second, then the net force on the mass is one newton. You may push with your hand on a set of bathroom scales and the scales record the size of the force. Measuring force with a set of scales or some other force-measuring instrument is consistent with the fundamental scheme of measuring mass and acceleration and using the relationship $F = ma$.

Perform this experiment. Hold a pencil and a coin the same distance above the floor and release them at the same time. From the sounds made when each hits the floor you probably conclude that they struck the floor at the same time (Fig. 2.3). But if you do the same experiment using a coin and a small scrap of paper, the coin clearly strikes the floor before the paper. You might conclude that the coin falls faster because it is more massive. But if you were to use a coin and a pad of paper (which obviously is more massive) dropped with its surface parallel to

FIGURE 2.3 Try dropping two objects and listen for the sounds as they hit the floor. You should find, as this drawing demonstrates, that air strongly influences a sheet of paper but weakly influences a pencil. In the absence of the air, both paper and pencil would accelerate at the same rate.

the floor, you would see that the coin wins again. Observation of the flights of the dropped objects reveals that the air affects the motion. It is not easy to contrive a set of experimental conditions to remove the air surrounding the objects being dropped. In a moderately sophisticated laboratory, it is possible to evacuate a reasonably large container; on the moon, which has no atmosphere, the Apollo astronauts dropped objects in an air-free environment. Under these conditions two objects dropped simultaneously take the same time to travel a given distance. In many situations even though the air is present its effect is minimal and can be neglected. Under this assumption gravity is the only force acting on an object. When gravity is the only force acting, experiment demonstrates that the acceleration is the same for all masses. This acceleration is given the symbol g and is numerically equal to 9.8 meters per second per second. That is to say, the velocity changes by 9.8 meters per second for every second that the object falls. The force of gravity on an object is called its weight. Thus, using Newton's second law, the weight of an object is simply its mass multiplied by the acceleration due to gravity. In symbols, $W = mg$. The connection between mass and weight is now clear. They are proportional but not equal. Weight is a force and in the metric system of units is expressed in newtons. In our everyday world, the unit of weight is the pound. For comparison, one pound is equivalent to 4.45 newtons. If you weigh 110 pounds, then your metric weight is

$$W = 110 \, \cancel{\text{pounds}} \times \frac{4.45 \text{ newtons}}{\cancel{\text{pound}}}$$
$$= 490 \text{ newtons},$$

and your mass is

$$m = \frac{W}{g} = \frac{490 \text{ newtons}}{9.8 \text{ meters per second}^2},$$
$$= 50 \text{ kilograms}.$$

Any two objects having mass, such as the earth and you, exert attractive gravitational forces on each other. The moon and the earth pull on each other and this gravitational pull influences tidal motion (Chapter 11). Our sun and our moon, and our sun and the earth, mutually attract because of gravity. Throughout the universe, gravitational forces are constantly exerting pulls on pairs of objects. The gravitational force between two masses m and M depends on the product of m and M. Additionally, the force decreases as the separation of the two masses increases. If you double the separation, the force is only one-fourth as much. If d meters separate the center-to-center positions of two objects of m kilograms and M kilograms, the gravitational force betwen the two can be written

$$F = 6.67 \times 10^{-11} \frac{mM}{d^2} \text{ newtons.*} \tag{2.7}$$

* The notation 6.67×10^{-11} means 0.0000000000667. This powers of ten notation is discussed in Appendix C.

According to Newton's second law of motion, an object cannot change its velocity, that is, it cannot accelerate unless a net or unbalanced force acts on it. Thus a runner or a car cannot begin forward motion from a stop unless something forces it forward. There is nothing in contact with the runner or car that can provide this force except the ground. The runner learns to push backwards on the ground and the ground responds by exerting a forward force on the runner. This reaction response to some instigated action is the essence of Newton's third law of motion. Newton's third law of motion states that if some object (label it *A*) exerts a force on another object (label it *B*), then B exerts an equal but oppositely directed force on A. Newton's third law of motion is put to use in myriad ways. As you sit in a chair or seat with no acceleration, the earth pulls down on you with a force equal to your weight (Fig. 2.4). You, in turn, cause an upward gravitational force on the earth. Were there no other forces on you, you would move downward like a pencil dropped from your hand. But as you push down on the chair, the chair pushes up on you and balances the downward gravitational force of the earth. When you decide to leave the chair, you push backwards on the chair and the chair responds with a forward force on you causing you to move out of the chair. Note that the two forces in Newton's third law of motion never act on the same object. You push backwards on the chair; that is a force on the chair. The chair pushes forward on you; this is the other force in Newton's third law of motion and it acts on you, not the chair.

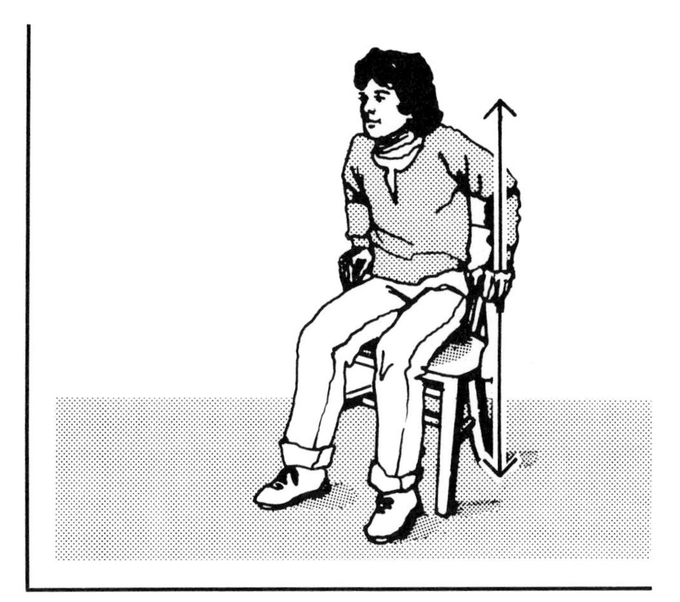

FIGURE 2.4 Several action-reaction pairs of forces can be identified in this drawing. The chair pushes up on the student's hand. The reaction force is her hand pushing down on the chair. This pair, as for all action-reaction pairs, acts on different objects. One force is on her hand, the other is on the chair.

2.5 FUNDAMENTAL FORCES

Every time something moving—a person running, an atom in a nuclear-fusion reactor, a smoke particle ejected by a coal-burning power plant—changes velocity, it accelerates, and some net force is responsible for the acceleration. Because of the countless number of objects experiencing accelerations, it might seem there would be a corresponding number of types of forces. But we find only four fundamental types of forces described as gravitational (or just gravity), electromagnetic, weak nuclear, and strong nuclear.* Energy is associated with each force, with each type playing a very important role in the societal use of energy. Water atop a dam has gravitational energy. Atoms have electric energy. Nuclei of atoms have electric energy and energies associated with both the weak and strong nuclear forces. As we proceed, we shall examine in some detail these four fundamental forces and their associated energies.

We cannot avoid gravitational and electric forces in our daily lives. Gravity exerts a downward force on a person standing on a floor that is balanced by

---

* One of the most exciting bits of contemporary research involves an effort to demonstrate that these four forces are special cases of a single unifying force.

an upward force exerted by the floor. That upward force is electric, arising from forces between atoms in the floor and atoms in the person's shoes.

2.6 WORK

A student does physical work when cleaning her room. She does mental work preparing for an exam. Whether physical or mental, work involves an effort directed toward the production of something. In the physical sense, effort is associated with force, and work is done when a force acts on an object as it moves through some distance. If the force does not change as the object moves, the numerical value of the work W is the product of the force F and the distance d moved.

$$\text{work} = \text{force} \times \text{distance},$$
$$W = Fd. \tag{2.8}$$

Work has the units of force times distance, that is, newton meters. One newton meter is called a joule, pronounced "jool" and given the symbol J. Thus the student pushing the broom 1.2 meters with a force of 20 newtons does 24 joules of work.

The physical definition of work retains much of the popular notion, but there is a subtle and important difference. No matter how much force (effort) is exerted, no work is done in the physics sense if the object does not move. A student may fret and sweat while tugging on the foot of her roommate in bed but if the roommate does not budge, no work is done.

It is important also to recognize that a force can act either in the same direction as the mass is moving or opposite to the direction in which the mass is moving (Fig. 2.5). We distinguish between these two physically different situations by calling the work positive if the force is in the direction of motion, and negative if the force is opposite to the direction of motion. The *net* work on an object is the algebraic sum of the works done by each force acting on the object. Suppose that a sled moves two meters eastwardly as a result of a girl pulling due east with a force of 100 newtons and a boy pulling due west with a force of 50 newtons. The girl does

$$+100 \times 2 = 200 \text{ joules of work},$$

the boy does

$$-50 \times 2 = -100 \text{ joules of work},$$

and the net amount of work is

$$+200 - 100 = +100 \text{ joules}.$$

2.7 ENERGY AND WORK

A day does not pass without our being involved in physical work. The forces that move us in walking and running perform work. Lifting food to our mouths and chewing food involves forces doing work. Some days we work more easily and more efficiently than on others. We might cast our activities in terms of our ability to do work. Because physical work involves forces and movement, we might ask, "Under what conditions does something have a capacity for moving an object through some distance?" A car in motion has this capacity because if it rams into the rear of a car stopped at a red light, the struck car will surely move some distance. The energy associated with masses in motion is called kinetic energy. Experience tells us that a big, massive car colliding with a stopped car will "do more work" on the stopped car than a less massive vehicle like a bicycle. Similarly, a fast-moving car will "do more work" on a

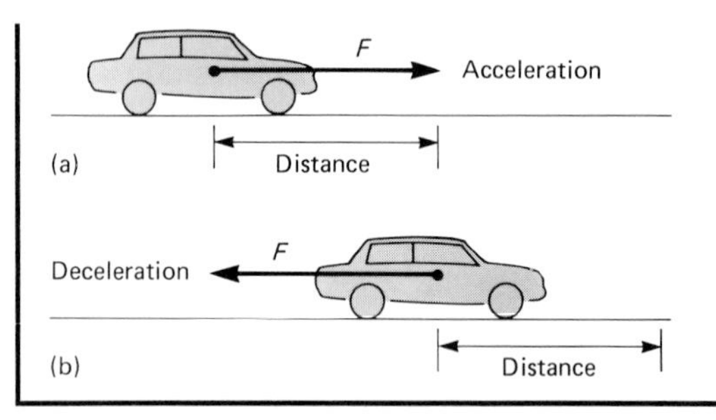

FIGURE 2.5 **(a)** Force acts in the same direction that the car moves tending to increase its speed. This is a case of positive work. **(b)** Force acts in a direction opposite to that in which the car moves tending to decrease its speed. This is a case of negative work.

stopped car than a slowly moving car of the same type. Thus a numerical evaluation of kinetic energy should reflect these observations. Formally, the kinetic energy K of an object having m kilograms of mass moving v meters/second is defined as one half of the product of mass and square of speed.

$$K = \frac{1}{2}mv^2. \tag{2.9}$$

There are some subtle reasons for writing the kinetic energy equation this way, but they are not essential for our discussion. Importantly, the equation shows that kinetic energy increases if either or both the mass and speed increase. Energy and work both are measured in joules. We can see this by examining the units for work and kinetic energy. Work is the product of force and distance. Force has units of kilogram meters/second and distance has units of meters. Thus work has units of kilogram meter^2/second^2. Kinetic energy is $\frac{1}{2}mv^2$. Mass is measured in kilograms and speed in meters/second. Thus kinetic energy has units kilogram meter^2/second^2, the same units as for work.

At some point in time an object such as a car may have speed v and kinetic energy $K = \frac{1}{2}mv^2$. Later on its speed may change to V so that its kinetic energy is $\frac{1}{2}mV^2$. Its kinetic energy will have changed by $\frac{1}{2}mV^2 - \frac{1}{2}mv^2$. The work-energy principle states that the net amount of work on the object between the initial and later times is equal to the change in kinetic energy. As an equation

$$W_{\text{net}} = \frac{1}{2}mV^2 - \frac{1}{2}mv^2. \tag{2.10}$$

When a hockey player hits the stationary 0.16-kilogram puck with a hockey stick and imparts to it a speed of 45 meters/second (100 miles/hour), its kinetic energy changes from zero to $\frac{1}{2}$ (0.16) $(45)^2 = 162$ joules. Thus the work done on the puck is +162 joules. A catcher receiving a 0.15 kilogram baseball traveling 40 meters/second (90 miles/hour) does negative work on the baseball because its kinetic energy changes from $\frac{1}{2}$ (0.15) $(40)^2 = 120$ joules to zero.

You see this principle in operation in many processes of interest to us. The kinetic energy of water flowing over a dam increases if its speed increases. This increase in kinetic energy is the result of work done on the water by the force of gravity. When a car is set into motion starting from rest, its kinetic energy increases. This is due to work done by a force in the direction of motion of the car. Likewise, when the car slows down, its kinetic energy decreases. This is due to (negative) work done on the car by a force acting in a direction opposite to the direction of motion.

2.8 POTENTIAL ENERGY

A car at rest has no kinetic energy, yet it is *potentially* capable of moving another object *if* it can be put into motion. The potentiality arises from the intrinsic energy of the gasoline that, when burned in the automobile engine, can be *converted* to energy of motion. Water at the top of a hydroelectric dam is capable of exerting a force on the rotor of a turbine *if* it is released. Before release, the energy is potential. Once released and set into motion, the water possesses kinetic energy (Fig. 2.6). There is potential energy in coal, oil, and gas because *if* ignited, each releases heat energy. There is also potential energy in the mass of the nucleus of the atom which can be released *if* an appropriate nuclear reaction is initiated.

The flow of water over a dam and onto the blades of a paddle wheel provides an ideal illustration of the connections between kinetic energy, potential energy, and work. Because it is easy to visualize the motion of an object that has some boundaries, let us follow the motion of a plastic bag of water moving in concert with the water (Fig. 2.7). At the top of the dam, the bag of water has kinetic energy ($\frac{1}{2}mv^2$) by virtue of its mass and speed, and potential energy (PE) by virtue of its position at the top of the dam. As the bag flows from the top of the dam to the bottom, work is done on it by the gravitational force. This force is the same as the weight $w = mg$. The work is just the product of the weight (mg) and the distance (h) through which it falls. Because the work

FIGURE 2.6 The potential energy of dammed water is converted to electric energy at the Fontana hydroelectric station located on the Little Tennessee River in North Carolina. This 480-foot high dam is typical of the many dams in the Tennessee Valley Authority. (Photograph courtesy of the Tennessee Valley Authority.)

is positive, the change in kinetic energy is positive. Thus the total energy at the bottom of the dam includes the kinetic energy it had at the top plus the work done.

$$\text{work} = mgh,$$

$$\text{change in kinetic energy} = \frac{1}{2}mv^2_{\text{bottom}} - \frac{1}{2}mv^2_{\text{top}},$$

$$mgh = \frac{1}{2}mv^2_{\text{bottom}} - \frac{1}{2}mv^2_{\text{top}},$$

$$\frac{1}{2}mv^2_{\text{bottom}} = \frac{1}{2}mv^2_{\text{top}} + mgh. \tag{2.11}$$

The bag of water arrives at the paddle wheel with the speed it has at the bottom of the dam. It exerts a force on the paddle wheel and does work (W) on it producing rotational motion. After it has passed the paddle wheel, its speed decreases because it has lost energy to the paddle wheel. The kinetic energy of the bag beyond the paddle wheel equals the kinetic energy it had at the bottom of the dam less the work done on the paddle wheel.

$$\frac{1}{2}mv^2_{\text{beyond wheel}} = \frac{1}{2}mv^2_{\text{bottom}} - W. \tag{2.12}$$

It is somewhat more difficult to visualize the energy transformations in an automobile engine, for example, by visualizing moving particles. The idea, though, is the same, and it is always revealing to follow the energy flow and transformations in any energy system.

2.9 ENERGY CONVERSION, CONSERVATION OF ENERGY, AND EFFICIENCY

We have dwelled at some length on the usefulness of energy in many facets of our society. When put to use, energy is kinetic. An automobile is useful when it is in motion. Heat energy involves the kinetic energy of atoms and molecules. Electricity, probably the most useful form of energy, results from the motion of electric charges. Useful energy is always derived by converting some other form of energy into the useful form through the use of a converting device. For example, heat energy is useful for heating a room in a home and can be derived by converting the intrinsic chemical potential energy in a fossil fuel and the oxygen in the air through burning in a furnace. An idealist might expect to get all the energy converted into the useful form. A realist suspects from the way things work in nature that you cannot expect to get all the energy converted into the useful form. The difference between the energy converted and the useful energy produced is diverted to other areas or uses. In physics, this idea is contained in the principle of conservation of energy or, equivalently, the first law of thermodynamics that reads—there is no loss of energy in any energy transformation. If 1000 joules of energy are derived from burning a certain amount of gasoline (the potential energy) in an engine, then there will still be 1000 joules after the burning. Perhaps only 300 joules were transformed to the useful form, i.e., mechanical energy of motion. The other 700 joules were diverted to other forms, which most often are not useful.

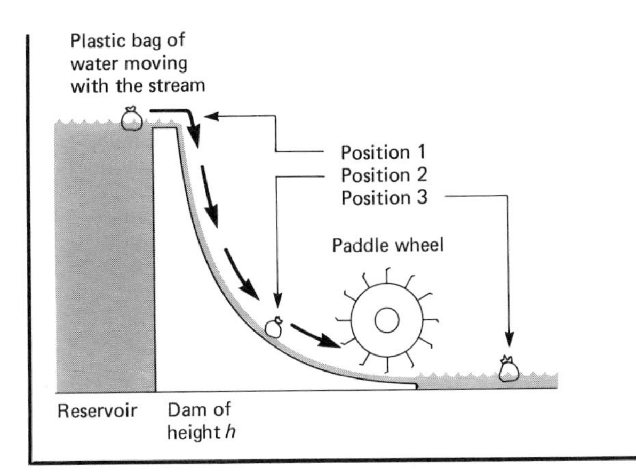

FIGURE 2.7 Energy transformations involved in water flowing over a dam.

POSITION 1. At the top of the dam, a bag of water has kinetic energy $\left(\frac{1}{2}mv_{\text{top}}^2\right)$ due to its motion, and potential energy (mgh) due to its position relative to the bottom of the dam.

POSITION 2. At the bottom of the dam, all of the potential energy has been converted to kinetic energy. The energy of the bag of water is all kinetic and includes the kinetic energy it has atop the dam plus the kinetic energy acquired by falling over the dam.

POSITION 3. Passing by the paddle wheel, the bag of water transfers energy (W) to the wheel. Its energy after passing the wheel is still all kinetic and is equal to its energy before impinging on the wheel minus the energy imparted to the wheel.

The conservation-of-energy principle provides an indispensable bookkeeping procedure. It is like a budget where energy rather than dollars is being dispersed. Like any budget, the numbers must total correctly.

Your first thought may be that this concept is inconsistent with concerns for energy supplies. After all, if there is no change in the total amount of energy, why worry about energy shortages? Just because there is energy present does not mean that it can be used or converted to some usable form. The basic conversion mechanisms divert energy to areas where it cannot be recovered. In particular, energy that is diverted to the environment as heat cannot be recovered, and the ultimate disposition of all energy is as heat. Even though the total amount of energy remains unchanged in any energy conversion, the energy ultimately ends up as heat that cannot be recovered.

As a practical matter we would like to know what fraction of the energy converted in any process gets into the useful form. This fraction is designated as the efficiency

$$\text{efficiency} = \frac{\text{useful energy out}}{\text{total energy converted}},$$

$$\epsilon = \frac{E_{\text{out}}}{E_{\text{converted}}}. \tag{2.13}$$

Because the useful energy is always less than the total energy converted, the efficiency is always numerically less than one. For example, if 300 joules of useful mechanical energy are derived from the conversion of 1000 joules of energy in gasoline, then the efficiency of the engine is

$$\epsilon = \frac{300}{1000} = 0.3.$$

This means that 30 units of useful energy are produced for every 100 units of energy converted. The energy unit is immaterial so long as it is the same for both the output of useful energy and the converted energy. The efficiency is expressed often as a percent. This simply means that you multiply the efficiency by 100. The engine in the previous example has an efficiency of 30%. Remember, in any numerical manipulation, the efficiency has to be expressed as a decimal or fraction. For example, the efficiency of an ordinary incandescent lightbulb is only 5%. The useful output from a light bulb is light energy, and the energy converted is electric energy. Thus if 5,000,000 joules of electric energy are converted by a light bulb, the amount converted to light energy is

$$\begin{aligned} \text{light energy} &= \text{total energy converted} \times \text{efficiency} \\ &= (5{,}000{,}000\,\text{joules}) \times 0.05 \\ &= 250{,}000\,\text{joules}. \end{aligned}$$

The remainder of the converted energy, 4,750,000 joules, emerges as heat energy. The reason that a light bulb tends to get very hot is because most of the energy converted by the bulb goes into heat.

The efficiencies of some important energy conversion processes are given in Table 2.1. Note that most of these are quite low. We will see that these low efficiencies have contributed to many of our problems resulting from conversion of energy.

The energy for our society, 93% coming from fossil fuels, is converted to useful forms at an overall efficiency of about 50%. The diverted energy enters the environment as heat. For every unit of useful energy, nearly one unit is wasted as heat. This means that if you can save one unit of useful energy, then two units of energy need not be converted. Thus the potential for saving energy is significant.

TABLE 2.1 Efficiencies of a few important energy conversion devices. A more complete list can be found in *Energy*, W. H. Freeman, San Francisco (1978).

Device	Energy conversion function	Efficiency (%)
Incandescent lamp	electrical to light	4
Fluorescent lamp	electrical to light	20
Automobile engine	chemical to thermal to mechanical	25
Steam turbine	thermal to mechanical	47
Home oil furnace	chemical to thermal	65
Charging a battery	electrical to chemical	72
Discharging a battery	chemical to electrical	72
Electric generator	mechanical to electrical	99

While the principle of conservation of energy allows you to account for each joule of energy converted, it says nothing about having deployed the energy in the optimum way. For example, if energy is needed to heat a home, the principle of conservation of energy does not help you select the best system for conversion. Nor does it tell you the optimum use of a given source of energy. For example, the energy derived from burning natural gas can be used in certain types of engines as well as for heating water and providing household heat. While the engine *must* have the high temperature provided by the burning gas, household heating does not. Thus a case may be made that the most efficient use of natural gas is not in household heating. We shall find that the second law of thermodynamics (to be discussed in Chapter 7) often provides a more realistic evaluation of energy use and energy converters.

2.10 POWER

Mountain climbing is a popular pastime. If a friend tells you he did a certain amount of work in scaling a height of 100 meters, you might be interested but not be particularly impressed. On the other hand, if he told you he did it in ten seconds you should take note because he did work at an impressive rate. The rate of doing work and the rate of converting energy are very important concepts. There may exist an enormous energy resource but if it cannot be extracted at a rate appropriate to societal needs, it is of little use. The rate of doing work or the rate of converting energy is called power.

$$\text{power} = \frac{\text{amount of work done (or amount of energy converted)}}{\text{time required to do the work (or convert the energy)}}$$

$$P = \frac{W}{t}. \qquad (2.14)$$

Measuring work and energy in joules and measuring time in seconds makes the units of power joules per second. One joule per second is called a watt, symbol W. An operating 100-watt light bulb uses electric energy at a rate of 100 joules per second. If the bulb were on for one hour, you would pay the electric power company for

$$100\,\frac{\text{joules}}{\text{second}} \times 1\,\text{hour} \times 3600\,\frac{\text{seconds}}{\text{hour}}$$
$$= 360{,}000\,\text{joules}$$

of electric energy.

Any energy unit divided by any time unit is a unit of power. Thus a 2500-calorie per day use of food is a power measure. While there are many energy units, we seldom encounter any power unit other than the watt or multiples like kilowatt (1000 watts), megawatt (1 million watts), and gigawatt (1 billion watts). The horsepower (hp) rating of an engine measures the rate at which an engine does work. One horsepower is equivalent to 746 watts. If an engine develops 200 horsepower, it is doing work at a rate of

$$200\,\text{horsepower} \times 746\,\frac{\text{watts}}{\text{horsepower}} = 149{,}200\,\text{watts}$$
$$= 149{,}200\,\frac{\text{joules}}{\text{second}}.$$

An electric motor on a small household appliance such as a food mixer produces only about $\frac{1}{6}$ horsepower. This equals about 120 watts.

To obtain a feeling for the size of the joule energy unit and the enormous range of energies we are exposed to, consider that

the sound energy in a whisper is about 0.01 joules,

the kinetic energy of a 1000 kilogram car traveling 55 miles per hour is about 300,000 joules,

the energy from burning a barrel of oil is about 6,000,000,000 joules,

the annual energy use in the United States is about 10^{20} or 100 billion billion joules, and

the daily energy input to the earth from the sun is about 10^{22} or 10,000 billion billion joules.

In terms of the rate at which work is done or energy is converted consider that

the human heart does work at a rate of about 0.01 joules per second,

a flashlight converts energy at a rate of about four joules per second,

a student in daily activity does work at a rate of about 90 joules per second or about 0.1 horsepower,

a horse can do work at a rate of about 800 joules per second,

an automobile develops about 100,000 watts of power, a large electric power plant produces about 1,000,000,000 watts of electric power, and the solar input to the earth (at the top of the atmosphere) amounts to about 10^{17} or 100 million billion watts.

2.11 OTHER COMMON ENERGY AND POWER UNITS

If the metric system of units were universal, then the units of joules for energy and watts for power would suffice. But as stated at the outset, we do not operate in a world where joules and watts are the last word. Thus to travel in these other "worlds" we must conform and convert units as required. Table 2.2 is provided to help in this chore. For example, units in column 1 are convertible to equivalent units by multiplying by the numbers given in columns 4 and 5. Thus, in joules, 2000 Btu are equivalent to

$$2000\,\text{Btu} \times 1055\frac{\text{joules}}{\text{Btu}} = 2{,}110{,}000\,\text{joules}.$$

The origin of many of these units will be explained as we proceed.

2.12 ENERGY CONVERSION SYMBOLISM

The analysis of the wide variety of energy conversion processes using the ideas put forth in this section is the essence of this book. It will be useful to depict an energy conversion process with the diagram in Fig. 2.8. The principle of conservation of energy allows us to write

$$\text{total energy converted} = \text{useful energy out} + \text{energy diverted}.$$

TABLE 2.2(a) Some special energy units.

Special energy unit	Study area of main use	Symbol	Equivalent in joules	Other useful equivalents
kilowatt-hour	electricity	kWh	3,600,000	3413 Btu, 860 kcal
calorie	heat	cal	4.186	
kilocalorie (food calorie)	heat	kcal	4186	1000 cal
British thermal unit	heat	Btu	1055	252 cal
electron volt	atoms, molecules	eV	1.602×10^{-19}	
kiloelectron volts	X-rays	keV	1.602×10^{-16}	1000 eV
million electron volts	nuclei, nuclear radiation	MeV	1.602×10^{-13}	1,000,000 eV
quintillion	energy reserves	Q	1.055×10^{21}	10^{18} Btu or a billion billion Btu
quadrillion	energy reserves	quad	1.055×10^{18}	10^{15} Btu or a million billion Btu

ENERGY EQUIVALENTS

1 gallon of gasoline = 126,000 Btu

1 pound bituminous coal = 13,100 Btu

1 therm = 100,000 Btu

1 cubic foot natural gas = 1030 Btu

1 42-gallon barrel of oil = 5,800,000 Btu

Variations of these energy equivalent values will appear in the literature.

The values quoted here are typical.

TABLE 2.2(b) Some special power units

Special power unit	Study area of main use	Symbol	Equivalent in watts
British thermal unit per hour	heat	Btu/hr	0.293
kilowatt	household electricity	kW	1000
megawatt	electric power plants	MW	1,000,000
gigawatt	total U.S. or world power production	GW	1,000,000,000
horsepower	automobile engines, motors	hp	746

The concept of efficiency assesses the ability of the system to convert energy to the useful form.

The diagram is not intended to picture the actual converting device. It is used to show the energy pathways and help with the energy bookkeeping. Although these diagrams are extremely useful and essential for our study, one should never lose sight of the complexity of an energy converting system such as an electric power plant and the engineering knowledge involved in producing energy at the rates required by our technological society.

A seemingly inescapable feature of energy conversion is the generation of by-products that produce troublesome environmental effects. The automobile engine in its capacity as a converter of chemical energy to mechanical energy produces carbon monoxide, hydrocarbons, and nitrogen oxides that in turn have produced undesirable effects in our cities. Coal-burning electric power plants produce sulfur oxides and fine particulate matter that can be dangerous to persons having respiratory ailments. The control of these emissions requires money and energy. Above all, incentives must exist for wanting to control these emissions. Chapters 5 and 6 are devoted to an assessment of these environmental effects.

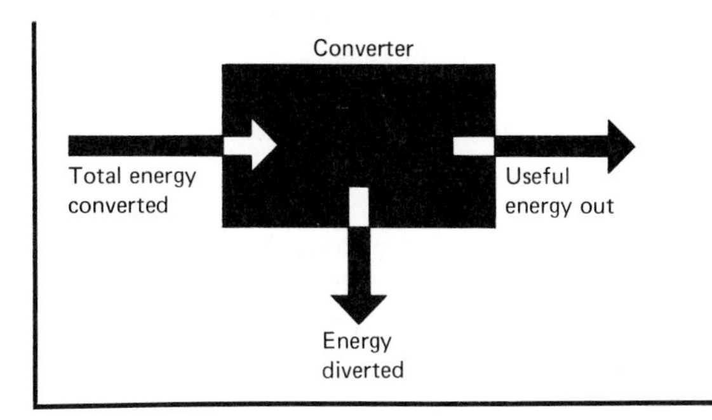

FIGURE 2.8 Schematic description of an energy conversion system. Regardless of the nature of the converter (engine, person, battery, etc.) the sum of the useful and diverted energies equals the total energy fed into the converter.

REFERENCES

PHYSICS OF ENERGY AND POWER

Any physics or physical science text will devote a substantial number of pages to the study of energy and power. Two such texts useful for expansion of the material presented in Chapter 2 are

Physics in Perspective, Eugene Hecht, Addison-Wesley, Reading, Mass., 1980.

Conceptual Physics, Paul G. Hewitt, Little, Brown, Boston, 1981.

CONVERSION OF ENERGY

The conversion of energy is covered in some detail in *Energy,* W. H. Freeman, San Francisco, 1978.

REVIEW

1. Express in words and as an equation the meaning of a rate.
2. Why is it important to include and keep track of units in an equation?
3. What is acceleration? Give two examples of units for acceleration?
4. Explain the meaning of an acceleration of two miles per hour per second.
5. What role do forces play in the motion of objects and systems?
6. Formulate Newton's three laws of motion in words and give two examples of the application of each law.
7. Even though energy and work have the same measuring units, how are they different?
8. Distinguish between kinetic energy and potential energy.
9. What is meant by energy conversion and energy conversion efficiency?
10. What is the physics meaning of "conservation of energy"?
11. Explain how energy differs from power.
12. Draw a generalized diagram for an energy converter.
13. What is a kilowatt? a megawatt?
14. Match the following quantities and units.

force	joules
heat	meters/second
speed	(meters/second) second
velocity	newtons
acceleration	joules/second = watts
work	
energy	
potential energy	
power	
kinetic energy	

15. Match the following quantities and expressions:

Newton's second law	$\frac{1}{2} mv^2$
work	$PE = mgh$
kinetic energy	$r = \frac{c}{t}$
rate	$v = \frac{d}{t}$
speed	$P = \frac{W}{t}$
power	$F = ma$
potential energy	Fd

16. If while traveling due north on an interstate highway, your speed changed from 50 miles per hour to 55 miles per hour in five seconds you could say (correctly)
    a) the acceleration of the car is one mile per hour per second.
    b) the velocity of the car is five miles per hour.
    c) the acceleration of the car is -25 miles per hour per second.
    d) the speed is unchanged.
    e) there is no net force on the car.
17. From the standpoint of different systems of units, a meter is to a foot as a newton is to a
    a) kilogram. b) pound. c) gram. d) joule.
    e) yard.
18. Newton's first law of motion states that an object in motion remains in motion in a straight line at constant speed unless acted upon by a net external force. Therefore
    a) an auto must use energy, even on a straight, level road, to overcome the force of air resistance and rolling friction.
    b) an auto traveling at constant speed over a straight, level road must use energy (gasoline) to overcome the force of gravity.
    c) Newton's first law does not hold for automobiles here on the earth's surface.
    d) an auto traveling over a straight, level road could, in principle, move forever with its engine off.
    e) none of the first four selections are correct.
19. In order to run, a woman learns to push on the ground supporting her. According to Newton's third law of motion, the reaction to this push is
    a) the force exerted on her joints by her muscles.
    b) mental anguish.
    c) the force exerted on her by the ground.
    d) the force of gravity pulling on her.
    e) the acceleration achieved by her.
20. Equal net forces are applied to a Cadillac and a Volkswagen. We would expect the Volkswagen to

experience the larger acceleration because
- a) it is less expensive.
- b) German workers make better cars.
- c) it has less mass.
- d) it has smaller wheels.
- e) it has more mass.

21. In a moment of exuberance, a student jumps from the floor to the top of the lecture table. A person in the audience might be heard to say (correctly)
    - a) the force of gravity has done work on the student.
    - b) the potential energy of the student has increased.
    - c) the floor pushed up on the student.
    - d) the student pushed down on the floor.
    - e) all the previous four statements are reasonable.

22. A 1000-kilogram car undergoes an acceleration of 0.4 meters per second per second. According to Newton's second law of motion, we would say
    - a) the net force on the car is 400 newtons.
    - b) the kinetic energy of the car is constant.
    - c) there is no work done on the car.
    - d) the reaction force is 1000 kilograms.
    - e) the net force on the car is 2500 newtons.

23. A car is traveling due west on an interstate highway. In order for the car to slow down there must be a force on the car that is directed ________ and the work done by this force is ________ .
    - a) west, negative
    - b) west, positive
    - c) west, zero
    - d) east, negative
    - e) east, positive

24. In physics, work is defined as
    - a) the product of mass and acceleration.
    - b) the quotient of distance and force.
    - c) the quotient of force and distance.
    - d) the rate of using energy.
    - e) the product of force and distance.

25. Water is pulled from the top of a dam to the bottom by the force of gravity. We would say
    - a) the work due to gravity is negative because the water moves in a downward direction.
    - b) the work due to gravity is zero because the water is unaccelerated.
    - c) the work due to gravity is negative because gravity pulls down on the water.
    - d) the work due to gravity is positive because the displacement of the water and the force of gravity are in the same direction.
    - e) work is not involved in this situation.

26. If the net work on a car is positive, we know
    - a) its kinetic energy decreases.
    - b) there is no net force on the car.
    - c) its kinetic energy increases.
    - d) there is no change in kinetic energy.
    - e) Newton's second law of motion has been violated.

27. If the brakes are applied in a moving automobile, the work done by the brakes is
    - a) negative, because the braking force applied and the motion are oppositely directed.
    - b) positive, because the braking force and motion are in the same direction.
    - c) converted to potential energy.
    - d) positive, because it is converted to positive energy in the form of heat.
    - e) none of the previous four selections are correct.

28. From the basic idea of kinetic energy, we expect the kinetic energy of a car to
    - a) increase as the mass of the car decreases.
    - b) decrease as the speed of the car increases.
    - c) not depend on the mass of the car.
    - d) increase if either or both the mass and speed increases.
    - e) depend only on the acceleration of the car.

29. Car A has a mass of 1000 kilograms and a speed of 60 kilometers/hour. Car B has a mass of 4000 kilograms and a speed of 30 kilometers/hour. The kinetic energy of car B is ________ the kinetic energy of car A.
    a) half b) twice c) four times d) three times
    e) equal to

30. Power is the rate of converting energy, and speed is the rate of changing distance traveled. Knowing that distance equals speed multiplied by time, then by analogy power multiplied by time equals
    a) energy. b) force. c) mass.
    d) acceleration. e) watts.

31. The work-energy principle in physics states that
    - a) work and energy mean exactly the same thing.
    - b) work is a capacity for expending energy.
    - c) the net work done on an object equals the change in its kinetic energy.
    - d) no energy converter can be 100% efficient.
    - e) force equals mass times acceleration.

32. A kilowatt-minute, though not a commercially used

unit, would be a unit of
   a) power. b) energy. c) force.
   d) acceleration. e) mass.

33. The conversion of energy in a certain system involves three energy conversion steps. If each individual efficiency is 80%, then the overall efficiency is
   a) 80%.
   b) the geometric mean which is 57%.
   c) 240%.
   d) 51.2%.
   e) zero.

34. In energy conversion processes, we can do a sort of "bookkeeping" to keep track of where energy is going. Which of the following is not interchangeable with energy, and so should not enter in this bookkeeping?
   a) potential energy b) kinetic energy
   c) work d) heat e) acceleration

35. In the energy converter below, the laws of physics allow you to say the energy input is ______ joules and the efficiency is ______.
   a) 500, 100% b) 1500, 50% c) 1000, 50%
   d) 1000, 100% e) 0, 0%

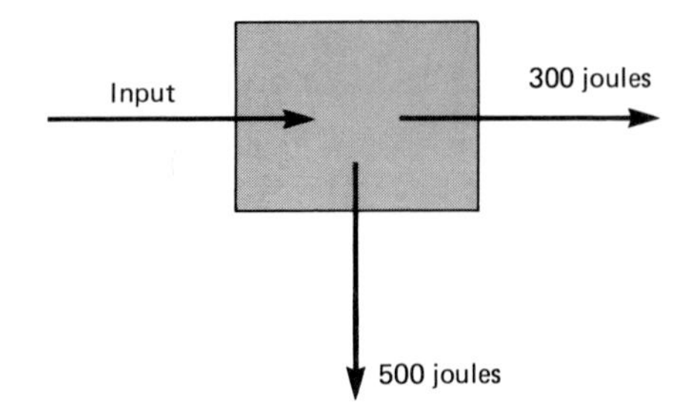

Answers

16. (a)	17. (b)	18. (a)	19. (c)
20. (c)	21. (e)	22. (a)	23. (d)
24. (e)	25. (d)	26. (c)	27. (a)
28. (d)	29. (e)	30. (a)	31. (c)
32. (b)	33. (d)	34. (e)	35. (c)

QUESTIONS AND PROBLEMS

2.1 MOTIVATION

1. A gram is a thousandth of a kilogram. How many grams are there in 12 kilograms? Devise a rule involving the decimal point for converting kilograms to grams. An ounce is a sixteenth of a pound. How many ounces are there in 12 pounds? Why can't you devise a rule involving the decimal point for converting pounds to ounces?

2. A consumer certainly would not argue against the importance of price units. It makes a lot of difference whether something costs ten dollars or ten cents per unit. Units for physical quantities are equally important. The work-energy principle equates the work done on an object to the change in kinetic energy. When applying this principle why can't you express the work in joules and kinetic energy in kilowatt-hours even though joules and kilowatt-hours are legitimate units of energy?

2.2 RATES

2.3 SPEED, VELOCITY, AND ACCELERATION

3. Some particles released from a smokestack settle toward the ground at a rate of 0.004 centimeters/second. How many hours does it take these particles to settle a distance of one kilometer? (One kilometer = 1000 meters, one meter = 100 centimeters.)

4. A modern coal-fueled electric power plant uses coal at a rate of about 4000 tons/day. Anticipating a curtailment of supplies, the company desires to have a two-month supply on hand. How much coal will be required?

5. Suppose your car obtains 15 miles per gallon of gasoline when it averages 45 miles per hour. You might also want to know how many gallons per hour the car uses at this speed. Determine this by combining the units, miles per gallon and miles per hour, in a way that gives units of gallons per hour.

6. An object falling freely in the absence of air resistance accelerates at a rate of 32 feet/second per second.
   a) Show that this corresponds to an acceleration of 35 kilometers/hour per second.
   b) A person stepping off a diving board five meters high takes about one second to reach the water. What is the person's speed in miles per hour when the water is reached if you neglect air resistance? (You may want to use this information: 39.37 inches = one meter and one kilometer = 0.62 miles.)

7. A train on the Bay Area Rapid Transit (BART) system has the ability to accelerate to 80 miles/hour in half a minute.
   a) Express the acceleration in miles per hour per minute.

b) Starting from rest, how fast does one of these trains move after 20 seconds?

8. An automobile burns gasoline at an average rate of 2.5 gallons per hour. How long will it take to deplete a 20-gallon supply of gasoline?

2.4 NEWTON'S THREE LAWS OF MOTION

9. If you had two identically shaped footballs, one filled with air, the other with sand, what differences in motion would you expect when they are kicked from a resting position on the ground with the same force?

10. Alvin Toffler in *Future Shock* writes, "The acceleration of change in our time is, itself, an elemental force." What analogies can you draw between this statement and Newton's second law of motion?

11. Shortly after a starting signal is given, a runner acquires a velocity. Newton's second law of motion requires a force on the runner to produce the acceleration. Identify the force, keeping in mind that it must act in the direction of the acceleration.

12. A pound is a measure of force. A gram is a measure of mass. It is customary to compare on the label of a can of food the contents in pounds and in grams. Why is this comparison inconsistent with the notions of force and mass?

13. Imagine two persons tugging on opposite ends of a rope. Pick out some action and reaction forces in this system.

14. A baseball acquires energy from the force of a bat hitting it. What is the reaction force to the force exerted on the ball?

15. Water at the top of a dam is pulled downward by a gravitational force. What is the reaction force to the force exerted on the water?

16. Knowing that one pound is equivalent to 4.45 newtons, express your weight in newtons. Why is it unlikely that the newton, the metric unit of weight, will catch on as a popular measure of body weight?

17. A cheerleader at a basketball game holds his female counterpart in a stationary position over his head. What is the net force on the girl and what forces contribute to the net force? How much work does he do?

18. A can of food is labeled as one pound (454 grams). What is its weight in newtons?

19. A 1200-kilogram car is cruising at a constant speed of 22 meters/second (50 miles/hour).
    a) What is the net force on the car?
    b) The car begins losing speed at a rate of one meter/second per second. What is the size and direction of the net force on the car?

2.7 ENERGY AND WORK

20. A person tugging feverishly on a rope attached to a car fails to move the car. Why has no work been done but energy has been expended?

21. In order to maintain a car at constant speed on a straight, level highway it is necessary to depress the accelerator pedal on the car. Why is the *net* work done on the car zero?

22. A child loses hold of her helium-filled balloon and it rises rapidly. Identify a force on the balloon doing positive work and a force on the balloon doing negative work.

23. Give some examples of negative work.

24. Work must be done on a car to reduce its speed from 50 to 40 miles per hour. Why is this work negative? What is the disposition of the kinetic energy when the speed is changed from 50 to 40 miles per hour?

25. Why does a compressed spring have "a capacity for doing work"? Why does it possess energy?

26. If the speed of a car doubles, what change is there in its kinetic energy?

27. Water at the top of a dam has potential energy by virtue of its position. If the height of the dam doubles, the potential energy doubles. Why won't the speed of the water after it flows to the bottom of the dam also double if the height of the dam doubles?

28. A person riding a bicycle at 12 miles per hour uses about 25 food calories for each mile of travel. How many miles of cycling at this speed are required to use up 1000 food calories of energy? Make some judgment as to the effectiveness of cycling for losing weight knowing that a person requires about 2500 food calories per day.

29. If gold is trading at $338 per ounce or 845 West German marks per ounce, how many dollars are equivalent to one West German mark? If the energy content of gasoline is 126,000 Btu per gallon or, equivalently, 133,000,000 joules per gallon, how many joules are equivalent to one Btu?

30. An air conditioner removes 5000 Btu of heat for every

kilowatt-hour of electrical energy used to operate it. Does the air conditioner run at a profit in the sense of removing more energy than is required to run it?

31. An 1100-kilogram automobile loses speed at a rate of 0.5 meters/second per second. What is the net amount of work on the automobile as it moves a distance of 800 meters?

32. A hockey puck having a mass of 0.16 kilograms skids to a stop after having been given a speed of 40 meters/second. Determine the change in kinetic energy and the net work done on the puck.

2.8 POTENTIAL ENERGY

33. An aerosol spray can used to hold household items such as deodorants, is a potential energy source. What are some other examples of potential energy in a household?

34. A roller coaster is a popular amusement park ride in which an open railroad-type car is released from an elevated position on a curved, undulating track. Discuss the potential and kinetic energy transformations that take place during the travel of a roller coaster. What are some other amusement park rides involving energy transformations?

2.9 ENERGY CONVERSION, CONSERVATION OF ENERGY, AND EFFICIENCY

35. What energy transformations are involved when a nail is driven into a board with a hammer?

36. From the time a bowler picks a bowling ball from a rack beside the bowling lane to the time the ball returns to the rack, the ball is involved in several energy transformations. Trace these transformations from the beginning to the end of the trip.

37. Water impinging on a paddle wheel imparts energy to the wheel. Why is it that all of the energy of the water cannot be transferred to the paddle wheel?

38. A car having a mass of 1000 kilograms starts from rest from the top of a hill 100 meters high. At the bottom of the hill the car is moving 30 meters/second. The energies involved in this situation are kinetic $\frac{1}{2}mv^2$, potential (mgh), and friction (call it FR).
    a) Write an equation involving the energies involved.
    b) Determine the energy due to friction.

39. A car atop a hill has potential energy mgh. Starting from rest at the top of the hill, it acquires kinetic (energy $\frac{1}{2}mv^2$) at the bottom. All the initial potential energy does not go into kinetic energy because some energy is lost through frictional forces.
    a) Show that the efficiency for converting potential energy to kinetic energy is $v^2/2gh$.
    b) Evaluate the efficiency for a hill 50 meters high if the speed is 20 meters/second at the bottom of the hill.

40. A typical oil furnace for a house has an efficiency of about 65% for converting the chemical energy of oil into heat for the house. How many Btu of heat are delivered to a house by an oil furnace burning 100 gallons of oil? (Refer to Table 2.2(a) for the energy content of a barrel of oil.)

41. A steam turbine often is used to drive an electric generator that produces electric energy. Although this involves two distinct energy conversion processes, it can be thought of as a single system that converts heat energy to electric energy. If the efficiencies of the engine and generator are 50% and 90%, respectively, show that the overall efficiency of the combined system is 45%. A straightforward way of doing this is to "feed" one unit of energy into the engine and follow it through the system seeing how much electric energy can be derived from it.

42. A steam engine converts heat energy to kinetic energy. Typically, it takes two Btu of heat energy to produce one Btu of kinetic energy.
    a) Draw an energy diagram like that shown in Fig. 2.8 which characterizes this typical steam engine.
    b) Assuming that in a given operation the engine produces 1,000,000 Btu of energy, determine the input energy and diverted energy and label these on your diagram.
    c) What is the efficiency of this engine?

2.10 POWER

43. A typical household air conditioner is rated as 8000 Btu. This rating alludes to the quantity of heat that it is capable of removing from a room. Although time is not mentioned explicitly in this rating, why is time important as far as the performance of the air conditioner is concerned?

44. Why would power expressed in megawatts multiplied by a year be a unit of energy?

45. How many joules of energy are delivered in one hour by an engine developing 10 horsepower?

46. A person applies a 200 newton vertical force to lift a box through a distance of one meter.
    a) How many joules of work are done?
    b) How much power is developed if the work is accomplished in four seconds?
    c) How does this power compare with the power developed by a motor on a household food mixer? (See Section 2.10.)

2.11 OTHER COMMON ENERGY AND POWER UNITS

47. Burning a pound of coal produces about 13,000 Btu of energy. How many joules of energy are produced by burning a pound of coal?
48. A person derives energy from food at the rate of about 2500 food calories per day. Show that this is equivalent to a consumption rate of about 100 watts.
49. Heat energy for a steam engine that drives an electric generator is normally expressed in British thermal units (Btu) but the electric energy output is normally expressed in kilowatt-hours (kWh). Knowing that a kWh equals 3,600,000 joules and 1 Btu equals 1055 joules, show that a kWh equals 3413 Btu.

3

Fossil Fuels

3.1 MOTIVATION

At the turn of this century a student probably worked and slept in a room heated by burning a prehistoric remnant of organic matter. So it is today and probably will be at the close of this century. But in 1901 the remnant would have been coal and today it is likely to be natural gas or oil. Fossil fuels have dominated our energy resources for over a century, and they could well be dominant for another century even though their abundances and availabilities are changing dramatically. Industries dependent on the use and supply of fossil fuels strive to satisfy consumers and to provide controls for unwanted by-products of combustion. It is essential that we understand the role of the fossil fuels and strive to use them in the most efficient and environmentally acceptable manner.

This area in Wyoming was once a coal mine pit. Land reclamation laws require restoration of the areas following removal of coal. (Union Pacific photograph from the Department of Energy.)

3.2 PROVEN AND UNPROVEN RESERVES

We read often that the United States reserves of oil and natural gas will be exhausted by the year 2000. A student in 1920 may well have read that the reserves would not last beyond 1980. These statements need careful scrutiny because they require an understanding of the meaning of reserves and depletion.

A proven energy reserve is not necessarily one in which a fuel has been gleaned from the earth and set aside for use as consumers see fit. Usually a proven energy reserve consists of an energy commodity in the ground ready for extraction. Detailed geological analyses, such as borings to see if oil or coal is present, have been done at the proven reserve sites. The energy commodity is available at current prices and extractable with current technology. Clearly, proven energy reserves will increase if new discoveries are made. American crude oil reserves increased substantially when the east Texas oil fields were discovered in 1930 and when oil was discovered in northern Alaska in 1968 and became a proven reserve in 1970. But proven reserves also increase if economic conditions make it attractive to recover theretofore economically unattractive resources, or if the removal technology improves. An energy reserve never gets depleted in the sense that the last ounce of fuel is removed. Rather there is a point in time when it is no longer economically attractive to mine or drill.

Apart from known proven reserves there are several kinds of less certain, unproven reserves. Some unproven reserves are inferred from geological studies and exploration experience. Other unproven reserves exist in areas where the technology for extracting the energy commodity has not been developed. Conventional oil removal technology averages only 30% extraction of the oil from an established well. The 70% remaining in the wells is an unproven oil reserve. If oil removal technology improves or if the price of oil improves sufficiently, these unproven reserves advance to proven reserves and the oil interests look to them for supplies. Another important unproved oil reserve is the huge amount of oil that might be gleaned from the shale deposits in the western United States if the technology for removing the oil can be developed in an economically and environmentally acceptable manner.

3.3 COAL

Solar energy for heating homes or water generally provokes thoughts of solar collectors mounted on rooftops. We could just as easily classify the heat from burning coal, oil, or natural gas as solar energy because the energy liberated has its origin in collected solar radiation. But the solar energy for making the fossil fuels did not fall on the earth this year or even a century ago. It was energy that prompted the growth of plants and animals hundreds of millions of years ago. From this living matter came the fossil fuels. The story is interesting and fundamental to understanding the dilemma we face in contemporary energy problems.

The matter in our living environment takes three forms—gas, liquid, and solid. A gas is distinguished from liquids and solids by being easily compressed. Thus you inflate a balloon and readily squeeze the balloon with your hands. Liquids and solids are not easily compressed and a solid can be sliced into pieces. Changes between these forms take place all around us. Solid ice melts into liquid water. Liquid water evaporates into gaseous water vapor. These two transformations involve an input of energy. Ice forms on a pond when energy is removed from the water as heat flows from the water to the cold air around it. When ice melts, the energy absorbed is literally stored in the water and the water becomes an energy reservoir.

Photosynthesis is another important process involving energy and matter. In photosynthesis gaseous carbon dioxide,* liquid water, and solar energy interact to produce solid carbohydrates and gaseous hydrogen (Fig. 3.1). Sugar and starch are carbohydrates. So is cellulose, the principal structural material of wood. Like water, carbohydrates

* Carbon dioxide is a common gas that we will learn more about later. It is released when a person exhales, for example.

FIGURE 3.1 The arctic poppy orients its cup of petals toward the sun to collect energy for its growth by photosynthesis.

are an energy reservoir having captured and converted the energy from the sun. We might construct a word equation for photosynthesis as

$$\text{carbon dioxide} + \text{water} + \text{solar energy} \rightarrow \text{carbohydrates} + \text{oxygen}$$

where the arrow is read as "to form." When this process is reversed by combining the carbohydrate with oxygen, the energy stored on formation is liberated and so is carbon dioxide and water. This happens when wood is burned in a fireplace or when a carbohydrate in food we have eaten is decomposed by our bodily functions to provide energy for living and working. Because oxygen is involved in the decomposition, the process is termed oxidation. It can also be thought of as burning whether or not a flame is involved. We burn wood to produce heat and we burn carbohydrates in our bodies to produce energy.

When a plant dies, it decomposes and, in the presence of oxygen, reverts to its initial constituents. Thus in the long run there would be no net accumulation of plant material if oxidation conditions exist. But nature provides some oxygen-deficient conditions in which plant matter is able to accumulate. But the imbalance between formation through photosynthesis and decomposition through oxidation is so slight that it takes millions of years to amass significant amounts of material. Coal is the end product of the accumulation of plant matter, and while its formation requires millions of years, its depletion takes only hundreds of years.

The coal burned today had its origin in swamps that existed from one to 440 million years ago. Giant ferns and other plants that died in these swamps were covered with water that limited their exposure to oxygen and stifled oxidation. Bacterial action decomposed the matter in a way that removed oxygen from the fallen debris. Under pressure from the weight of the water, the decomposed matter layered producing peat, a brownish solid material (Fig. 3.2). In Ireland peat is extracted, dried, and burned as a fuel, albeit a fuel of low quality. Significant deposits

FIGURE 3.2 Peat is a common mulch for gardens. You can usually see plant remnants in peat and feel the moisture in it.

of peat are found in the United States but the peat in them is used mostly for a gardening material. However, there are some efforts to exploit peat as a fuel.

As geological conditions changed, sand accumulated on the buried plant material, sandstone or shale formed, and the overcovering water flowed away. The weight of the solid overburden compacted the buried material squeezing water from it and forcing decomposition in an oxygen-deficient environment. As time progressed, lignite, a solid brown material, was formed and then subbituminous coal, bituminous coal, and anthracite (Fig. 3.3). Each of these types of coal is composed primarily of varying proportions of carbon and hydrogen in a very complex arrangement. Additionally, there is included a wide variety of elemental impurities, notably sulfur, that produces environmentally damaging sulfur oxides when burned. Lignite and anthracite are roughly 30% and 90% carbon, respectively. The carbon content of bituminous coals ranges between 30% and 90%. The moisture content decreases from lignite to anthracite. The heating value ranges from about 8300 Btu per pound for lignite to about 15,000 Btu per pound for quality bituminous and anthracite coals. Most United States coal is bituminous. Only 1% is anthracite. Layers or seams of coal range from an inch to upwards of 400 feet thick and are generally covered with tens or hundreds of feet of sandstone, shale, or other solid overburden. To be economically profitable a seam must be more than 24 inches thick. United States coal is located in three broad regions referred to as Appalachia, the Central Region, and the Rocky Mountain Region (Fig. 3.4). About 75% of the high quality bituminous coal is located in Appalachia and the Central Region. Lignite and subbituminous coal is most often found in the Rocky Mountain Region.

FIGURE 3.3 A close look at a piece of coal reveals a subtle beauty that American Indians once exploited in trinkets. Unlike peat, its much younger counterpart, coal shows little evidence of moisture. (AMAX Coal Company photograph courtesy of the National Coal Association.)

Coal is recovered from the ground from either tunnellike structures leading to the deposits or by stripping the overburden and exposing the coal seam. Both methods are highly mechanized and rather spectacular (Fig. 3.5). Strip mining accounts for about 60% of all the coal mined. Strip mining is performed in areas in which the coal layers lie reasonably close to the earth's surface. Beneath the dirt surface there generally lies layers of rock that must be loosened by detonating explosives. A strip of the rock-laden overburden, called spoils, is removed by mechanical shovels and formed into a row of piles. The exposed strip of coal is dug out, loaded into trucks, and processed for the marketplace. When a strip in the coal seam has been exhausted, a parallel strip is exposed. The overburden from the second strip is dumped into the space vacated by the coal extracted from the first strip. In hilly regions the spoils are removed from the sides of the hills to expose the coal seam. The coal is then removed by screwlike augers that bore into the coal seam from the side. When the overburden is too thick to be removed economically, tunnellike structures (shafts) are built so that equipment and miners can reach the coal seam and ultimately bring the coal to the surface. Historically the accumulation and abandonment of the overburden in strip mining led to severe environmental degradation because of acrid solutions evolving from water interacting with the spoil piles. Environmental concerns led to passage of the Surface Mining Control and Reclamation Act in 1977 requiring a coal

FIGURE 3.4 In the United States, coal is spread over three broad regions called Appalachia, the Central Region, and the Rocky Mountain Region. Most of the low-sulfur content, subbituminous coal is in the Rocky Mountain Region. The high-sulfur content, high-heat content coal is concentrated in Appalachia. Significant deposits of lignite are found in the upper Great Plains area of the Central Region.

mining company to restore areas mined after 1977 to their initial conditions as nearly as possible (Fig. 3.6).

Coal is America's most abundant fossil fuel in terms of total energy availability. Taking a proven reserve estimate of 164 billion tons (Table 3.1) and comparing it with a 1981 production rate of 0.82 billion tons per year, the amount of available coal is impressive. Thus it is tempting to expand coal use beyond its 21.7% contribution in 1981. But considering that 81.9% of this coal was used to produce electricity and the remainder to make coke (8.4%), for industrial uses (8.8%), and for other uses (0.9%) the options for use are limited. Expansion in the electric utilities sector will result from replacement of oil-fired and gas-fired systems and growth in demands for electricity. As the technology for converting coal to liquid and gaseous fuels progresses, demands for coal will increase. But rapid expansion is thwarted by two important considerations. First, the production of sulfur oxides, nitrogen oxides, and carbon dioxide on burning poses serious environ-

FIGURE 3.5 Coal is being extracted from a seam some 70 feet thick. A single scoop of the shovel fills a waiting truck with nearly 50 tons of coal. Some feel for the size of the operation can be obtained by comparing the size of the personnel with other features in the photograph. (AMAX Coal Company photograph courtesy of the National Coal Association.)

FIGURE 3.6 Environmental laws require that strip-mined regions be restored for productive use. This area has been graded and reseeding is in progress. (Photograph courtesy of the National Coal Association.)

mental threats. Second, the coal industry is very transportation-dependent and much of the desirable low-sulfur coal is in the Rocky Mountain Region where environmental concerns exist for disturbing pristine areas.

3.4 PETROLEUM

Like coal, petroleum is a remnant of organic matter. Unlike coal, petroleum

1. may be solid (for example, kerogen, the oil-producing material in shale), liquid (crude oil), or gaseous (natural gas),
2. is formed mostly from creatures rather than plants,
3. is chemically different although both coal and petroleum are composed mostly of hydrogen and carbon.

Because hydrogen and carbon are bound with oxygen in plants and animals, petroleum formation requires conditions for separating oxygen from dead organisms. A commonly accepted scheme views countless dead marine creatures accumulating on the bottom of large bodies of water forming the petroleum source bed. Sandstone, a rather porous rock, forms over the source bed, exerting pressure on the organic accumulation. Heat from the pressure and heat from the earth's interior enhance the

TABLE 3.1 United States proven reserves and 1981 production rates for oil, natural gas, and coal. Data for proven reserves are from "1981 Annual Report to Congress," U.S. Department of Energy (May 1982). Production rates are from "Monthly Energy Review," U.S. Department of Energy (1982). Proven reserves for 1973 and 1976 are from Department of Energy sources. The fact that the reserves are declining is significant and is likely characteristic of the estimates provided by any source. A historical account of the proven reserves of natural gas and oil is shown in Figs. 3.9 and 3.10.

Fossil fuel	Proven reserves			Production rate in 1981
	1980	*1976*	*1973*	
Oil (billion barrels)	29.8	30.9	35.3	3.13 billion barrels per year
Natural gas (trillion cubic feet)	199	216	250	20.2 trillion cubic feet per year
Coal (billion tons)	475			0.820 billion tons per year

petroleum formation. Petroleum is forced out of the source bed and into the sandstone and escapes unless some peculiar geological disturbance forms a trap (Fig. 3.7). There are several kinds of traps but three are notable—fold, salt dome, and fault (Fig. 3.8). In a fold, the petroleum-bearing sandstone lies between two nonporous layers of rock, the totality of which is folded into a humplike structure. Pressure on the sandstone squeezes the petroleum into the hump where its movement is halted by the nonporous stone cap on the hump. A salt dome trap is produced by the upward squeezing of salt from deep-down salt formations. A fault trap results when the petroleum-bearing formation is fractured into two segments, followed by movement of the segments along the fracture. Petroleum migrating toward the surface of the earth in a sandstone layer finds its movement blocked by a nonporous rock.

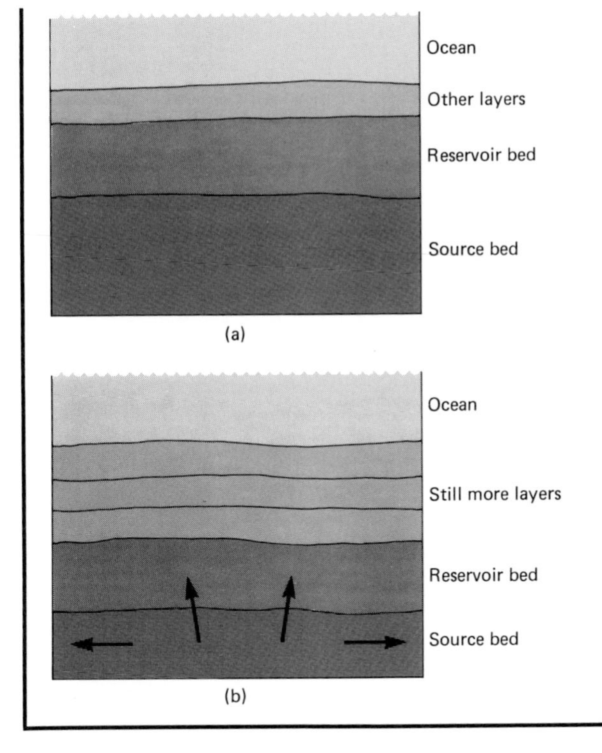

FIGURE 3.7 **(a)** Marine creatures accumulate in a source bed consisting of silt and mud. A reservoir bed of porous material such as sandstone forms over the source bed. **(b)** Pressure from the weight of material over the source bed squeezes petroleum into the reservoir bed.

Petroleum is pumped from the trap through a pipe in a drilled hole of upwards of 22 inches in diameter. Considering that traps are generally no more than a few miles on a side and are obscured by more than a mile of overlying material, including water in offshore regions, locating traps requires impressive techniques. In addition to relying on their experience, petroleum explorers also rely on subtle measurements of the earth's magnetism, concentrations of important rock types, and analysis of earthquakelike wave disturbances (seismic waves) purposely created in the earth. One of four drillings in 1980 produced a commercial petroleum well and only one in fifty identified a significant new reserve. The average well depth was 5870 feet, more than a mile, the deepest was 26,518 feet, nearly five miles. Gleaning petroleum from the harsh, north-shore environment of Alaska is indicative of the difficulty in locating new petroleum sources. But this is a fact of life because it is believed that 33% of undiscovered petroleum is in Alaska and another 32% is located in offshore ocean regions. Twelve states are considered to be major petroleum producers but in 1980 Texas, Louisiana, California, and Alaska accounted for 76% of the crude oil and Texas, Louisiana, Oklahoma, and New Mexico are credited with 70% of the natural gas recovered.

Figures 3.9 and 3.10 show United States crude oil and natural gas reserves beginning in 1950. Alaskan discoveries are responsible for the sharp increase in both crude oil and natural gas in 1970. But even when Alaskan oil became available in 1978, the dependence on foreign imports to satisfy demands did not vanish. Deregulation of the price of natural gas and crude oil spurred exploration activity but it is unlikely that the United States will ever meet all its petroleum needs through drilling. This country must either continue to import crude oil or develop alternatives. Importing is unattractive not just because the major producers are in politically unstable nations but also because worldwide resources in general are not overly abundant (Fig. 3.11).

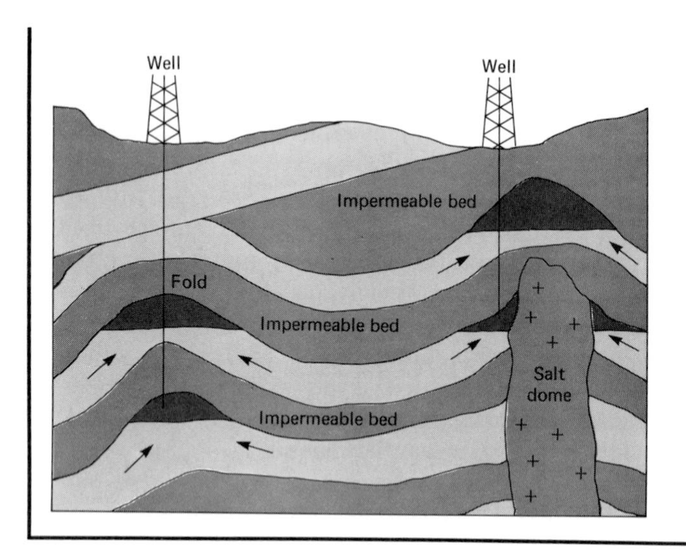

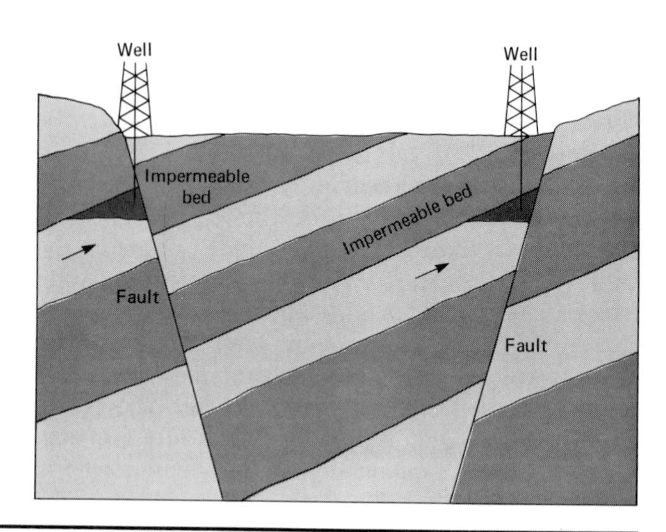

FIGURE 3.8 Sketches of three important petroleum traps. Each trap requires a mechanism for halting petroleum movement in the porous sandstone reservoir bed. In a fold, a nonporous rock such as limestone caps the fold. A salt dome utilizes the impermeability property of rock and salt. In a fault trap, an impermeable layer slides along a fracture in the reservoir bed and prevents upward migration of petroleum.

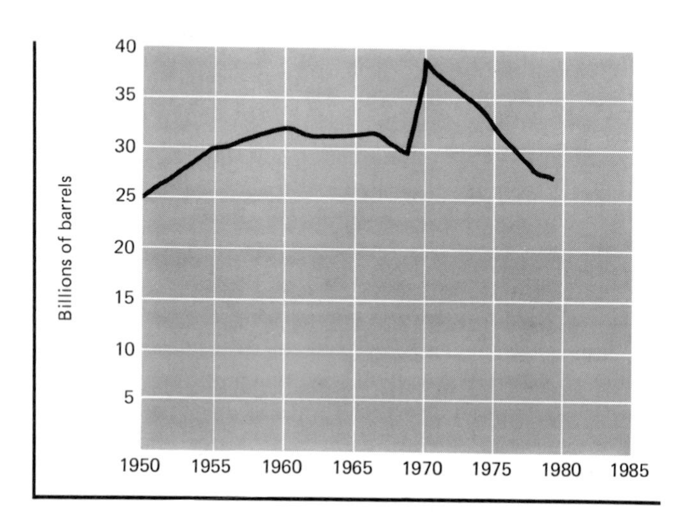

FIGURE 3.9 Proven United States crude oil reserves expressed in billions of barrels. The sharp increase in 1970 is a result of discovering petroleum on the north shore of Alaska. The reserves have declined steadily since the Alaskan discovery. The data are from the *1981 Annual Report To Congress*, Energy Information Administration, United States Department of Energy. They are presented in Table A.6.

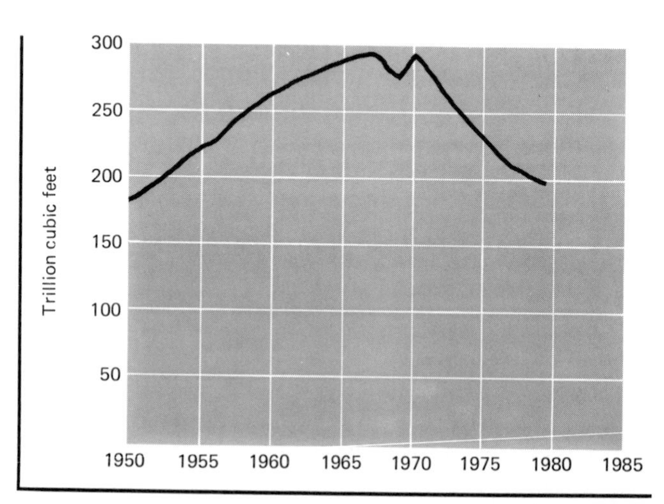

FIGURE 3.10 Proven United States natural gas reserves expressed in trillions of cubic feet. Although discoveries in Alaska produced a noticeable increase in 1970, the reserves have since declined steadily. The data are from the *1981 Annual Report to Congress*, Energy Information Administration, United States Department of Energy. They are presented in Table A.6.

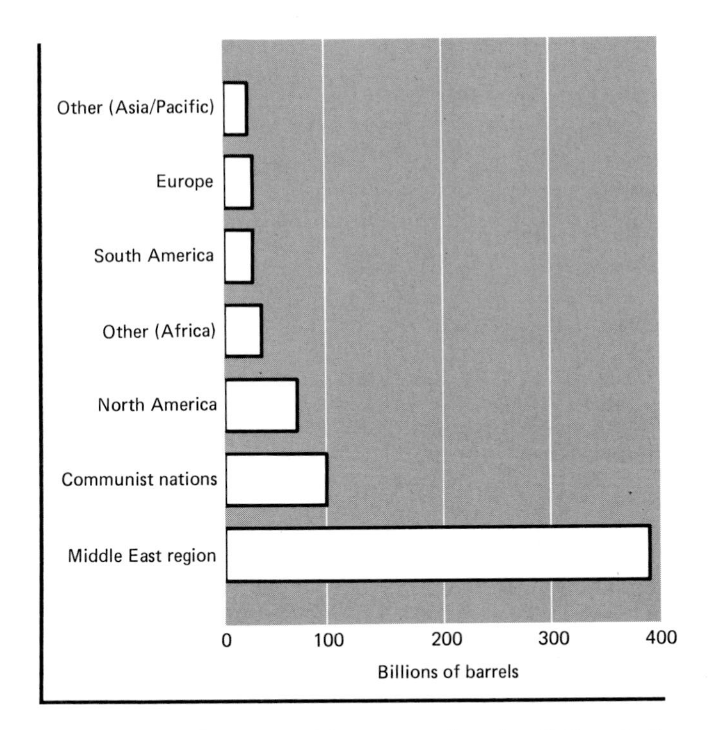

FIGURE 3.11 Proven world reserves of crude oil. Over half the reserves are in the Middle East region. The data are from *The Oil and Gas Journal,* December 31, 1979.

3.5 OIL SHALE AND TAR SANDS

Shale is a layered, rock structure formed from the deposition of fine-grained particles on the bottoms of ancient seas. Oil shale trapped a wide variety of organic material that developed ultimately into a solid fossil remnant called kerogen. When oil shale is heated to temperatures in the range 800–1000° F, the kerogen breaks down into a hydrocarbon liquid that is very similar to crude oil. This shale oil can be refined like crude oil to produce gasoline, diesel fuel, heating oil, and so on. High-grade shale produces 25–35 gallons (about 0.6 of a barrel) of shale oil for each ton of oil shale processed. In perspective, a cube of shale 2.5 feet on a side weighs about a ton. Visualize the size of a gallon of milk for the size of a gallon of shale oil. The vast Green River Formation in Colorado, Utah, and Wyoming is the richest oil shale formation in the United States. Because of the large extent of the deposit and the variation in kerogen content, estimates of recoverable oil vary from 500 billion to a trillion barrels. Bear in mind that proven United States crude oil reserves are measured in tens of billions of barrels, not the hundreds of billions number used for shale oil estimates. But assessing the shale oil resource and converting it to a refinable product are two very different issues. Shale oil recovery faces technological, environmental, economic, and social roadblocks that may delay indefinitely its becoming a significant proven energy reserve.

Generally, shale oil recovery involves mining, although some schemes envision recovery without mining or at least with limited mining. In a mining operation, shale is removed from the ground either by strip or shaft mining, crushed, and heated in a vessel where the shale oil is recovered. In-place recovery requires explosive fracturing of the formation so that the shale can be heated in place. A combination of in-place recovery and mining starts with a shaft–mine operation to remove sufficient amounts of shale. Having opened caverns in the shale deposit, fracturing and in-place recovery follows. To produce 100,000 barrels of crude oil a day in a strip-mining operation requires processing 500,000 tons of shale a day, a mammoth operation! The waste which fluffs to a larger volume than that occupied by the shale presents a disposal problem. Processing requires significant quantities of water, a scarce commodity in the regions where the best shale is located. Possible pollution of streams by runoff from the wastes is also of concern. Because the industry would be very labor intensive, living communities and services are required for the work force and their families. The sociological considerations for establishing communities in remote areas may be the biggest deterrent to exploitation of oil shale resources. Although the shale deposits and the basic techniques for shale oil recovery have been known for more than half a century, no significant commercial production has developed.

Tar is a thick, oily, petroleum liquid used as a binder for sand and gravel in some types of roads. Sandstone containing natural tar in its porous structures is termed tar sand. High-temperature steam can force the tar from the sandstone. Then it can be refined into liquid petroleum. Like the yield from a high-quality oil shale, the yield from tar sand is about 25 gallons per ton of material processed. Substantial tar sand resources are found in North America but those in the United States are limited. The richest tar sand deposits occur in northern Alberta, Canada. Estimates of the amounts of extractable oil are in the range of billions of barrels. Limited commercial exploitation of the Canadian tar sands has taken place, but a significant production of crude oil is hampered by unfavorable economics.

3.6 EXPONENTIAL GROWTH OF ENERGY USE

Change is a part of life. Students change majors. The travel time from New York to San Francisco has changed from a few days by train to a few hours by plane. The number of automobiles on the highways more than doubled between 1950 and 1970. Unfortunately, some things change at exponential rates that can lead to disaster if unchecked. Energy consumption and population growth typify exponential changes.

There are several ways of examining exponential behavior. The following description works well for our purposes. If a quantity doubles in value at regular intervals of time, it is said to be increasing exponentially. To illustrate, suppose you opened a bank account with a very modest deposit of one cent. Then each day for a month a benefactor adds to your account twice the previous deposit. Thus the first day two cents are added, the second day four cents, the third day eight cents, and so on (Fig. 3.12). At week's end you are not wealthy but if the benefactor sticks to the agreement for 30 days, you would have over $20,000,000! A small beginning led to a highly profitable ending. Note that each deposit is just one cent more than the total of the account just prior to the deposit. Thus when the 1,073,741,824 cents were added on the 30th day, the total prior worth of the account was 1,073,741,823, or one cent less. Essentially half the total accumulation comes from the final deposit. When something like a bank account changes exponentially, very large changes occur in short intervals of time once a significant amount of that quantity has accumulated. Eventually, maintaining exponential changes becomes unrealistic.

	Deposit	Accumulation
Opening		1 cent
Day 1	2 cents	3 cents
Day 2	4 cents	7 cents
Day 3	8 cents	15 cents
•	•	•
•	•	•
•	•	•
Day 29	536,870,912 cents	1,073,741,823 cents
Day 30	1,073,741,824 cents	2,147,483,647 cents

FIGURE 3.12 The Exponential Bank Account. The account opens with a one-cent deposit. Each successive deposit is twice the previous one. After 30 deposits, over $20 million accumulates. Half the accumulation resulted from the final deposit.

Quantities can decline exponentially as well as grow exponentially. You may win $16,384 in a lottery and decide to spend $1 the first day, $2 the second day, $4 the third day and so on. The account does not lose much value those first few days but on the 14th day you spend $8,192 and your windfall is reduced to $1. Both increasing and decreasing exponential changes are very important in energy considerations.

Working with the doubling time simplifies exponential accounting. However, the exponential rate is often given rather than the doubling time. But don't despair. The mathematics of exponential behavior leads to a useful relation between doubling time and the exponential rate. The doubling time is determined by dividing 69.3%, or 70% for approximate figuring, by the rate expressed as a percentage per unit time. For example, if something is growing exponentially at a rate of 3.5% per year, the approximate time for that something to double is

$$\text{doubling time} = \frac{70\%}{3.5\%/\text{year}} = 20 \text{ years.} \qquad (3.1)$$

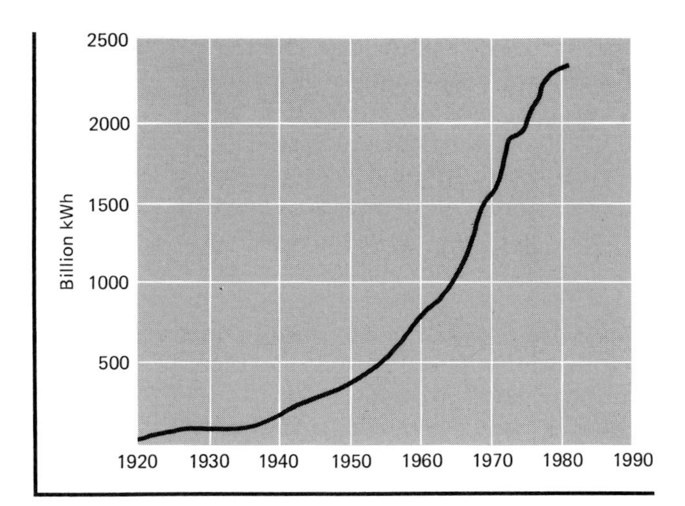

FIGURE 3.13 United States annual production of electric energy. The energy units are in billions of kilowatt-hours (kWh). A kWh is the energy used in ten hours by a hundred-watt desk lamp. The data are from the Department of Energy. They are presented in Table A.7.

If the rate were 3.5% per minute, the approximate doubling time is 20 minutes.

The growth of electric energy production in this country is a classic example of exponential behavior. Beginning in 1935, electric energy production doubled about every ten years (Fig. 3.13). A calculation dramatizing the growth of electric energy use determines the time required for the United States to be covered by electric power plants if the production continues to double every ten years. In 1981, all of the electric energy in the United States could have been produced by 400 large plants. Each plant could be built on a square of land about 1000 feet (about 0.2 miles) on a side so that all 400 plants require about 16 square miles, representing the area encompassed by an average size city. The contiguous United States is about 3000 miles long and 1000 miles wide so that its total area is about 3 million square miles. But if the growth of electric energy were to double every ten years, it would take about 18 doublings of 180 years for electric power plants to completely cover the United States. This cannot happen, but until 1973 the trend was that way.

As noted, the change in any doubling period of an exponential process exceeds the prior accumulation by an amount equal to the initial amount. Thus if an exponential growth starts with x units the change and prior accumulation differ by x units. This means that between 1963 and 1973 more electric energy was used than the entire period up to 1963. This frightening fact points to the burden placed on the electric energy industry to supply the energy if the growth maintains the exponential trend.

Total energy consumption betwen 1950 and 1973 grew at a rate of about 3.2% per year. Subsequently this rate has fluctuated. The future course is speculative but the rate probably will not return to the pre-1973 value of 3.2% per year. Whatever the growth, the rate is significant and has important implications for the lifetime of the fossil fuels. To illustrate, consider the following example.

If the amount of oil in a given reserve were actually twice the original estimate, and if the oil is taken at an exponential rate, does that mean the lifetime of the reserve is doubled because there was twice the anticipated reserves? No, it does not! Consider the time it takes to deplete to one billion barrels a reserve thought to contain eight billion barrels if the amount taken doubles every five years starting with one billion barrels during the first five-year period. After 15 years, seven billion barrels would be taken and one billion barrels remain (Fig. 3.14). Convince yourself that if the reserve contained 16 billion barrels rather than eight billion barrels, it would take 20 years to reduce the reserve to one billion barrels. Hence twice the amount of oil in the reserve increases its lifetime

	Remaining	Production
Start	8	0
5 years	7	1
10 years	5	2
15 years	1	4

FIGURE 3.14 Depletion of an eight-billion barrel crude oil reserve at a hypothetical exponential rate of 14% a year.

by only five years or one doubling period. This point is an extremely important aspect of exponential behavior. There may be large uncertainties in estimates of the reserves, but it makes only small variations in the lifetime if they are being depleted exponentially. It also means that although the Alaskan oil discovery may yield 50 billion barrels, it doesn't add significantly to the lifetime of the reserves if consumption continues to grow exponentially. One of the most difficult decisions facing our society is whether we really need to continue this exploitation of energy resources.

Not everyone will agree on the estimates for lifetimes of energy resources, but all must accept the fact they are finite. When you continually draw from a finite source, the useful lifetime is limited. On a smaller scale this is exactly what happens when a gold deposit is worked. Following discovery gold is easy to recover. More mining equipment and miners are brought in, the production rate rises, and the local economy flourishes. As the deposit nears depletion, gold is harder to find and the production rate declines. Equipment and miners gradually move out tending to further decrease the production. Time passes and the production rate fades away. Perhaps it does not vanish completely because a single miner may stick it out making a livelihood from the scattered gold remnants. Nevertheless, the local economy dies and a ghost town remains. In the same sense, oil production is not likely to ever come to a complete halt. Hence the notion of a lifetime is a little misleading. But as the reserves deplete and it becomes harder and harder to find commercial wells, the production rate will decline and fade away. M. King Hubbert, a highly regarded geophysicist who has spent a large portion of his life estimating oil reserves, believes that oil production in the United States has peaked.* Hubbert reasons that the exponential rise seen in oil production will be followed by an exponential decline. Oil production declined from 1970 to 1976. Increased production from the Alaskan fields subsequently produced a slight increase that leveled off and remained essentially constant from 1978 through 1981. If Hubbert is correct and production eventually declines as anticipated, 80% of the United States reserves of oil will be used in a 65-year period, less than the average lifetime of a person (Fig. 3.15).

The pattern of energy use is symbolic of many other facets of our society that are changing exponentially. Population in many countries grows exponentially. There have been periods of exponential population growth in the United States but, while the population is still growing (Fig. 3.16), the rate is less than exponential. In fact the birthrate in the United States has reached the level needed for zero population growth.* One might suspect that the growth in energy use reflects only the growth in population. However, this is not the case as is

* See, for example, "Energy and the Environment," John M. Fowler. In Lon C. Ruedisili and Morris W. Firebaugh (eds.), *Perspectives on Energy,* Oxford University Press, New York, 1982.

* See, for example, "The Populations of the Developed Countries," Charles F. Westoff, *Scientific American* **231,** 3 (September 1974): 108.

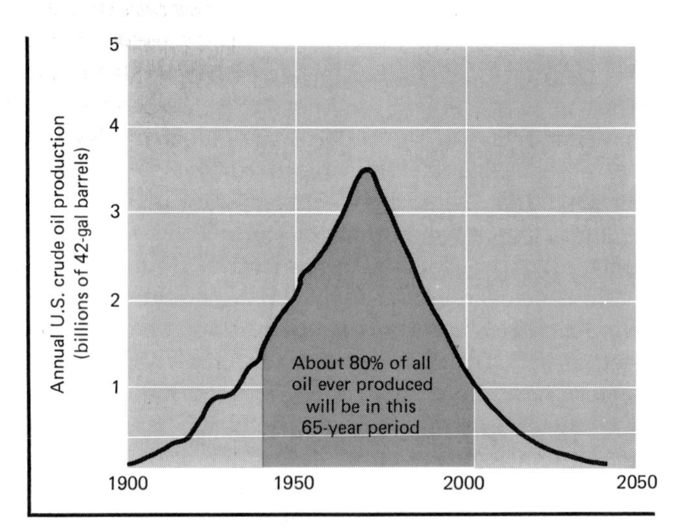

FIGURE 3.15 A possible way that consumption of crude oil will progress in time. The heavy black area represents production of 80% of the anticipated total lifetime production. The time period for 80% production is about 65 years—roughly the lifetime of a United States citizen.

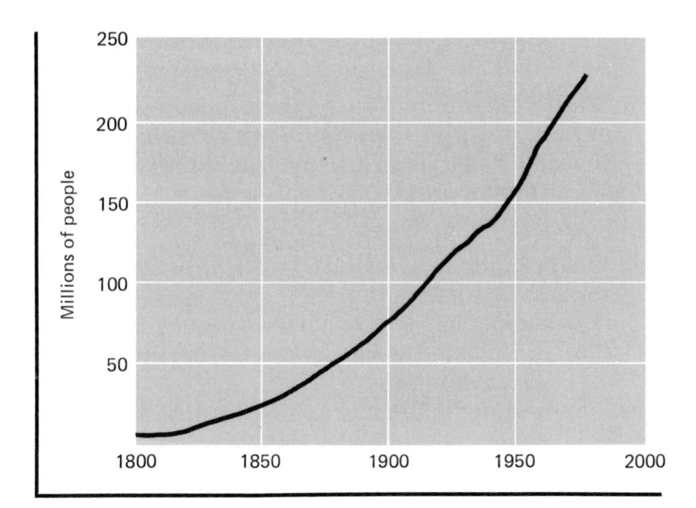

FIGURE 3.16 United States residential population expressed in units of millions of people. The data are from *Statistical Abstracts of the United States,* 1981 edition.

illustrated in Fig. 3.17, which shows energy use per person. Realize that even if energy use per person is constant, energy consumption increases so long as the population increases. We are demanding more energy to run our automobiles, to air-condition our homes, to manufacture our convenience items, etc. The growth of many of these convenience items, such as aluminum cans, nonreturnable bottles, and plastic containers, is nearly exponential, and this growth has aggravated the disposal problems and placed more demands on energy sources for their production. The exponential depletion of minerals and metals, such as lead, zinc, and iron, is also of great concern. Exponential exploitation of finite resources can lead only to disaster.

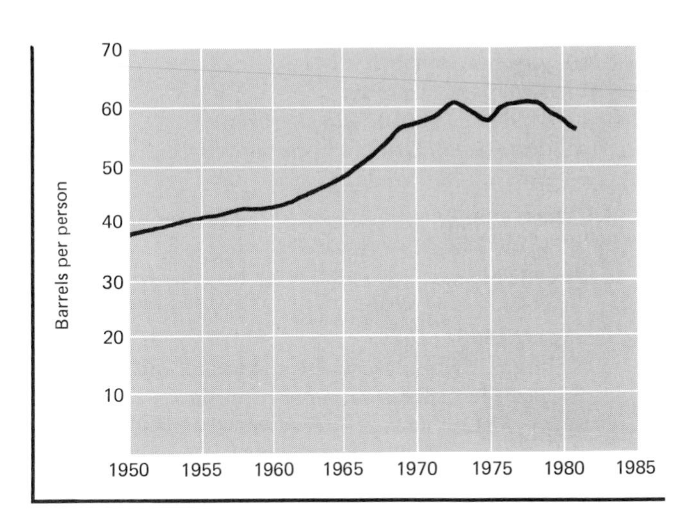

FIGURE 3.17 United States annual per capita total energy consumption. The leveling off in per capita energy use around 1973 is produced by an overall decrease in total energy use. The energy data are from the Department of Energy and the population data are from the *Statistical Abstracts of the United States,* 1981 edition.

REFERENCES

FORMATION AND USE OF FOSSIL FUELS

The origin of fossil fuels, including oil shale and tar sands, and extraction methods are discussed in *Man and His Geologic Environment,* David N. Cargo and Robert F. Mallory, Addison-Wesley, Reading, Mass., 1974.

"The Origin of Petroleum," John M. Hunt, *Oceanus* **24,** 52 (Summer 1981).

"Toward a Rational Strategy for Oil Exploration," H. William Menard, *Scientific American* **244,** 55 (January 1981).

"The Chemistry and Technology of Synthetic Fuels," Alan Schriesheim and Isidor Kirshenbaum, *American Scientist* **69,** 536 (September–October 1981).

EXPONENTIAL GROWTH OF ENERGY USE

"World Oil Production," Andrew R. Flower, *Scientific American* **238,** 42 (March 1978).

"Forgotten Fundamentals of the Energy Crisis," Albert A. Bartlett, *American Journal of Physics* **46,** 876 (1978).

"Life Expectancy and Population Growth in the Third World," Davidson R. Gwatkin and Sarah K. Brandel, *Scientific American* **246,** 57 (May 1982).

"Coal," *The Worldbook Encyclopedia.* This seventeen-page article can be obtained by writing the National Coal Association, Coal Building, 1130 Seventeenth Street NW, Washington, D.C. 20036.

A Guide to Petroleum Exploration and Production, a 30-page booklet, can be obtained from the Exxon Corporation, 1251 Avenue of the Americas, New York, N.Y. 10020.

"Understanding Nonrenewable Resources Supply and Behavior," Douglas R. Bohi and Michael A. Toman, *Science* **219,** 927 (25 February 1983).

REVIEW

1. Describe the process of photosynthesis.
2. Write a short discourse on the formation of coal and petroleum.
3. Describe three types of petroleum traps.
4. Distinguish between proven and unproven reserves.
5. Which is the most plentiful of the fossil fuels?
6. What are some of the difficulties involved in estimating the lifetime of the fossil fuel reserves?
7. Why is it impractical to entirely deplete an energy resource?
8. What is meant by exponential growth? Why is it of vital importance in energy considerations?
9. Define doubling time. How is doubling time related to growth rate?
10. What are tar sands and oil shale? Why are they of interest?
11. Pick the equation below that describes photosynthesis.
    a) carbohydrate + carbon dioxide + energy → oxygen + water.
    b) water + oxygen + energy → carbon dioxide + carbohydrate.
    c) carbon dioxide + water + oxygen → energy + carbohydrate.
    d) carbon dioxide + water + carbohydrate → energy + oxygen.
    e) carbon dioxide + water + energy → carbohydrate + oxygen.
12. A certain photosynthesis process requires one unit of energy in combining water and carbon dioxide to form a carbohydrate and oxygen. When the process is reversed and the carbohydrate and oxygen combine, we can expect
    a) one unit of energy will have to be added to the carbohydrate and oxygen.
    b) two units of energy to be liberated because each side of the photosynthesis equation has two types of material products.
    c) no energy to be released.
    d) one unit of energy to be released.
    e) that water and carbon dioxide will not be formed when the carbohydrate and oxygen combine.
13. When a carbohydrate decomposes by interacting with oxygen, we might characterize this as burning even though a flame is not involved because carbon dioxide and water are produced either way.
    a) true b) false
14. Energy has to be absorbed by ice in order for the ice to melt. This energy
    a) escapes immediately to the surroundings.
    b) is converted to heat in the ice and escapes to the environment.
    c) is stored internally in the water making the water a reservoir of energy.
    d) is converted to nuclear energy in the atoms of the water.
    e) The statement is wrong. Energy does not have to be absorbed by the ice.
15. For any fossil fuel to form, there is required
    a) an oxygen-deficient environment.
    b) an environment rich in sulfur.
    c) a complete lack of sunshine.
    d) a plentiful amount of nitrogen.
    e) an environment rich in carbon dioxide.
16. Which of the groups listed below fall into the category of petroleum?
    a) gasoline, kerosene, wood
    b) coal, peat, uranium
    c) crude oil, natural gas, kerogen
    d) distillates, diesel fuel, carbohydrates
    e) coal tar, tar sands, steam
17. Of those energy supplies listed below, the one used most in electric power plants is
    a) petroleum b) natural gas c) wind
    d) trash e) coal
18. Coal is a result of
    a) slight imbalances in the production and decomposition aspects of photosynthesis.
    b) converting nuclear energy into chemical energy.
    c) concentrating carbon dioxide released in the decay of plants and animals.
    d) sedimentation in prehistoric oceans.
    e) cosmic rays from outer space.
19. Order coal, peat, and lignite in the time order in which they are formed by natural processes.
    a) coal, peat, and lignite
    b) lignite, coal, and peat

c) peat, coal, and lignite
d) peat, lignite, coal

20. It would not be completely inappropriate to classify fossil fuels with solar energy because
    a) visible light is produced when the fuels burn.
    b) the formation of the fuels required solar energy in the photosynthesis process.
    c) solar energy is required to make the fuels burn.
    d) solar energy is used to dry the fuels before burning.
    e) both solar energy and fossil fuels can be used to produce synthetic oil.

21. The type of underground oil deposit sketched below is termed a
    a) fault b) fold c) nonconformity
    d) salt dome e) malformation

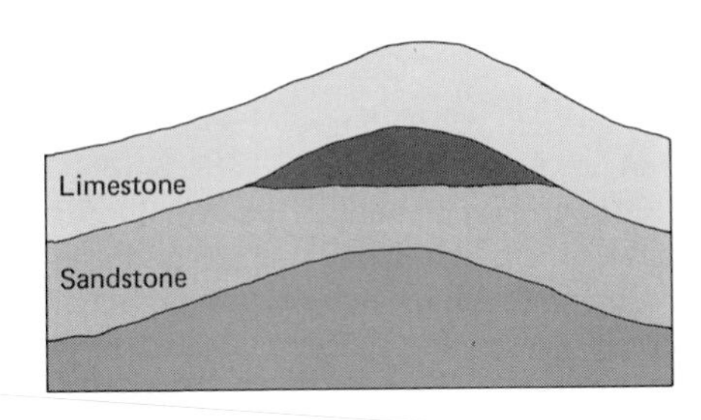

22. When you examine the depletion of a natural resource such as petroleum, it is fair to say that
    a) there will come a time when every drop of petroleum will have been removed from the earth.
    b) economics plays no role in determining how long the resources last.
    c) the resource is never completely removed from the earth but economics determine when it is no longer profitable to extract it.
    d) technology will always provide means for finding new reserves.
    e) there are no limits of any resource because they are constantly being replenished by natural processes.

23. Of those energy supplies listed below, the one in most plentiful supply in the United States is
    a) petroleum. b) natural gas. c) wind.
    d) trash. e) coal.

24. Shown below are five graphs of the rate of use of an energy reserve as a function of time. Pick the one that best describes the use of a fossil fuel reserve.

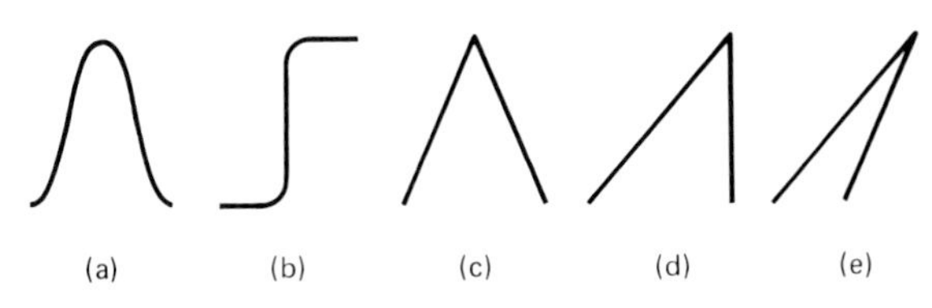

25. Our proven reserves of oil are 30 billion barrels. This means that
    a) when 30 billion barrels are used, we will have no more oil to use.
    b) at a consumption rate of 1.5 billion barrels per year, the reserves will be depleted in 45 years.
    c) depending on the economy and technology, we could still have proven reserves of 30 billion barrels 20 years hence.
    d) we have about a one-year supply of oil.
    e) we have little worry about where our oil for transportation will come from in the next decade.

26. Oil shale is a material
    a) containing a tarlike substance intermixed with rocks.
    b) having little potential for producing oil.
    c) found primarily in the Appalachian Mountains.
    d) that is a by-product of the oil-refining industry.
    e) containing a fossil remnant called kerogen.

27. If the population of a country grows at an exponential rate of 2% per year, then the population will double in about
    a) 35 years. b) 2 years. c) 10 years.
    d) 45 days. e) one century.

28. The world population for a series of dates is presented below.

1850	1 billion
1930	2 billion
1960	3 billion
1976	4 billion

An examination of these data reveals
    a) the growth of population is exponential because it doubles every so often.
    b) the growth of population is not exponential because the doubling time changes.
    c) the doubling time has increased as time progressed.

d) the growth of population is both exponential and exothermic.
e) there is no way of determining from these data whether or not the population growth is exponential.

29. In the January 22, 1980 issue of the *Cincinnati Enquirer,* the Gulf Corporation ran a full-page ad relating to the extraction of oil from shale. They announced that there is more oil to be garnered from North American shale deposits than from the Middle East oil fields. Just knowing that this oil could possibly be extracted from shale, it follows that
   a) the shale deposits should be classified as a proven reserve.
   b) we can expect significant oil production within the year.
   c) economics may never allow any of the shale oil to be recovered.
   d) our worries about oil supplies are over.
   e) the price of crude oil will drop on the basis of the announcement by the Gulf Corporation.
30. An increase in the price of a certain energy resource tends to increase the proven reserves.
   a) true   b) false

Answers

11. (e)	12 .(d)	13. (a)	14. (c)
15. (a)	16. (c)	17. (e)	18. (a)
19. (d)	20. (b)	21. (b)	22. (c)
23. (e)	24. (a)	25. (c)	26. (e)
27. (a)	28. (b)	29. (c)	30. (a)

QUESTIONS AND PROBLEMS

3.1–3.5 USE AND FORMATION OF FOSSIL FUELS

1. Discuss some of the problems in estimating fossil fuel reserves.
2. A significant amount of oil can be derived from the shale formations in the western United States. Until at least 1982, why weren't these shale deposits considered to be a proven reserve?
3. The first drilling for oil in the continental shelf off the coast of the eastern United States began in 1978. Does this area constitute a proven oil reserve?
4. When the price of a fuel increases how does this tend to increase the proven reserves of the fuel?
5. Figure 1.1 shows that, on a percentage basis, coal use peaked in 1910. When was the coal use half the 1910 amount? If petroleum and natural gas use follows the coal use pattern, when will the petroleum and natural gas use be one half the maximum use?
6. Suppose you had a fixed amount of money, say $1 million, and you started spending this at an increasing rate. Why must this rate eventually peak and about how much money would you expect to have left when the rate peaks? What is the relation between this example and the depletion of a finite resource?
7. Suppose that estimates of the recoverable oil in an offshore region vary from 180 million barrels to 650 million barrels. At the consumption rate of 5.8 billion barrels per year, about how many years would this offshore oil last?
8. The recovery of oil from shale requires considerable amounts of water. Why might this be a deterrent to exploitation of the shale deposits in the western United States?
9. A hypothetical shale deposit yields one barrel of oil per ton of material processed. If the deposit yields 500,000 barrels of oil per day, what is the daily amount of shale processed? If this shale is transported by railroad cars each carrying 50 tons, how many carloads are required each day?
10. What are some environmental trade-offs to be made if the oil shale deposits in the western United States are to be developed?
11. A cube of shale 10 feet on a side weighs about 70 tons. How many cubes could be formed from 350,000 tons of shale? If a system recovering oil from shale processed 350,000 tons of shale a day and the shale was laid end to end in cubes 10 feet on a side, how far would the blocks extend?

3.6 EXPONENTIAL GROWTH OF ENERGY USE

12. From 1935 to about 1975 the electric energy in the United States grew at an exponential rate of 7% per year. What was the doubling time? Does the doubling time agree with that quoted in the text? The doubling time for oil consumption in the same period was 20 years. What was the growth rate?
13. What are some rates that have a direct bearing on our environment?
14. The sizes of some of the numbers encountered are difficult to comprehend. For example, about 30 million barrels of oil were produced domestically each day in 1981. A barrel is roughly half the height of a

person, say three feet. If 30 million barrels were laid end to end, how would the total length compare with the distance from New York to San Francisco (about 3000 miles)? This should give you an idea of the magnitude of the problem of developing a source of oil that produces a million barrels a day.

15. The following experiment is one that you can do to illustrate exponential growth. Take a sheet of $8\frac{1}{2}'' \times 11''$ paper, typically 0.003 inches thick, and cut it into two equal pieces. Then take the two pieces and cut them into four equal pieces. Continue this process of doubling the number of pieces a total of eight times. Are you surprised that the final stack is about one inch thick? Calculate how thick the stack would be if you continued for a total of 25 steps. This is a good example of how a system grows if it doubles at regular intervals.

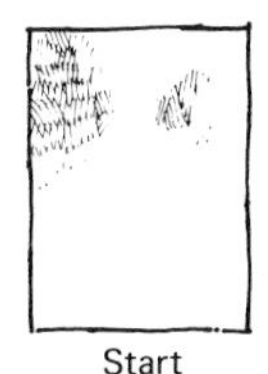
Start

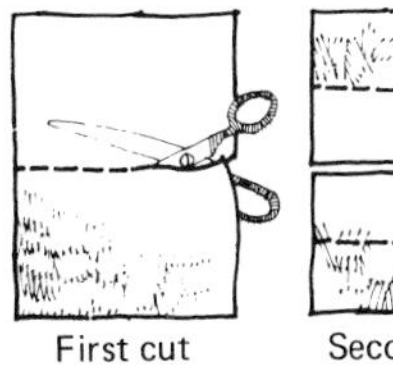
First cut

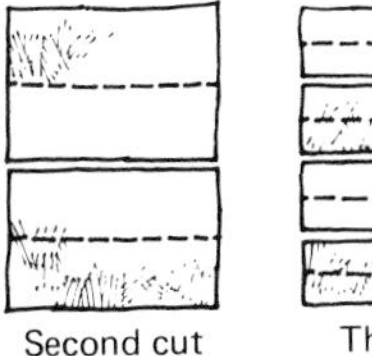
Second cut

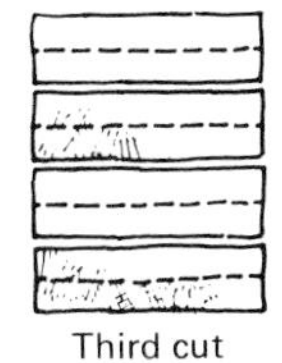
Third cut

16. A familiar quote states that "Half of *all* the energy consumed by human beings since the birth of Jesus has been consumed within the last century." Knowing that worldwide energy consumption increases at an exponential rate of about 3%/year, does this quote make sense?

17. In two days a moth caterpillar can consume an amount equal to 70,000 times its birth weight. Assuming that the caterpillar consumes at an exponential rate, show that about 16 doubling periods are required. What would be the doubling time?

18. At present the rate of electric energy production in the United States is only a millionth of the rate of solar energy flow to the earth. If the rate of electric energy production continues to double every ten years, how long will it be until the rate equals the rate of solar energy flow to the earth?

4

Electric Energy

4.1 MOTIVATION

It is a spectacular sight to see water lazily approach Niagara Falls and then be accelerated downward by the force of gravity, delivering much of its kinetic energy to the rocks at the bottom of the falls. Above the falls another impressive event attracts few sightseers. There, water is quietly shuttled into a large pipe under the city of Niagara and thence to water turbines downstream from the falls. The water turbines spin electric generators, and electricity is channeled through wires to consumers. When you watch a river, you sense the motion of its water and its kinetic energy. However, you see nothing in motion and no evidence of energy when you view wires leading from electric generators. But energy is clearly there. We all learn at an early age to plug motor-

The metallic cables that transmit electric energy from a generating station to a distribution center often extend over hundreds of miles of countryside. They bring the most versatile energy of all to our homes and industries. (Photograph by David Falconer. Courtesy of the EPA.)

driven appliances into electrical outlets and almost magically motion and energy are visible. Even though nothing visible moves in electrical wires submicroscopic electrons move in response to electric forces exerted on them. Electric energy acquired by electrons is converted to a wide variety of other types of energy including heat, light, sound, and motion. Because of its versatility and the relative ease of its production and transmission, electric energy is a precious commodity.

Electricity accounted for one-third of the gross energy consumption in 1981. About two-thirds of this energy used in electric power plants was lost in the conversion to electricity and in the transmission to consumers. This means about three units of energy are converted to deliver one unit of electric energy to a consumer. Thus every unit of electric energy saved by a consumer means that three units of energy need not be converted. Table 4.1 itemizes electric energy production according to type of energy converted. The decline in electricity produced from burning petroleum is notable. Electric utilities turned increasingly to clean-burning oil in the early 1970s. But as oil became more expensive in the latter part of the 1970s, restrictions were placed on burning oil in newly developed systems. Electric utilities then turned to coal and uranium (Table 4.1) for energy to pick up the slack created by the oil situation. Hydroelectric energy in terms of kilowatt-hours produced remained essentially constant but the percentage contribution declined from 14.6% in 1973 to 11.4% in 1981. Although hydroelectric production of electricity has several attractive features, expansion of hydroelectric facilities is thwarted by a lack of suitable sites and by environmental restrictions. Electricity sales grew by about 24% between 1973 and 1981. Distribution of electricity changed very little in this period. The residential, commercial, and industrial sectors purchased 33.5%, 23.8%, and 39.0%, respectively, of the electric energy sold in 1981.

Some time or other all of us have to purchase light bulbs, as well as many other electrically operated devices, and many of us have to pay for their operation. On a light bulb you can read notations such as 120 volts and 100 watts. On the utility bill you see kilowatt-hours or maybe an abbreviation, kWh. In newspapers you may read that electric utilities have problems in meeting consumer demands. The purpose of this chapter is to provide you with an insight into volts, watts, kilowatt-hours, and the generation of electricity so you may become a more informed citizen.

4.2 ELECTRIC FORCE AND ELECTRIC CHARGE

If you were to see a child's balloon resting on a floor, you could analyze its motionless condition by saying: The earth pulls downward on the balloon with a gravitational force. Responding to pressure from the balloon, the floor pushes upward with an equal but oppositely directed force. The net force is zero, therefore the balloon is in equilibrium. You might pick up the balloon, rub it on your hair, touch it to the wall and see the balloon cling peacefully to the wall (Fig. 4.1). Now explain. Certainly, the net force on the balloon is zero, otherwise it would fall. The

TABLE 4.1 Electric energy production according to type of energy converted. Production is expressed as a percentage of the total electric energy produced which is given in billions of kilowatt-hours. The data are from the Department of Energy.

	Coal	Petroleum	Natural gas	Nuclear	Hydroelectric	Other	Total
1973	45.6%	16.9%	18.3%	4.5%	14.6%	0.1%	1861
1976	46.3%	15.7%	14.5%	9.4%	13.9%	0.2%	2038
1979	47.8%	13.5%	14.7%	11.4%	12.4%	0.2%	2247
1981	52.4%	9.0%	15.1%	11.9%	11.4%	0.3%	2295

(a)

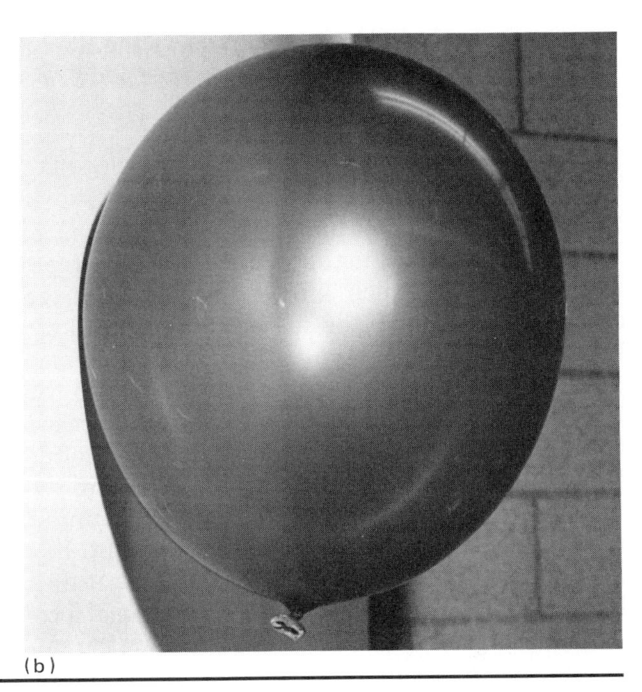
(b)

FIGURE 4.1 (a) This student's hair is attracted to a balloon that was previously rubbed against her hair. (b) The balloon filled with helium gas would normally rise rapidly in the air but after it has been rubbed on a person's hair, it is held fast by electric forces to the wall.

earth still pulls downward with a gravitational force. Something must pull upward with an equal but oppositely directed force. But what? Had you not rubbed the balloon on your hair, the balloon would not adhere to the wall. Therefore the upward force must be associated with the rubbing.

This force merits further investigation, and the balloons are our tools. Take two inflated balloons and with threads suspend them an inch or so apart. Probably the threads are vertical unless you rubbed the balloons on something. Rub both balloons on your hair and remove your hands and you see the balloons exert mutual forces that tend to separate them. Move your head near the balloons and you see that your hair is not repelled by the balloons but attracted to them. These observations are a consequence of electric forces and the property of charge acquired by the balloons and hair. Because of their mass, two objects exert mutually attractive gravitational forces. If these objects have charge in addition to mass, they exert a mutual electric force that may tend to pull the objects together or tend to push them apart. If the electric force is attractive, we say the charges are unlike, and if they repel, we say they are like. We distinguish unlike charges from each other by calling one positive and the other negative. The charge on a balloon rubbed on your hair is negative. Any charge repelled by the balloon is also negative; any charge attracted is positive.

The mass of an object is normally measured with something like a pan balance (Fig. 2.2) but in principle mass could be determined using the law of gravitation (Eq. 2.7) and measuring the gravitational force between two objects separated a known distance. There is no "pan balance" to measure charge, but charge can be and is measured by determining the electric force between two charged objects separated a known distance. It is not essential to dwell on the mechanics but it is important to know the unit of charge is the coulomb represented by the symbol C. Electric charge, electric force, and energy associated with the electric force play roles ranging from the structure of atoms to the liberation of energy in either a nuclear power plant or a coal-fired plant.

4.3 ATOMIC STRUCTURE

Visualize an ice skater moving smoothly in a nice, neat circle. Attached to the skater's waist is a rope, the other end of which is held by a person at the center of the circle. If the skater attempts to skate in a larger circle, she is pulled inward and confined to move in a circle that has a radius equal to the length of the rope. The language of physics describes this condition as bound. The skater is bound to the circular path by a binding force provided by the rope pulling on her waist. Bound systems, similar in principle, are common and very important.

An orbiting satellite and the earth constitute a bound system held together by the gravitational force between the two (Fig. 4.2). At any instant, the satellite tends to shoot off in a straight line but gravity pulls continually toward the earth and the satellite executes circular motion about the earth. Conceptually, an atom is much like a system composed of the earth and satellites revolving about it. In the atom, electrons are satellites and a nucleus plays the role of the earth (Fig. 4.2b). Just as the earth is much more massive than a satellite, a nucleus is much more massive than an electron. The mass of the nucleus is confined to a volume approximating a sphere having a radius of about 0.000000000000001 (1×10^{-15}) meters. The closest electron is about 0.0000000001 (1×10^{-10}) meters from the nucleus. In perspective, if the radius of the nucleus were about the thickness of a dime, the nearest electron would be about the length of a football field away. Like the earth–satellite system, electrons and the nucleus experience the attractive gravitational force existing between any masses. However, this force is much too weak to hold the electron to such a small region. The primary binding is from the electric force between negatively charged electrons and a positively charged nucleus.

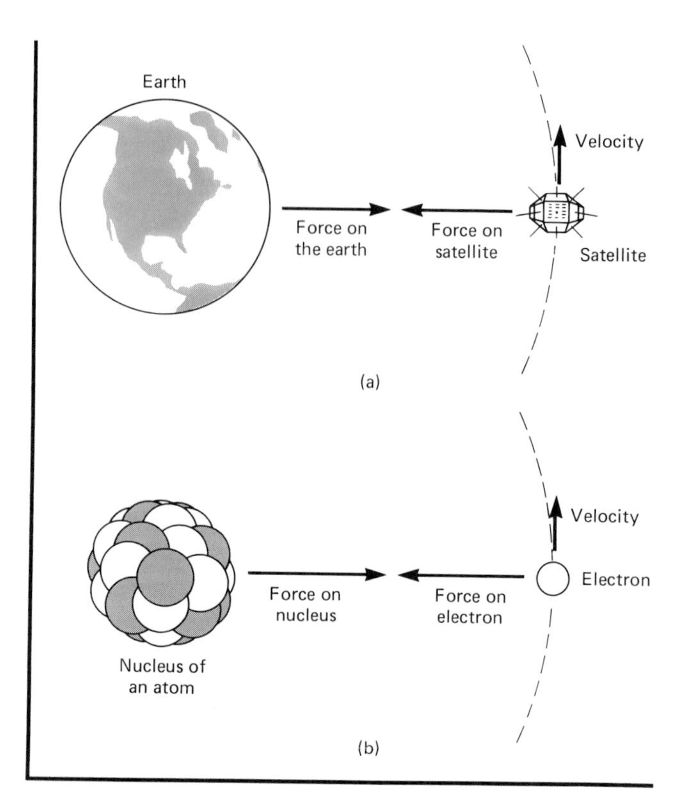

FIGURE 4.2 (a) An orbiting satellite and the earth. The earth and the satellite are bound together by a gravitational force. The force of gravity continually pulls inward on the satellite and keeps the satellite progressing in a circular orbit. **(b)** A nucleus and an electron. The nucleus and electron are bound together by an electric force. The electric force continually pulls inward on the electron and keeps the electron progressing in a circular orbit.

In a normal atom the total negative charge of the electrons balances the charge of the nucleus and the atom is said to be electrically neutral. The number of electrons in a neutral atom is called the atomic number. The atomic number distinguishes different atomic species. For example, carbon has six electrons and oxygen has eight electrons. Most electrical wires in homes are made of copper. Each atom of copper has 29 electrons. As the atomic number increases, the mass of the atom also increases. This is caused by the additional electrons and by an increase in the mass of the nucleus. It is possible to have atoms with the same number of electrons, but with nuclei of different masses. These are called

TABLE 4.2 The first 18 atoms in order of increasing number of electrons and mass. The mass is that of the most abundant isotope and is based on a scale of exactly 12 for the most abundant form of carbon.

Atom	Chemical symbol	Number of electrons	Approximate relative mass
Hydrogen	H	1	1
Helium	He	2	4
Lithium	Li	3	7
Beryllium	Be	4	9
Boron	B	5	11
Carbon	C	6	12
Nitrogen	N	7	14
Oxygen	O	8	16
Fluorine	F	9	19
Neon	Ne	10	20
Sodium	Na	11	23
Magnesium	Mg	12	24
Aluminum	Al	13	27
Silicon	Si	14	28
Phosphorus	P	15	31
Sulfur	S	16	32
Chlorine	Cl	17	35
Argon	Ar	18	40

isotopes and later we will see the origin of their mass differences. Because isotopes have the same number of electrons, they share the same chemical characteristics. Table 4.2 lists the first 18 atoms in the order of increasing number of electrons. The approximate relative mass corresponds to the most abundant isotope of a given atom.

This model of the atom allows us to interpret the experiment with the balloons. The rubbing process initiated a transfer of electrons from the hair to the balloons. The balloons acquire electrons and become negatively charged; the hair loses electrons and is left with a deficiency of negative charge and, therefore, an excess of positive charge. The balloons repel each other because they both have a net negative charge.

4.4 ELECTRICITY

Standing near a busy highway out of view of the traffic but within hearing distance of it, you could listen and count the number of cars passing by in a measured time interval and express the rate in units like cars per minute. If each car had two passengers, then multiplying the car rate by two yields the number of people passing by per minute. Electrons moving through wires are analogous to cars moving on a highway. You can't see the electrons but in principle you can stand at the edge of the wire and count the electrons passing by in a measured time interval, and express the rate in units of electrons per second. Because each electron carries the same amount of charge, multiplying the electron rate by the charge per electron yields the rate at which charge passes by in units of coulombs per second. The rate of charge flow is called electric current, or current, for short. As an equation

$$\text{electric current} = \frac{\text{amount of charge flowing through a wire}}{\text{time required for the flow}}.$$

$$I = \frac{Q}{t} \tag{4.1}$$

Recording charge (Q) in coulombs and the time interval (t) in seconds yields coulombs per second as the units of electric current (I). A coulomb per second is called an ampere, symbol A. If 100 coulombs of charge flow by some position in a wire in ten seconds, we say the curent is ten amperes. The electric current in a lit flashlight bulb is about one ampere. If the bulb were on for ten minutes the total charge passing through the bulb in ten minutes is

$$1\,\frac{\text{coulomb}}{\text{second}} \times 10\,\text{minutes} \times 60\,\frac{\text{seconds}}{\text{minute}}$$

$$= 600 \text{ coulombs.}$$

Because a wire is composed of atoms that have positive charges in their nuclei, and of electrons that have negative charge, electric current could conceivably involve both positive and negative charges. However, the nuclei form the solid structure of the conductor and cannot move unless destructive forces are operative. Only the outermost electrons of an atom are free to contribute to an electric current in a wire because they are weakly bound to the positively charged nucleus. The closer the electron is to the nucleus, the harder it is to remove it from the atom. In a copper wire, only about one of the 29 electrons possessed by each copper atom contributes to an electric current.

A force must be exerted on the electrons to cause them to flow through a wire just as a force must be applied on water to keep it moving through a pipe. Electric forces do work on electrons just as work is done on an automobile pushed some distance by a person. It is the role of an electric power company to supply electrical "pressure" for moving electric charges. This electrical "pressure" is called potential difference* and is defined in terms of the amount of work done by electric forces on a charge as it moves between two positions. As an equation

$$\text{potential difference} = \frac{\text{electrical work on the amount of charge moved}}{\text{amount of charge}},$$

$$V = \frac{W}{Q}. \tag{4.2}$$

The idea of potential difference is very abstract. You never see the charges let alone what is exerting the electric forces. But you can employ the same idea for the gravitational force by replacing electrical work with gravitational work and charge with mass in the definition of potential difference. Potential difference has units of joules per coulomb. A joule per coulomb is called a volt, symbol V. If 100 joules of work are done by electrical forces on one coulomb of charge when it moves between two positions, the potential difference between the two positions is 100 volts. The potential difference between the two openings in the electrical outlets in a house is effectively 115 volts. Batteries for automobiles generally provide a 12-volt potential difference between the terminals of the battery. If two coulombs of charge moved through a 12-volt potential difference provided by a car battery, electric forces do 24 joules of work on the two coulombs of charge.

Be sure you understand this notion of a potential difference. An everyday analogy may help. If you were to tell a friend a building is 500 feet high, he would probably assume you mean the difference in "height" between the top and bottom is 500 feet. Clearly, though, if the building were on top of a mountain, its height would be more than 500 feet relative to the base of the mountain. Potential differences, like heights, are also relative. When you say that the potential difference across an electrical outlet in a house is 115 volts, you mean that one of the contacts is 115 volts relative to the other contact. This other contact in a household electrical socket is at the same potential as any other part of the room, including yourself. You can touch this contact and not get an electrical shock because there is no difference in potential between you and that

---

* We use the term potential difference initially to stress the involvement of two positions. Voltage, a less formal word describing the same idea, replaces potential difference as we proceed.

A light bulb glows when connected to the "hot" terminal of an electrical outlet and a cold-water faucet. This is because the pipe connecting the faucet passes through the earth, thereby completing the electrical circuit. Electrical shock occurs when a person touches the "hot" terminal while standing on the earth or something connected to the earth. Because of the possibility of electrical shock, you should not attempt the demonstration shown in this photograph.

contact. This contact is called the ground because it is fastened to a metal stake driven into the ground (earth). Generally you can see this ground wire running down an electric utility pole holding wires bringing electricity to a house. The other contact in an outlet is referred to as "hot." Electrical shocks occur when you touch both ground and the "hot" contact.

When an appliance like a lamp is plugged into an outlet and the switch turned "on," an electrical circuit is said to be completed. It is complete in the sense that a complete (or closed) path is provided for the flow of electrons. It is entirely analogous to connecting a hose to a pump that pulls water out of the bottom of a container and returns it to the top (Fig. 4.3). The electric power company uses an electric generator (Section 4.8) for the electric pump. The wire plays the role of the water hose and the switch functions as a water faucet (valve).

4.5 ELECTRICAL RESISTANCE

In the water circuit (Fig. 4.3), the hose resists the flow of water through it. Pinch the hose and the flow decreases and we say the resistance of the hose increased. Similarly, a wire resists the flow of the electrons. Just as some hoses are better conductors of water than others, some wires are better conductors of electrons than others. This resistance depends not only on "obstructions" in the path of the electrons but also on the size of the conductor. If two similarly constructed hoses of different lengths are connected to the same water tap, the flow in gallons per minute is greater out the open end of the shorter hose. Replacing these two hoses with ones having the same length but different diameters shows that the flow is greater out of the hose with the larger diameter. You may have made these observations with garden hoses. Electrical resistance, like hose resistance, increases as the conductor length increases, and decreases as the conductor diameter increases.

In line with our thinking about the electric current in a wire connected to an outlet, we should expect the current to increase if the potential difference (the electrical "pressure") increases but should decrease if the electrical resistance increases. Accordingly electrical resistance of a wire or an

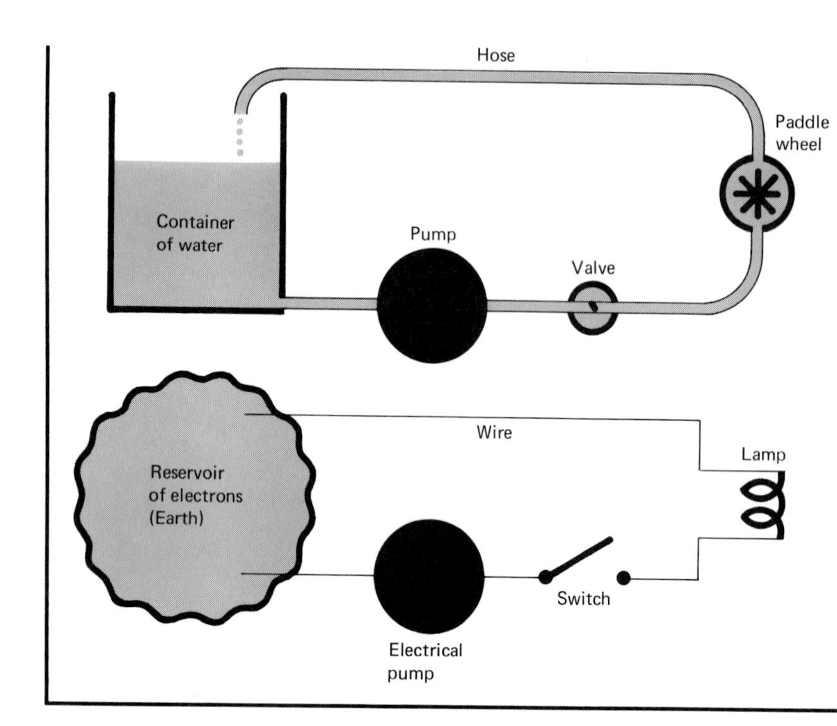

FIGURE 4.3 Schematic comparison of a fluid circuit and an electrical circuit. The water pump forces water to circulate through the water circuit. The electrical pump forces the free electrons to circulate through the electrical circuit. The sum total of electric charge never changes in the process just as in the absence of leaks the sum total of water never changes in the water circuit.

electrical appliance is determined numerically from the equation

$$\text{resistance} = \frac{\text{potential difference across the element measured in volts}}{\text{current in the element measured in amperes}},$$

$$R = \frac{V}{I}. \tag{4.3}$$

Resistance has units of volts divided by amperes. This unit is called an ohm. If a 12-volt battery causes 1.5 amperes of current when connected to a bulb, we say the bulb has a resistance of 12 volts ÷ 1.5 amperes = 8 ohms. The relation $R = V/I$ is called Ohm's law when R does not depend on the size of the current. Common things like wires, toasters, and irons obey Ohm's law. Transistors doing marvelous things in calculators do not obey Ohm's law. In a household where the potential difference is fixed at 115 volts, the current in an appliance depends on its resistance. As the resistance decreases, the current increases. The energy used to overcome the electrical resistance appears as heat. For example, the heat liberated in a toaster is produced by electrons overcoming the electrical resistance provided by the wire elements of the toaster.

4.6 ELECTRIC POWER AND ELECTRIC ENERGY

Like water flowing over a dam, electrons flowing in a conductor acquire kinetic energy as a result of the work done on them. This is another example of the work–energy principle (Section 2.7). Electrons convert this energy in a variety of ways. In a lamp the energy is converted to heat and light. In an electric motor, the energy is converted to mechanical rotational energy. The rate at which moving charges convert energy is called electric power.

$$\text{power} = \frac{\text{energy converted measured in joules}}{\text{time required for conversion measured in seconds}},$$

$$P = \frac{W}{t}. \tag{4.4}$$

Because electric energy, like mechanical energy, is expressed in joules and time is in seconds, electric power is expressed in joules per second, or watts. (See Section 2.10.) Many electrical devices use power in the thousands-of-watts range and electric power plants generate power in the millions-of-watts range. You will often see the terminology, kilowatts and megawatts, symbolized kW and MW, meaning thousand and million watts, respectively.

Any conductor will have some resistance, and there will be some energy loss in the form of heat when it carries a current. If the conductor is very long so that the resistance is large, these losses can be significant. This situation occurs when conductors are used to transmit electric power from a generating facility to a customer. We will say more about this in Section 4.10.

At this point you should understand that if a wire is connected to a potential difference provided by a battery or a household outlet, a current is established in the wire. There are four quantities of interest—voltage, current, and resistance related through $V = IR$, and power related to voltage and current by $P = VI$. Two of these four quantities need to be known in order to determine the other two. The table below shows the equations needed.

When the quantity in this column is unknown,	then use these equations
voltage, V	$V = IR$ and $P = I^2R$
current, I	$I = \frac{V}{R}$ and $P = \frac{V^2}{R}$
resistance, R	$R = \frac{V}{I}$ and $P = VI$

To illustrate, suppose a kitchen toaster requiring 1000 watts is connected to a 115 volt outlet for two minutes. Line 3 in the table guides us for this situation. The current produced is

$$I = \frac{P}{V} = \frac{1000 \text{ watts}}{115 \text{ volts}} = 8.7 \text{ amperes.}$$

The resistance of the toaster is

$$R = \frac{V}{I} = \frac{115 \text{ volts}}{8.7 \text{ amperes}} = 13.2 \text{ ohms.}$$

The energy converted in kilowatt-hours is

$$\text{E} = 1 \text{ kilowatt} \times 2 \text{ minutes} \times \frac{1 \text{ hour}}{60 \text{ minutes}}$$
$$= 0.033 \text{ kilowatt-hour,}$$
$$= 0.033 \text{ kWh.}$$

Companies that provide the voltage for forcing electrons through our appliances are often referred to as power companies. You might suspect, then, that they are paid for power. But they are not. They are paid for energy. If this is not obvious, a little thought should convince you. Power is the rate of using energy. A 100-watt light bulb consumes energy at the rate of 100 joules per second regardless of how long it is turned on. Clearly though, the cost of operation depends on how long the bulb is lit. Using Eq. (4.4), it follows that energy, power, and time are related by

$$\text{energy} = \text{power} \times \text{time,}$$
$$W = Pt.$$

So any unit of power multiplied by a unit of time yields a unit of energy. Watts (or kilowatts) and hours are units of power and time. Thus the watt-hour (or kilowatt-hour, the unit for paying the utilities company) is a unit of energy. It is a peculiar unit because the units of time (hours) do not cancel the time units of seconds in power.

$$W = P(\text{watts}) \times t(\text{hours})$$
$$= P\left(\frac{\text{joules}}{\text{second}}\right) \times t(\text{hours}).$$

The arrow emphasizes that units of hours do not cancel the units of seconds.

Table 4.3 lists the power requirements for several household appliances. Note that those involving the generation or removal of heat require the most power. These devices tend to have low resistance.

TABLE 4.3 Power requirement (in watts) of some common household appliances.

Appliance	Watts
Cooking range (full operation)	12,000
Heat pump	12,000
Clothes dryer	5,000
Oven	3,200
Water heater	2,500
Air conditioner (window)	1,600
Microwave oven	1,500
Broiler	1,400
Hot plate	1,250
Frying pan	1,200
Toaster	1,100
Hand iron	1,000
Electric space heater	1,000
Hair dryer	1,000
Clothes washer	500
Television (color)	330
Food mixer	130
Hi-fi stereo	100
Radio	70
Razor	14
Toothbrush	7
Clock	2

The energy consumed by a 1100 watt (1.1 kilowatts) toaster operating for one hour would be $W = 1.1$ kilowatts $\times$ 1 hour $=$ 1.1 kWh. If electric energy costs 9 cents per kWh, it would cost 9.9 cents to operate it each hour.

4.7 MAGNETIC FORCE AND MAGNETISM

We have discussed electric current and the idea that an electrical "pressure" is required to establish the current but have said nothing about how the power company generates the pressure nor how it is transmitted to homes and factories. To proceed we must grasp some fundamentals of magnetic force and magnetism.

Magnets are as useful as they are fascinating. They hold notes on the kitchen refrigerator. Homeowners use magnets to hold quilted window insulation in place. For centuries travelers have relied on magnetic compasses to find their way. Magnetic effects of interest to us are attributed to electric charges but the charges must be moving. It is easy to demonstrate this. Just watch a magnetic compass in a car behave erratically when you drive under overland electric transmission lines. Or more formally, take a straight piece of wire, orient it in a north–south direction, and place the compass on top of the wire. With no electric current in the wire, the compass needle points north and lies parallel to the wire. But if a current is in the wire, the compass needle rotates and points perpendicular to the wire (Fig. 4.4). The electron current does something to exert a magnetic force on the compass needle.

When you experiment with toy magnets, you find their magnetic effects are concentrated near the ends. For centuries, these concentrated magnetic regions have been called poles. The earth possesses magnetic poles located near the north and south geographic poles. If a pole of a magnet is placed near the poles of another magnet, it will be attracted toward one pole and repelled by the other pole (Fig. 4.5). We describe the fundamental difference between the two poles of a magnet by calling one N and the other S and say an N pole is that pole of a freely suspended magnet, such as a compass, that points toward the geographic north pole. Thus the tip labeled N on a compass needle is an N magnetic pole. Any other pole repelled by an N pole is itself an N pole, and any pole attracted by an N pole is an S pole.

You may have used a magnet to pick up paper clips or nails having some iron content. An iron-based paper clip exhibits no magnetic properties when in the vicinity of a similar paper clip. But bring the clips near a magnet and they are attracted to the magnet. Pull the paper clips away and if they exhibit magnetic effects, we say they are magnetized. A useful model views unmagnetized iron as having a vast number of tiny atomic magnets oriented at random. The effect of an atomic magnet pointing in one direction is canceled by another atomic magnet pointing exactly opposite. Consequently, the randomness of the large number of atomic magnets produces no net magnetic effects (Fig. 4.6).

When a magnet is brought near unmagnetized iron (Fig. 4.7), the poles of the atomic magnets expe-

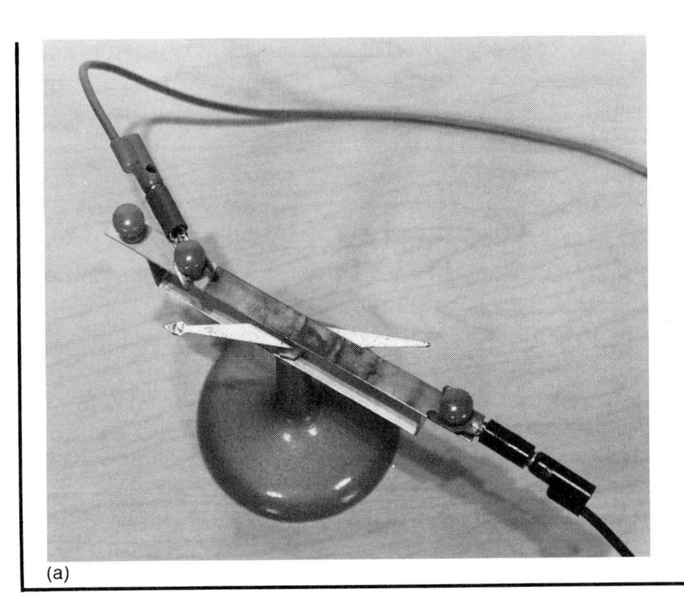

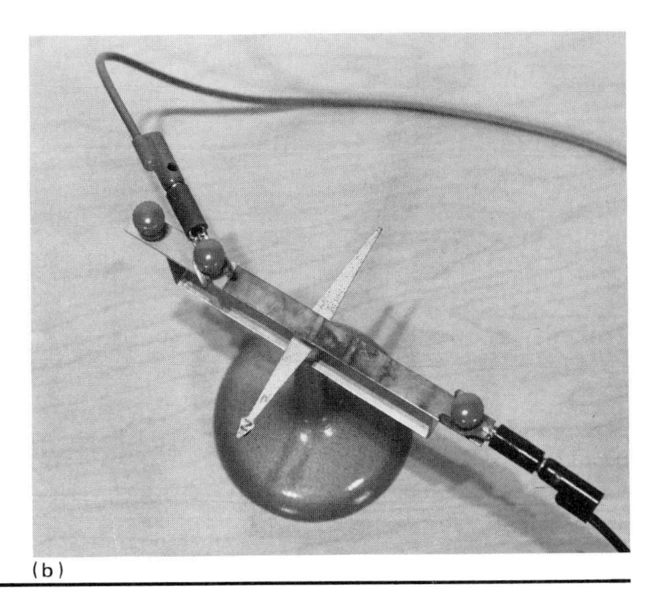

FIGURE 4.4 (a) A ribbonlike, metal strip is located above the compass needle. With no electric current in the strip the needle aligns with its N pole pointing toward the north geographic pole. **(b)** With sufficient current in the strip, the needle orients perpendicular to the strip because of a magnetic field created by the electric current.

rience a magnetic force and align. In Fig. 4.7 the repulsive force between S poles causes atomic magnets to align with their S poles away from the S pole of the magnet. Inside the iron, adjacent N and S poles neutralize their magnetism. But at one end, there is an accumulation of N poles and at the opposite end there is an accumulation of S poles. The

If these two poles attract each other,

then these two poles will repel each other

FIGURE 4.5 A pole of a magnet is attracted to one of the poles of a nearby magnet but is repelled by the opposite pole in the nearby magnet.

unmagnetized iron is now magnetized and is attracted to the magnet that induced the magnetism (Fig. 4.7). In some types of iron, the magnetism remains long after the inducing magnet is removed.

Magnetized bits of iron make tiny compass needles. Sprinkle these bits on a plastic plate covering a magnet and almost magically they align in a rather striking pattern (Fig. 4.8). The space around the magnet seems to be filled with something instructing the bits of iron to align. Accordingly we say there is a magnetic field around the magnet.

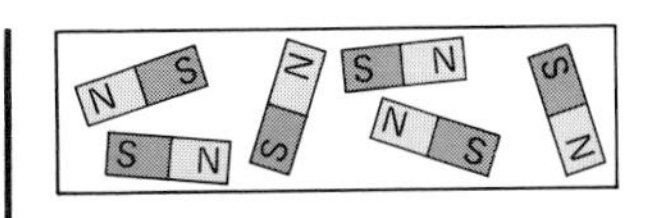

FIGURE 4.6 An unmagnetized bar of iron is viewed as containing a vast number of randomly oriented tiny magnets. The magnetism of any one magnet is canceled by neighboring magnets having different orientation.

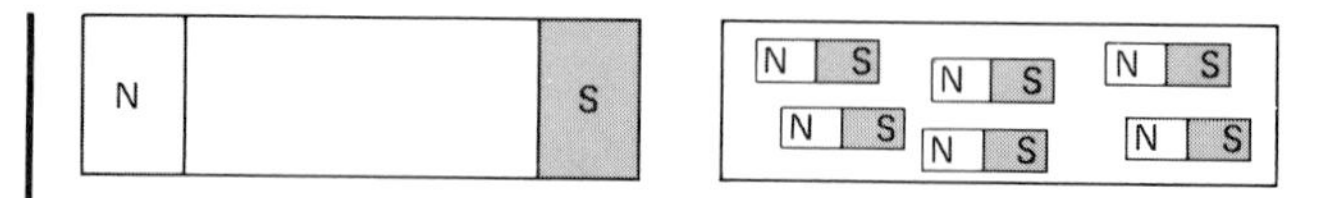

FIGURE 4.7 When unmagnetized iron is brought near a magnet, magnetic forces align the tiny magnets in the iron, leaving the iron with N and S poles.

The imaginary lines along which the iron bits align are called magnetic field lines. The direction assumed by the N pole of a bit of iron determines the direction of the magnetic field line at the position of the bit. Although there is nothing material about magnetic field lines, it is meaningful to visualize a magnetic field as being filled with magnetic field lines. If a wire moves in a magnetic field, we say the wire "cuts" the magnetic field lines. Think of this visual picture as an aid to predicting the effect of a moving charge in a magnetic field.

A TV picture becomes distorted noticeably when a magnet is brought near the screen. The distortion is a result of a magnetic force on electrons speeding toward the TV screen. As a general principle, any moving charge "cutting" magnetic field lines experiences a magnetic force. If a wire moves in a magnetic field (Fig. 4.9) so as to "cut" the magnetic field lines, all charges in the wire experience a force by virtue of moving in a magnetic field; but only the free electrons acquire motion in the wire. The motion of these electrons gives rise to an electric current in the wire. Work done by the agent moving the wire accounts for the energy acquired by the electrons. In a commercial electric power company, the work is provided by a steam turbine or a water turbine coupled to an electric generator. While we have illustrated the production of the current by holding the magnet fixed and moving the wire, a current would also develop if the wire were held fixed and the magnet moved. As long as the wire "cuts" the magnetic field lines, a current results.

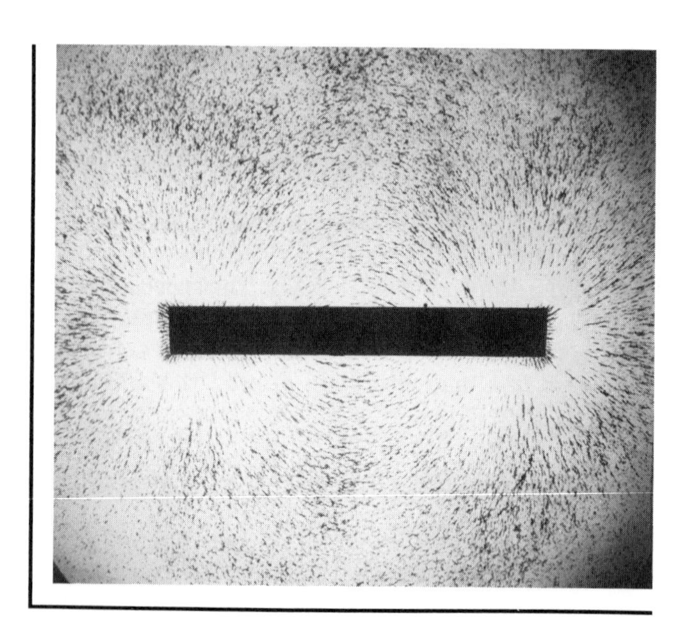

FIGURE 4.8 Tiny bits of iron sprinkled near a magnet align along the magnetic field lines of a magnet. Here you see the projection of the outlines of iron bits sprinkled on top of a glass plate resting on top of a bar magnet.

If the wire were moved in a direction opposite to that shown in Fig. 4.9, the electron current changes direction. Cutting magnetic field lines with an oscillatory motion produces a current whose direction alternates. Such a current is called an alternating current symbolized ac. The current produced by a battery (Fig. 4.10a) is called direct cur-

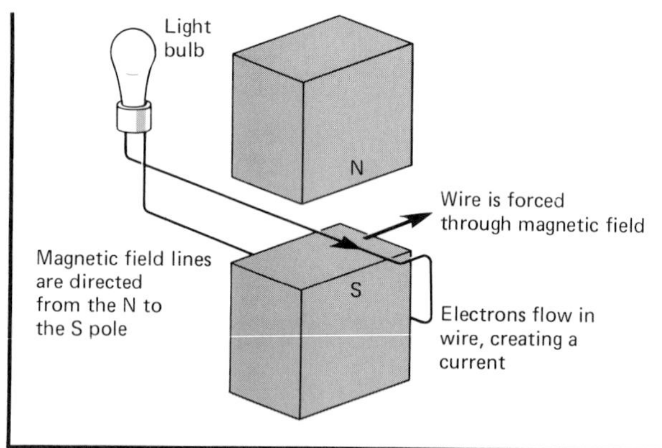

FIGURE 4.9 A current is established in a wire when it moves in a magnetic field so as to "cut" the magnetic field lines. If the wire were moved directly from the N to the S pole, no magnetic field lines are cut and, therefore, no current results.

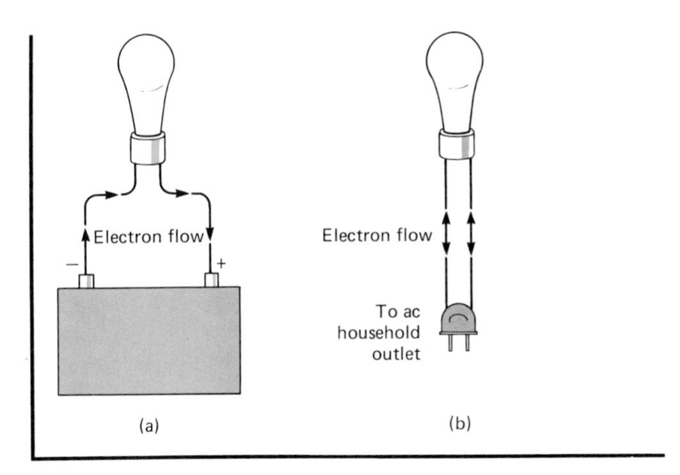

FIGURE 4.10 (a) When a device such as a light bulb is connected to a battery, electrons flow out of the negative (−) terminal, through the device, and into the positive (+) terminal. This type of current is called direct current and is denoted dc. **(b)** When a device such as a light bulb is connected to an ac household outlet, electrons oscillate back and forth in the conductors.

rent, symbolized dc, because the charge flows in one direction only.

4.8 ELECTRIC GENERATORS

The ac voltage generated by a commercial power plant or a generator on a bicycle is not produced by oscillating a single wire in a magnetic field, but the voltage derived has ac character. A practical generator employs a rotating coil with many turns of wire. To understand the principle let us look at a single loop of a coil as shown in Fig. 4.11. The metallic slip rings are connected electrically to the ends of the wire loop. The brushes are metal contacts touching the slip rings mounted on the shaft of the rotating coil. The coil, slip rings, brushes, and connected electrical device form an electrical circuit in which charges can flow. While this scheme involves circular motion of a loop of wire, it is essentially two parallel wires moving in opposite directions because only the sides labeled A and B are able to cut the magnetic field lines extending from the N pole to the S pole of the magnet. The two portions of the coil labeled A and B always move in opposite directions. Thus the current is always in opposite directions in the two wires. The maximum current results when the wires move perpendicular to the magnetic field lines. There is no induced current when the wire moves parallel to the magnetic field lines. When the portion of the loop designated A in Fig. 4.11 is at the top of its rotational path, it is moving horizontally to the right. Because the magnetic field lines are directed to the right from the N to the S pole, the wire is instantaneously moving parallel to the magnetic field lines. At this instant there is no current in the loop. As the loop rotates, the wires begin "cutting" the magnetic field lines and a current develops in the wire. When the portion of the loop designated A has moved down to the position shown in Fig. 4.11(b), it instantaneously is moving perpendicular to the magnetic field lines. In this position the electron current is a maximum and has the direction shown by the arrows on the wire loop. The electron current decreases to zero when the coil reaches the position shown in Fig. 4.11(c). As the coil rotates further, the electron current again increases but it also changes direction. It reaches a maximum when the coil achieves the position shown in Fig. 4.11(d). Finally the electron current drops to zero when the coil returns to the initial starting position. A plot of electron current versus position of the coil, or equivalently time since the position depends on time, is shown in Fig. 4.11(e).

When a toaster, for example, is plugged into an ac outlet in a house, the current produced in the wires of the toaster continually changes in size and direction. Nevertheless, the current causes the toaster wires to warm. Otherwise we would not be able to toast bread. A household voltage is usually quoted as 120 volts even though we know that the voltage is continually changing in size. This 120-volt rating means that the voltage is as effective at producing heat as a 120-volt battery, for example, connected to the toaster. The 120-volt ac label on a light bulb is the effective value of an ac voltage needed to power the bulb.

A generator that provides electricity for a headlamp on a bicycle functions very much like the one illustrated in Fig. 4.11. Mechanical energy is provided by the bicycle wheel rubbing against and rotating a shaft connected to the coil of the gener-

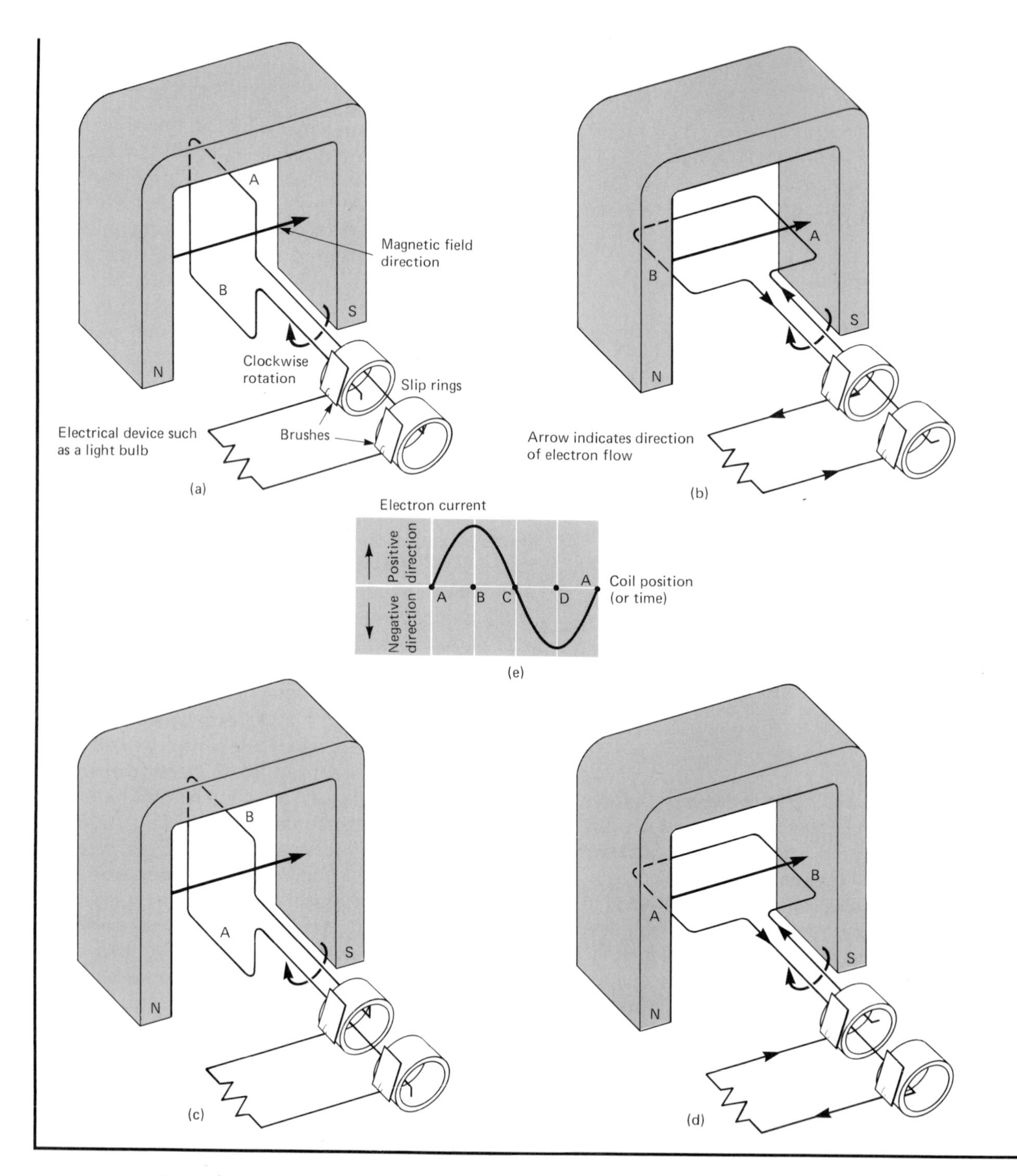

FIGURE 4.11 Principle of operation of an ac generator. A current develops in the rotating coil when it moves through the magnetic field created by the magnet. The slip rings rotate with the coil and make electrical contact with the brushes that are connected to an electrical device such as a light bulb or an engine. **(e)** Plot of induced current in the rotating coil of an ac generator. Positive and negative currents are used to distinguish the two directions of charge flow. The letters (*A, B, C, D*) on the horizontal axis refer to the coil positions designated by parts (a), (b), (c), and (d).

ator. The magnetic field is provided by a magnet. A generator like this produces about five watts of electric power. While the generator in a commercial electric power plant is enormously larger in physical size and often produces a billion watts of electric power, the basic physical principle is the same. The commercial generator uses a steam turbine for the input mechanical energy and a massive stationary coil with a circulating electric current to provide the magnetic field.

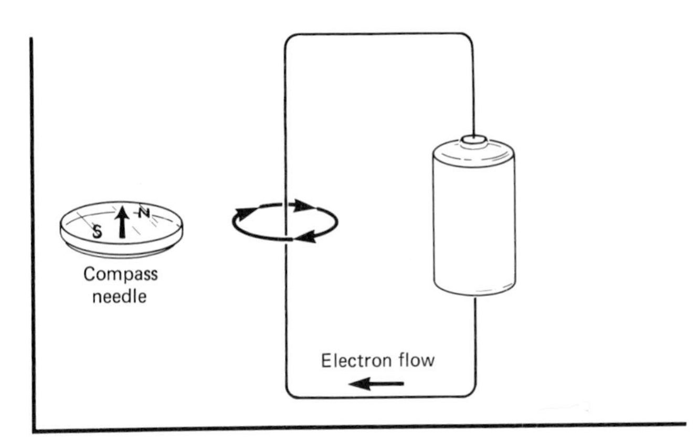

FIGURE 4.12 An ordinary compass used to determine the presence and direction of a magnetic field around a current-carrying wire. If the direction of the current is changed, the compass needle will point in the opposite direction.

4.9 TRANSFORMERS

An ac generator and a battery both deliver energy to such things as light bulbs connected to them. If a light bulb, for example, requires a voltage that does not match the voltage of a battery, it is relatively difficult to use the battery to power a device that produces a suitable voltage for the bulb. However, if ac voltages are involved, it is relatively easy to tailor voltages using an apparatus called a transformer. Transformers play very important roles in the transmission of electric power and in tailoring voltages for industrial and household uses. The transformer principle is straightforward.

A magnetic compass on top of a straight wire connected to a battery orients perpendicular to the wire (Fig. 4.4). Changing the battery connections reverses the current direction as well as the direction assumed by the compass. Reversal of the compass direction signals a change in the direction of the magnetic field lines (Fig. 4.12). If the battery that produces a current in only one direction is replaced by an ac voltage, the current changes direction periodically and the magnetic field lines change direction accordingly. If a loop of wire is placed in this changing magnetic field, an alternating current develops in it because of the changing feature of the magnetic field lines. It is just as if magnetic field lines were put in motion by a moving magnet. This principle is used in transformers to change the size of an ac voltage. The little black box you often see connected to an electrical outlet and to a calculator to recharge the calculator's battery is a transformer that reduces the line voltage from 115 to 9 volts.

Transformers are often seen near the top of utility poles supporting wires bringing electricity to a home. An actual transformer uses coils of wire (Fig. 4.13) rather than single wires as in the example. The winding to which the ac voltage is connected is called the primary winding. The other winding is called the secondary winding. The iron core guides the magnetic field lines from the primary winding to the secondary winding. Experiment reveals that the ratio of the primary and secondary voltages is equal to the ratio of the number of turns of wire in the primary and secondary windings of the transformer. In symbols

$$\frac{V_s}{V_p} = \frac{N_s}{N_p}. \tag{4.6}$$

Thus the size of the voltage on the secondary is determined by the turns ratio of the transformer and the voltage applied to the primary. For example, if there are five times as many secondary turns as primary turns, the voltage on the secondary will be five times the voltage applied to the primary. It would be to our benefit if at the same time the current in the secondary were also five times the current in the primary. This would mean that the power in the secondary would be 25 times greater. How-

(a)

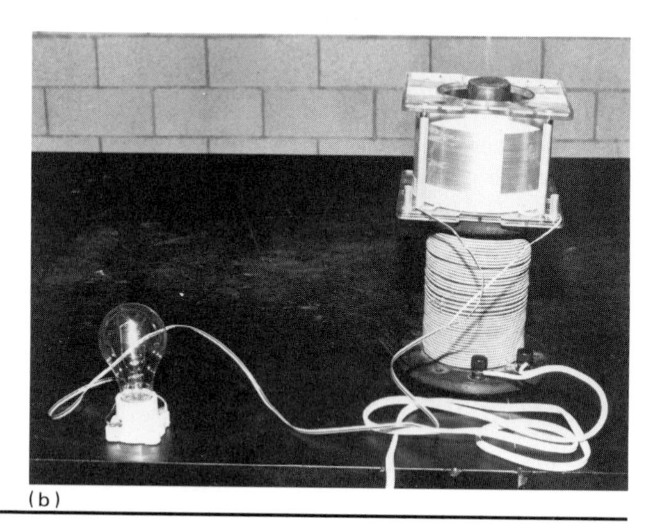
(b)

FIGURE 4.13 Essentially, a transformer consists of two coils of wire, the primary and the secondary, and an iron core to guide a changing magnetic field produced by current in the primary coil into the core of the secondary coil. In (a) the secondary coil that has a light bulb connected to the ends of the coil is isolated from the iron core of the primary coil that has its ends connected to an electrical outlet. There is no current in the secondary coil and the bulb is not lit. In (b) the secondary coil is placed over the iron core. A changing magnetic field within the secondary coil induces an electric current that causes the bulb to light.

ever, this would violate the principle of conservation of energy; you cannot get more power out of the system than you put in. Ideally you can get the same power out that you put in. The efficiency for energy transfer is about 99% in a well-designed transformer. Thus if the voltage on the secondary is five times the primary voltage, the secondary current is one-fifth the primary current, assuming 100% efficiency for the transformer.

4.10 TRANSMISSION OF ELECTRIC POWER

The voltage from a generator in an electric power plant is about 10,000 volts. A very large generator may deliver one billion watts of electric power. From the relation between power, voltage, and current ($P = VI$), it follows that the current is about 100,000 amperes in the wire leading from the generator. Heat produced in the wires depends on the current and resistance ($P = I^2R$). A large current produces substantial heat in a transmission line whose resistance is large because its length may be hundreds of miles long. The power company can maintain the same power in the transmission lines and reduce the heat losses by placing a transformer between the generator and the transmission lines (Fig. 4.14). The transformer steps up the voltage to as much as 75 times the 10,000-volt generator output. The current in the transmission lines drops accordingly. Transmission losses are reduced because the heat produced depends on the current and resistance ($P = I^2R$). Energy saved this way in transmission is energy available to consumers. At a home where the power is used, the voltage is stepped down to 230 volts by a series of transformers. By tapping the secondary of the transformer at the midpoint of the windings (Fig. 4.15), 115 volts are obtained between the center tap and either of the outer connections of the secondary. The center tap becomes the ground connection (see Section 4.4). The 230-volt connection provides energy for large appliances such as electric stoves, water heaters, clothes dryers, and air conditioners. The 115-volt connections are used for such things as light bulbs, mixers,

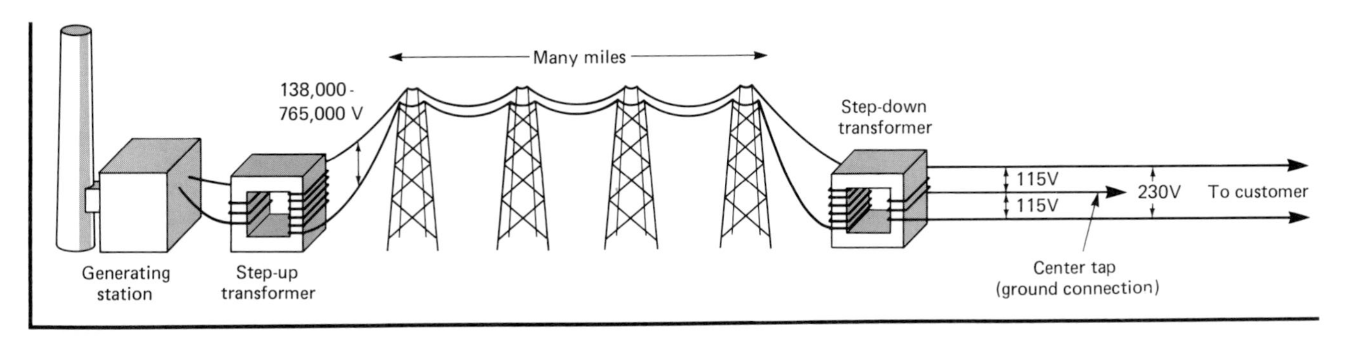

FIGURE 4.14 Elements of an electric power transmission system.

vacuum cleaners, and toasters. Homeowners often use small transformers to reduce 115 volts to 6 volts for operating door bells and door chimes. In addition to the 115 and 230 voltages used in a home, a factory will often use still higher voltages for some operations.

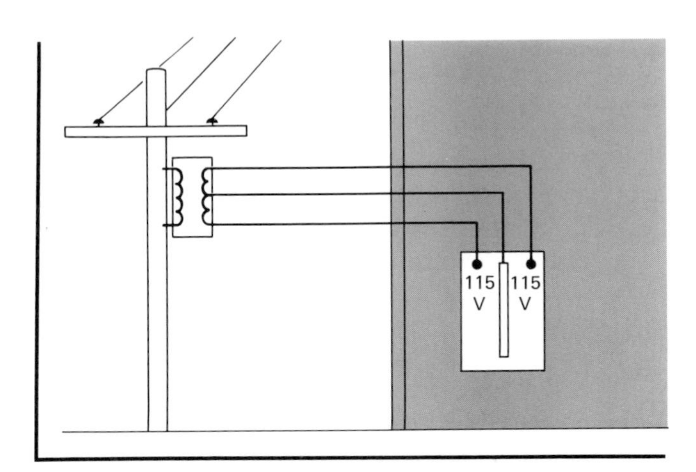

FIGURE 4.15 The manner of delivering electric power to a household. Somewhere away from the house there is a transformer usually mounted near the top of a utility pole. Electrical connections are made to the ends and midpoint of the secondary coil of the transformer. Wires from these three connections lead to a circuit-breaker box in the house. The consumer connects wires to the center terminal and either of the two outer terminals to obtain 115 volts, and the consumer connects to the two outer terminals to obtain 230 volts.

4.11 MEETING CONSUMER DEMANDS FOR ELECTRICITY

Examine a bicycle equipped with a small electric generator for powering a headlight and you see two wires leading from the generator to the light. An on–off switch on the headlight completes the electrical circuit when light is desired. As long as the switch is off, there is no current in the light and a rider does no work against electric forces. But turn the switch on and a current develops. Then the rider must do extra work to supply the energy delivered to the lamp. The energy is not stored in the generator and delivered on request. Electric energy is produced on demand. If the cyclist requires no light and keeps the switch off, the generator delivers no energy anywhere. An automobile uses a battery to store energy and deliver energy to headlights and other units on demand. A form of electric generator called an alternator replenishes energy to the battery. Although this scheme works very well for an automobile, it is impractical for electric power plants to store large amounts of energy in batteries. Consequently, a large electric power plant produces electricity on consumer demand. If there is no demand, the generators shut down. Once switches are turned on in factories and homes, the electrical connection triggers

1. the production of heat by burning coal or fissioning uranium,

2. the vaporization of water (boiling) to produce steam, and
3. the injection of steam to a turbine to spin a shaft coupled to an electric generator.

The utility must be prepared to meet demands that vary throughout the day, the week, and the year. A representative hourly demand is shown in Fig. 4.16. The demand is sectioned into three parts termed base load, part time, and peak. Base load demand is met with large units having slow response. Usually these units are fueled by coal or uranium. They may be large hydroelectric plants. Base load electricity is the most economical of the three categories, and if the demand were constant only base load electricity would be needed. The demand labeled part time is generally provided by the same large units but because of the difficulty of slow response, electricity produced in the part-time mode is about two to three times more expensive than when the units are in the base load operation. Peak demands requiring fast response are met with oil- and gas-fired turbogenerators, hydroelectricity, pumped-storage hydro systems, purchases from other utilities and, in a few isolated cases, wind generators and solar-powered systems. Peak electricity may cost three to four times more than base load electricity.

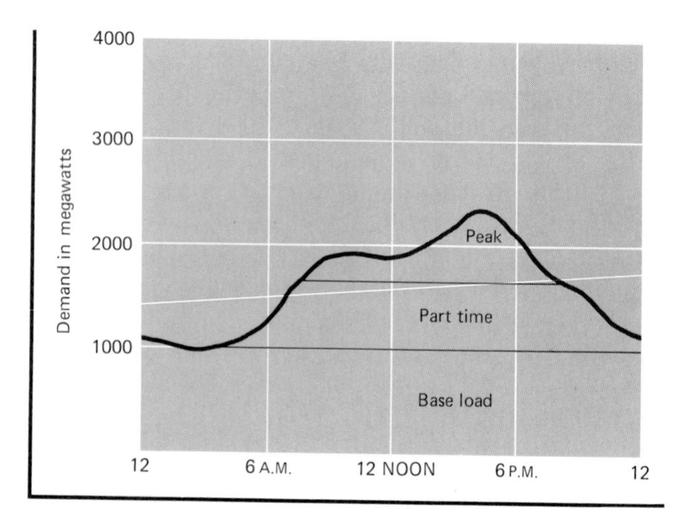

FIGURE 4.16 Hourly demand for electricity.

Economically it is in the interest of both the electric utility and the consumer to make an effort to "flatten out" the demand curve. As a result, a utility sometimes offers a cheaper late nighttime rate for electricity.

4.12 ENERGY AND POLLUTION MODEL OF A FOSSIL-FUEL ELECTRIC POWER PLANT

The steam turbine is the major source of mechanical energy for commercial electric generators in the United States. Water vaporized by the heat generated from burning fossil or nuclear fuels is forced onto blades attached to the shaft of the turbine (Fig. 4.17). It is somewhat like blowing air onto a pinwheel. With this basic model in mind, we can expand the system to that shown in Fig. 4.18. Several of the major components of this process are identified in the commercial facility shown in Fig. 4.19. These diagrams serve as a flowchart for energy and are useful for obtaining quantitative information. So let us "dump" 1000 pounds of coal (1000 pounds of coal would occupy a cubic space about three feet on a side) into the hopper and see how much electric energy comes out and what by-products evolve along the line.

To optimize burning, coal is pulverized to a powder. The pulverized product includes unburnable particles (fly ash) and sulfur along with the burnable carbon-based content. When burned, the sulfur combines with oxygen to form gaseous sulfur oxides. Unless controlled, the sulfur oxides and particulates escape out the smokestack. Thus assessing the power plant performance requires a knowledge of the energy content, sulfur, and unburnable content of the coal. These data are rather easy to come by* but there is a wide variation in quality.

A "typical" coal from the vast West Virginia coalfields might contain 72% carbon and 2.5% sulfur. For each pound burned, this coal produces about 13,000 Btu of heat, 0.05 pounds of sulfur oxides, 0.1 pound of fly ash, and 2.6 pounds of carbon dioxide if all the carbon is converted to carbon dioxide. Let

---

* *Handbook of Chemistry and Physics,* CRC Press Inc., Boca Raton, Fla.

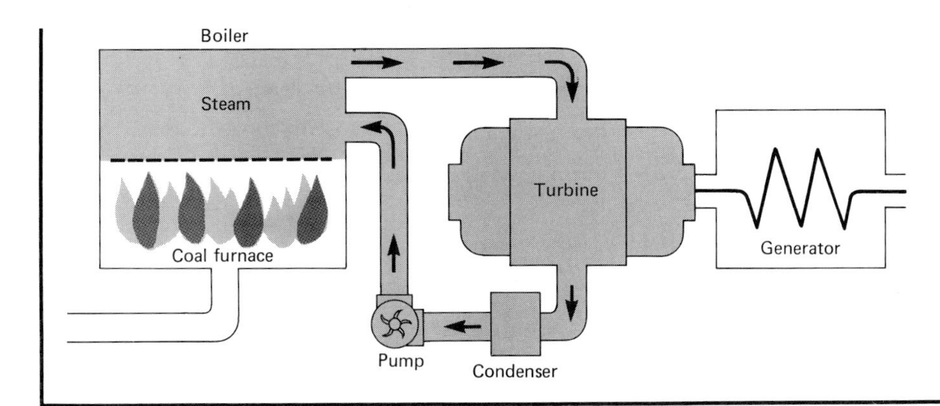

FIGURE 4.17 Schematic representation of the generation of electricity. High pressure steam impinges on the blades of a turbine producing rotational mechanical energy. The turbine rotates the generator shaft and the mechanical energy is converted to electric energy. The condenser converts the steam back into water and the pump circulates the water through the boiler.

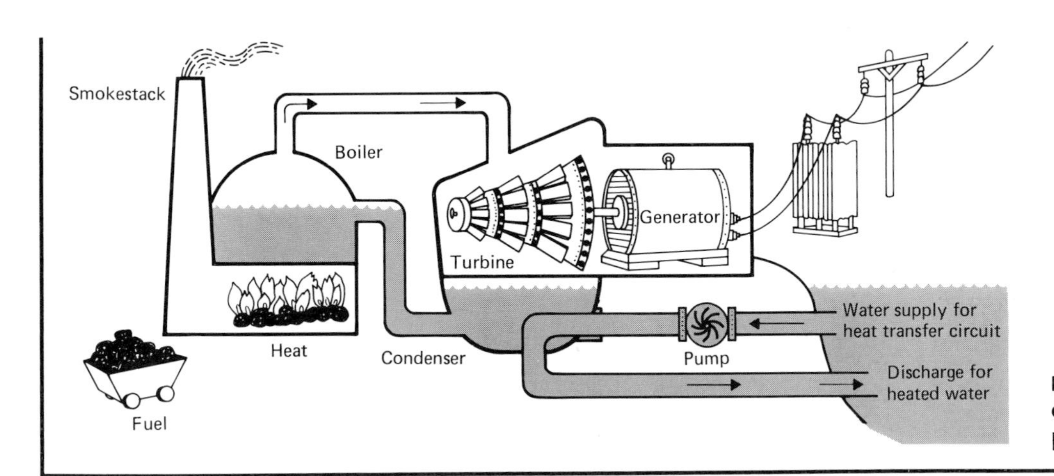

FIGURE 4.18 Major energy components of an electric power plant.

FIGURE 4.19 The Cincinnati Gas and Electric Company Beckjord Station located on the Ohio River. Several of the components shown schematically in Fig. 4.18 are identified in the photograph.

us now follow the 1000 pounds of coal through the power plant assuming the boiler, turbine, and electric generator have energy conversion efficiencies of 88, 47, and 99%, respectively.*

Boiler

$$\text{Energy available at the boiler} = \text{energy available per pound of coal} \times \text{number of pounds}$$
$$= \frac{13{,}000\ \text{Btu} \times 1000\ \text{pounds}}{\text{pound}}$$
$$= 13{,}000{,}000\ \text{Btu.}$$

$$\text{Energy into the turbine} = \text{energy in the boiler} \times \text{energy conversion efficiency of boiler}$$
$$= 13{,}000{,}000\ \text{Btu} \times 0.88$$
$$= 11{,}440{,}000\ \text{Btu.}$$

The energy that doesn't enter the turbine escapes, or is released, mostly through the smokestack.

$$\text{Energy out the smokestack} = \text{energy into boiler} - \text{energy into turbine}$$
$$= 13{,}000{,}000 - 11{,}440{,}000$$
$$= 1{,}560{,}000\ \text{Btu.}$$

The sulfur oxides and carbon dioxide are produced when the coal is burned. The amount of sulfur oxides produced when 1000 pounds are burned is

$$\text{sulfur oxides (lbs)} = 1000\ \cancel{\text{lbs of coal}} \times 0.05 \frac{\text{lbs}}{\cancel{\text{lb of coal}}}$$
$$= 50\ \text{lbs.}$$

These oxides do not necessarily escape out the smokestack if there is some mechanism installed to extract them. Modern power plants are equipped to remove about 90% of the sulfur oxides produced. We will assume that 5 pounds go out the smokestack and into the atmosphere.

The amount of carbon dioxide produced is

$$\text{carbon dioxide (lbs)} = 1000\ \text{lbs coal} \times 2.6 \frac{\text{lb carbon dioxide}}{\cancel{\text{lb}}}$$
$$= 2600\ \text{lbs.}$$

The carbon dioxide is vented to the atmosphere. The fly ash is simply an unburnable product in the coal.

$$\text{Fly ash} = \text{pounds of coal} \times \text{pounds of fly ash for each pound of coal}$$
$$= 1000 \times 0.1$$
$$= 100\ \text{pounds.}$$

Most power plants have facilities for removing about 99% of the fly ash before it goes out the smokestack. Hence 1% goes out the smokestack.

$$\text{Amount leaving smokestack} = 100\ (.01)$$
$$= \text{one pound.}$$

Turbine

$$\text{Energy out the turbine and into generator} = \text{energy into turbine} \times \text{energy conversion efficiency of turbine}$$
$$= 11{,}440{,}000\ (0.47)$$
$$= 5{,}377{,}000\ \text{Btu.}$$

The low value for this efficiency is probably the most difficult thing to accept in this discussion. However, there is a practical limit to the efficiency of a steam turbine which is imposed by the physical laws of thermodynamics. This is discussed in detail in Chapter 7.

Condenser

$$\text{Heat energy rejected to condenser} = \text{energy into turbine} - \text{energy into generator}$$
$$= 11{,}440{,}000 - 5{,}377{,}000$$
$$= 6{,}063{,}000\ \text{Btu.}$$

This heat energy goes into the condenser cooling water.

* The efficiencies used in these calculations were taken from "The Conversion of Energy" by Claude M. Summers in the book *Energy*, W. H. Freeman, San Francisco (1979). The definition of efficiency is given in Section 2.9.

Generator

$$\text{Energy out of generator} = \text{energy into generator} \times \text{energy conversion efficiency of generator}$$
$$= 5{,}377{,}000\,(0.99)$$
$$= 5{,}323{,}000 \text{ Btu}$$
$$= 5{,}323{,}000 \text{ Btu} \left(1055 \frac{\text{joules}}{\text{Btu}}\right) \cdot \frac{1}{3{,}600{,}000} \frac{\text{kWh}}{\text{joules}}$$
$$= 1560 \text{ kWh.}$$

Thus this 1000 pounds of coal would produce enough electric energy to run a 1560-watt air conditioner for 1000 hours.

With these numbers, the model of Fig. 4.18 can be quantified as in Fig. 4.20.

This procedure accomplishes several things.

1. Numerical results are obtained for a given set of conditions. This gives a feel for the emissions from a typical power plant.
2. The origin of the pollutants in the energy conversion process has been established.
3. Most importantly, a model applicable to any system of this type has been deduced. Only the numbers are different for another system of the same type.

The magnitude of the pollution cannot be fully appreciated because the element of time has not been considered. A 1000-megawatt unit operating at capacity would use the 1000 pounds of coal in about five seconds. This means that about 10,000 tons of coal are used per day and this would produce 500 tons of sulfur oxides, 26,000 tons of carbon dioxide, and 1000 tons of particulates.

REFERENCES

ELECTRIC ENERGY AND POWER

Extensive discussions of electricity, electric energy, and electric power are given in most physics and physical science texts. Two useful texts for expansion of the material presented in Chapter 4 are *Physics in Perspective,* Eugene Hecht, Addison-Wesley, Reading, Mass., 1980; and *Conceptual Physics,* Paul G. Hewitt, Little, Brown, Boston, Mass., 1981.

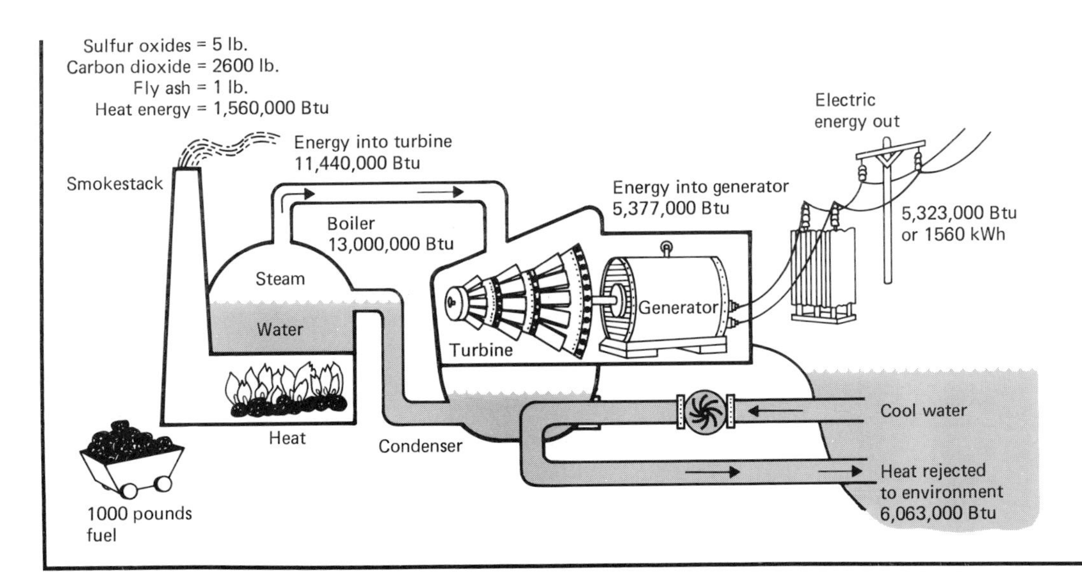

FIGURE 4.20 Typical outputs for conversion of 1000 pounds of coal into electric energy.

REVIEW

1. Devise an experiment to deduce whether or not a balloon has a net charge.
2. What type of energy is converted by an electric generator? How efficient is the generator in performing its stated function?
3. How is an electric current similar to the flow of water in a pipe?
4. What are the units of electric current, potential difference, power, and energy?
5. Make a sketch of an electrical circuit that includes a household outlet, a switch, and a light bulb.
6. Why are electrical devices producing heat more expensive to operate than household appliances like mixers producing mechanical motion?
7. How does electric power differ from electric energy?
8. What is a kilowatt-hour?
9. Distinguish between alternating current and direct current.
10. What physical principle involving electric charges and a magnetic field is employed in the production of an electric current by an electric generator?
11. Describe in words the function of a transformer. Why is a transformer useful in the transmission of electric power?
12. Why is electric power transmitted at very high voltages?
13. Name the by-products of significance produced by electric power plants using coal for input energy.
14. What is the most *inefficient* energy conversion device in an electric power plant?
15. Match the equations with the appropriate term and name the quantity going with each symbol.

Ohm's law	$I = \frac{Q}{t}$
voltage	$P = VI$
electric current	$I = \frac{V}{R}$
power	$V = \frac{W}{Q}$
power loss	$P = I^2R$

16. In a certain house, a 100-watt light bulb is normally on for 30 hours each week. If electricity costs 8¢/kWh and, for conservation reasons, it is decided to keep the bulb lit only half as much, then the savings per week would be
    a) 12¢ b) 6¢ c) \$60 d) \$120 e) 60¢
17. For a device that obeys Ohm's Law, if you halve the voltage across the device, then you
    a) double the current in the device.
    b) halve the current in the device.
    c) double the resistance of the device.
    d) quadruple the resistance of the device.
    e) do none of the above.
18. Homeowners pay their electric utilities bills in units of cents per
    a) watt.
    b) kilowatt.
    c) joule.
    d) Btu.
    e) kilowatt-hour.
19. A typical light bulb is rated at 100 watts. This rating means
    a) the bulb requires 100 watts of energy.
    b) the bulb converts 100 joules of energy each second it is on.
    c) the bulb could also be rated as 1 kilowatt.
    d) the bulb converts 100 watts of energy each second it is on.
    e) if the bulb is on 10 hours, it will use 1000 kilowatt-hours of electric energy.
20. When the switch in an electric circuit is in the off position, the current in the circuit is zero. Therefore, we would say the electrical resistance of the switch is
    a) zero.
    b) very large, essentially infinite.
    c) 115 ohms.
    d) impossible to determine without knowing the voltage.
    e) irrelevant. Only wires have electrical resistance.
21. An electric iron draws ten amperes, and operates on a 120-volt electric line for two hours. It used 2.4 kWh of energy and required the same power as twelve 100-watt light bulbs operating at the same time.
    a) true b) false
22. A certain 50-watt light bulb costs 50¢ and, on the average, lasts for 100 hours before burning out. Another longer-lived 50-watt bulb costs \$1.50 and lasts for 250 hours.
    a) The \$1.50 bulb is clearly a poorer buy.
    b) The \$1.50 bulb is clearly a better buy.
    c) The two bulbs are equivalent buys.

d) There is insufficient information to decide which bulb is better.
e) Both bulbs will use the same amount of energy in their lifetimes.

23. For water to flow through a pipe, there must be a water pressure difference between the ends of the pipe. For electric charge to flow through a wire there must be a (an) _____ difference between the ends of the wire.
    a) electrical b) current c) potential
    d) power e) charge

24. Different types of light bulbs are connected to a battery. The bulb having the least electrical resistance will produce the
    a) least amount of current.
    b) greatest amount of current.
    c) least amount of energy.
    d) least amount of electric power.
    e) least amount of potential difference.

25. A bird sitting on an electric power line does not get electrocuted because
    a) unless it touches another wire, there is no potential difference across the bird.
    b) its electrical resistance is too high.
    c) its electrical resistance is too low.
    d) the current in the wires is ac rather than dc.
    e) it wears rubber shoes as suggested by a student.

26. A night-light uses five watts of power. If electricity costs 10¢/kWh, the daily cost of continuous operation is
    a) 50¢ b) 0.5¢ c) $1.20 d) $5 e) 1.2¢

27. Electric current is generated essentially by forcing a large number of wires to move through a magnetic field.
    a) true b) false

28. A small electric generator is operated by a crank turned by a student. The student must do work in order to generate electricity because
    a) the magnetic field produced by the current tends to aid the magnetic field needed for the generation process.
    b) of the first law of thermodynamics.
    c) of Newton's third law.
    d) magnetic forces tend to oppose the motion provided by the student.
    e) of the electrical resistance of the wire.

29. Electric energy that your bedside clock uses during the night
    a) comes from electric energy stored by the power company during the day.
    b) is probably provided by solar powered generators.
    c) comes from batteries.
    d) is produced by generators operating during the night.
    e) probably costs more on a kWh basis than the energy used to run the clock in the daytime.

30. When electric power is transmitted from the generating site to a community, the company tries to keep the electric current small in order to
    a) keep energy losses small due to heating of the transmission wires.
    b) minimize magnetic energy losses.
    c) make the electrical resistance of the wires small.
    d) speed up the transmission of the electric power.
    e) protect birds that might perch on the transmission wires.

31. An electric power company uses transformers to
    a) transform ac currents to dc currents.
    b) increase the amount of electric power.
    c) prevent electrocution of birds sitting on the transmission lines.
    d) change the sizes of voltages.
    e) transform electric fields to magnetic fields.

32. Electric utilities, in an effort to reduce energy loss in transmission wires, employ _____ to step-up the _____ at the power plant to a very high level. (Similar devices are employed at the home to reverse this process.)
    a) transformers, voltage
    b) turbines, voltage
    c) transformers, current
    d) generators, voltage
    e) generators, resistance

33. In a commercial electric power plant, it takes about _____ units of energy to produce one unit of electric energy.
    a) 1 b) 2 c) 3 d) 4 e) 5

34. The commercial production of electric energy requires three energy conversion steps. If two of these steps have efficiencies of 90% and the overall conversion efficiency is 40%, then the efficiency of the third step is (pick the closest answer)
    a) 36% b) 50% c) 44% d) 90%
    e) 20%

35. Generation of electricity is a multistep process. Match the following converters with the particular energy

transformation accomplished by that converter. Assume a coal-burning electric power plant.

1. boiler	a. mechanical energy to thermal energy
2. steam turbine	b. chemical energy to thermal energy
3. generator	c. mechanical energy to electrical energy
	d. thermal energy to mechanical energy

a) 1-a, 2-b, 3-c
b) 1-b, 2-d, 3-a
c) 1-c, 2-b, 3-d
d) 1-b, 2-d, 3-c
e) none of the above are correct match-ups.

Answers

16. (a)	17. (b)	18. (e)	19. (b)
20. (b)	21. (a)	22. (a)	23. (c)
24. (b)	25. (a)	26. (e)	27. (a)
28. (d)	29. (d)	30. (a)	31. (d)
32. (a)	33. (c)	34. (b)	35. (d)

QUESTIONS AND PROBLEMS

4.2 ELECTRIC FORCE AND ELECTRIC CHARGE

1. Rubbing a glass rod with a piece of silk produces a net positive charge on the glass. Describe the mechanism by which the glass acquires the positive charge.
2. When there is little moisture in the air, it is common for a person to acquire a net electric charge (sometimes called static electricity) when getting out of a chair. How would you explain the acquisition of this charge?

4.3 ATOMIC STRUCTURE

3. A model airplane attached to a tethering cord flies in a circular path. In this sense it is bound to the controller like an electron is bound to the nucleus of an atom. What exerts the force on the plane that holds it in its circular path?

4.4 ELECTRICITY

4. For water to flow through a pipe, there must be a difference in water pressure between the ends of the pipe. What is the analogous situation for the flow of charge through a wire?
5. Figure 4.3 diagramming the circulation of water applies as well to the circulation of blood in the human body. What constitutes the pump, hose, valves, and container in the human system? How can fluid resistance occur in the human circulatory system?
6. Defend a friend who claims you have a 50% chance of not getting shocked if you touch one of the terminals of a household electrical outlet.
7. When a battery is connected to a small light bulb (Fig. 4.10), the direction of the flow of charge through the bulb depends on which terminal of the bulb is connected to the positive terminal of the battery. If you distinguish the two directions by calling one positive and the other negative, make a sketch of current and time if you changed the battery connections every second. How is this current similar to the ac current in a bulb connected to a household outlet?
8. An electrical shock results when two areas of a person's body touch the ground and "hot" leads of the electrical system in a household. The better the electrical connection, the more severe the shock. Knowing this, why is it especially dangerous to touch an appliance such as a hair dryer while in electrical contact with the water pipes in a bathroom?
9. If a current of one ampere exists in a wire, how many electrons flow past any position on the wire each second?
10. The lifetime of a battery depends on the amount of charge it delivers: the more the current, the shorter the lifetime. The quality of a battery is often specified by the number of ampere-hours (that is, amperes multiplied by hours) that it can provide.
    a) What physical quantity would an ampere-hour measure?
    b) If a battery is rated at 100 ampere-hours, how long would it last if it delivers 0.25 amperes?

4.5 ELECTRICAL RESISTANCE

11. Electrical resistance of a wire is figured by dividing potential difference and electric current. Using fluid analogies of potential difference and electric current, how could you compute the fluid resistance of a pipe?
12. Copper wires with large diameters have less resistance than copper wires of the same length but with smaller diameters. If the diameter doubles, why does the resistance decrease by a factor of four?
13. How does the current in a wire change if the potential difference between the ends is cut in half?

14. According to the equation $I = V/R$, current (I) decreases as resistance (R) increases. Why can a light switch in the off position be thought of as having very high (essentially infinite) electrical resistance?
15. The current in a light bulb is one ampere when connected to a 115-volt household electrical outlet. What is the resistance of the bulb and how much electric power does the bulb require?
16. When the voltage between the wire that leads to some device is 10 volts, the current in the device is one ampere. When the voltage is increased to 20 volts, the current increases to 2.5 amperes. Does this device obey Ohm's law? Explain.
17. A circuit breaker is a device placed in an electrical circuit to open the circuit like a switch when the current exceeds some given amount. In a house, circuit breakers for a line delivering power to small appliances are usually designed to open when the current exceeds 15 amperes. If the line voltage is 115 volts, what is the least resistance you could connect to a line and not have the circuit breaker open? How many 100-watt light bulbs could you connect to a line and not have the circuit breaker operate?
18. An electric toaster requires 1100 watts of electric power. How much electric current is required when the toaster is connected to a 115-volt household electrical outlet and what is the resistance of the toaster?

4.6 ELECTRIC POWER AND ELECTRIC ENERGY

19. A boy sliding down a rope from the top of a building converts gravitational potential energy into heat—his hands and the rope get warm. Charges flowing through a wire lose electric potential energy. Energetically, how is the electrical situation similar to the mechanical situation?
20. How does the wattage rating of a light bulb enter into figuring its cost of operation?
21. Electric power companies are paid on the basis of the number of kWh used by consumers. Would you find an appliance labeled in units of kWh? If not, what unit is used to rate an appliance?
22. A power company often reduces the generator voltage when demands for electric energy exceed the capacity for production. Recalling how power and voltage are related, why does reducing the voltage tend to conserve energy?
23. Electric stoves are usually designed to operate on 230 volts. If 115 volts were used on a stove designed to operate on 230 volts, does it mean that the power delivered drops by one half?
24. What is wrong with the physics in the following statement taken from a newspaper description of a new type of electronic device? "The device should save about 26,000 watts of electric power per hour."
25. What is the electric current in a 100-watt light bulb connected to a 115-volt outlet in a home?
26. A typical electric range requires 12,000 watts of power. How many amperes of current result when the range is connected to a 230 volt source?
27. A certain household clothes dryer requires 5000 watts of power. If electric energy costs ten cents per kWh, how much does it cost to run the dryer for 4 hours?
28. a) A typical household electric clothes dryer consumes electric energy at a rate of 6000 watts. Estimate the time required to dry a load of clothes and figure the total cost if the rate is ten cents/kWh.
    b) A night-light in a child's bedroom typically uses a 5-watt bulb. Estimate the cost per year if the bulb is left on continuously. Assume the rate is nine cents per kWh.
29. A certain 100-watt light bulb costs 75¢ and lasts for 120 hours. Another longer-lived 100-watt light bulb costs $1.20. How much longer would the more expensive bulb have to last to make it a "good buy"?
30. Selling electric energy is big business. To illustrate, assume that the electricity from a 1000-megawatt power plant is sold for 8¢/kWh. How much revenue is generated in a day's time if the plant operates at capacity the entire day?
31. Each of two electric clothes dryers requires 5000 watts of electric power. However, one costing $200 takes a half hour to dry a load of clothes and the other costing $150 takes one hour to dry an identical load. If electric energy costs 5 cents per kWh, how many loads would you have to dry with the more expensive model to make up the $50 difference in price between the two machines?
32. Light bulbs for home movie projectors typically have a lifetime of operation of 25 hours and a power rating of 500 watts. How much electric energy is used in the lifetime of one of these bulbs?

33. A certain 50-watt light bulb costs 50¢ and, on the average, lasts for 100 hours before burning out. Another longer-lived 50-watt light bulb costs $1.50 and has a lifetime of 250 hours. Is the longer-lived bulb a good buy? Explain.
34. A certain room is illuminated by 30 fluorescent lamps each having a light conversion efficiency of 10% and requiring 50 watts of electric power. A new 50-watt lamp is discovered that is 5% more efficient at producing light. How many of the new lamps would be required to produce the same amount of light as the 30 older-type lamps?
35. In the operation of a two-battery flashlight, the electric power delivered by the batteries is about five watts.
    a) Express the power in kilowatts.
    b) If the flashlight burns continuously, the batteries will "go dead" in about one hour. How many kilowatt-hours of electric energy are used in the lifetime of the batteries?
    c) If the batteries cost 50¢, how much per kWh did the energy cost? How does this cost compare with the cost per kWh from an electric power company?

4.7 MAGNETIC FORCE AND MAGNETISM

36. If the N pole of a compass points toward the north geographic pole, what kind of a magnetic pole is located near the north geographic pole? What merit would there be in calling one pole of a magnet positive and the other pole negative?
37. Automobiles are sometimes equipped with magnetic compasses for determining direction. Why might these compasses give erroneous readings when the car passes under electric transmission lines?

4.8 ELECTRIC GENERATORS

38. Many bicycles are equipped with a light powered by a small electric generator. The shaft of the generator is turned by allowing it to rub against one of the tires. Describe the basic operation of the generator. What energy transformations are involved in the operation of the generator? Why is the bicycle slightly easier to pump if the switch for the light is off?
39. A small electric motor produces rotational energy when connected to a flashlight battery. If the shaft of the motor is connected to the shaft of an identical motor, the second motor functions as a generator producing a potential difference between its coil leads. If a small bulb is connected to the leads, it will light up. Why isn't as much electric energy delivered to the bulb as was extracted from the battery to run the motor?
40. Figure 4.11 shows how a current develops in a coil of wire rotated in a magnetic field. Suppose the coil were held fixed and the magnets were rotated. Why would there still be a current in the coil?
41. A bicyclist cycling at a moderate speed expends about 25 food calories per hour. If an electric generator on the bicycle produces five watts of electric power when in operation, by how much will the bicyclist have to increase the power output (measured in food calories per hour) to maintain the same speed? (From Table 2.2, 1 food calorie = 4,186 joules and 1 joule per second = 1 watt.)

4.9 TRANSFORMERS

42. The input power and output power are identical in a perfect transformer. This is nearly the case for a practical transformer. What power losses are involved to make the output power smaller than the input power in a practical transformer?
43. If a flashlight battery is connected to the primary of a step-up transformer, a friend claims a voltage appears across the secondary as long as the battery is connected. Is he right or wrong? Why?
44. As an outdoor, visual exercise, see if you can identify near a house or building (1) a transformer, (2) a watt-hour meter, and (3) the ground wire near an electric utility pole.
45. Household door bells and chimes use a transformer to reduce the 115-volt house voltage to six volts. Does the primary winding contain more or fewer turns than the secondary in these transformers? If the secondary winding contains 1000 turns, how many turns are there on the primary?
46. The input power to a certain "perfect" transformer is 1000 watts. How much current is in the secondary if the secondary voltage is ten volts? From the information given, can you determine whether this is a step-up or a step-down transformer?
47. Using a step-up transformer, a student applies ten volts to the primary causing a current of two amperes in the primary. He reports that the potential difference is 20 volts and the current is two amperes in a light bulb connected to the secondary. What advice do you have for the student?

4.10 TRANSMISSION OF ELECTRIC POWER

48. Why are the wires used to transmit electric power separated by several feet when considerable space could be saved by placing them close together?

49. Why isn't a bird electrocuted by sitting on a conductor of an electric power transmission line?

50. A certain electric utility produces 120,000 kilowatts of power at 12,000 volts.
    a) How much current is delivered to the transmission lines?
    b) If the voltage is stepped-up to 240,000 volts for transmission, what is the current in the transmission line?

51. There is a trend toward increasing the voltage for transmitting electric power because heat losses decrease as current in the transmission lines decreases. Referring to Problem 50, if 10% of the power is lost when the 120,000 kilowatts of power produced are transmitted at 240,000 volts, how much power will be lost if it is transmitted at 480,000 volts?

4.12 ENERGY AND POLLUTION MODEL OF A FOSSIL-FUEL ELECTRIC POWER PLANT

52. Assume a 1000-megawatt power plant has an overall efficiency of 33⅓% and on the average it operates at 50% of capacity in a day's operation. Using coal having an energy content of 13,000 Btu per pound, show that a daily supply amounts to about 4700 tons. (Remember 1 Btu = 1055 joules and 1 watt = 1 joule/second.)

53. How many pounds of sulfur are contained in one ton of coal having 5% sulfur content?

54. Calculate the overall efficiency for the electric power plant model discussed in Section 4.12.

55. If a railroad car can transport 100 tons of coal and an electric generating facility requires 10,000 tons of coal per day, how many cars are required to supply coal for one day's operation? Comment on the importance of coal transportation facilities and long-term labor agreements.

56. A 1000-megawatt electric power output means electric energy is being produced at a rate of 1,000,000,000 joules per second. If the plant has a 33⅓% conversion efficiency, energy is derived from coal at a rate of 3,000,000,000 joules per second. Show that 1000 pounds of coal having a heat content of 13,000 Btu per pound burns in about four seconds in this plant.

57. a) How much heat is delivered by an 85% efficient boiler burning 5000 pounds of a typical coal? (See Table 2.2.)
    b) What happens to the energy that doesn't get into the turbine?

58. In 1900 about 20,000 Btu of heat delivered one kWh of electricity. Today it takes about 9000 Btu. What are the corresponding efficiencies? (Be careful of units.)

59. In the electric power plant model discussed in Section 4.12 it was assumed the coal came from a West Virginia coalfield. Redo the arithmetic assuming the coal came from Montana and had the following characteristics:

heat content = 10,000 Btu per pound
sulfur oxide production = 0.03 pounds per pound of coal
fly ash production = 0.12 pounds per pound of coal

5

Electric Power Plants and the Environment

5.1 MOTIVATION

The electricity that powers the light bulb of your reading lamp probably comes from a coal-burning electric power plant. Somewhere someone breathes air that contains products of the burning. Because those remnants have potential for damaging the human respiratory system and producing other detrimental environmental effects, a program for controlling power plant effluents has been underway since 1971.

The undesirable products from burning coal have their origin in nonhydrocarbon constituents. Sulfur compounds and unburnable ash are the most publicized but there are also concerns for nitrogen compounds and traces of mercury, lead, cadmium, selenium, nickel, arsenic, and radioactive radon gas. Burning the coal using air as a source of

These thick seams of coal in Wyoming contain sulfur and unburnable ash that lead to the production of sulfur oxides and particulates when the coal is burned. (United States Department of Energy photograph by Jack Schneider.)

oxygen produces sulfur oxides and nitrogen oxides, and pulverizing the coal makes particulate matter of the ash. The oxides complicate the environmental effects by participating in chemical reactions in the atmosphere. Carbon dioxide, although not biologically harmful, may contribute to a general warming of the earth through the so-called greenhouse effect (Section 8.7).

Regulation of emission limits for large coal-fired boilers were instituted in 1971. Stricter limitations followed in 1978. The effect on emissions is illustrated in Fig. 5.1. During the period shown, coal use rose by 33%. However, particulate emissions more than halved. Credit implementation of proven technology for this pronounced reduction. Sulfur oxides declined by about 16% which is very significant. But control of sulfur oxides is technically much more difficult than control of particulates, and the question of the best control method is very open. Emission limitations for nitrogen oxides are much less stringent and this is reflected in the nearly constant production seen in Fig. 5.1.

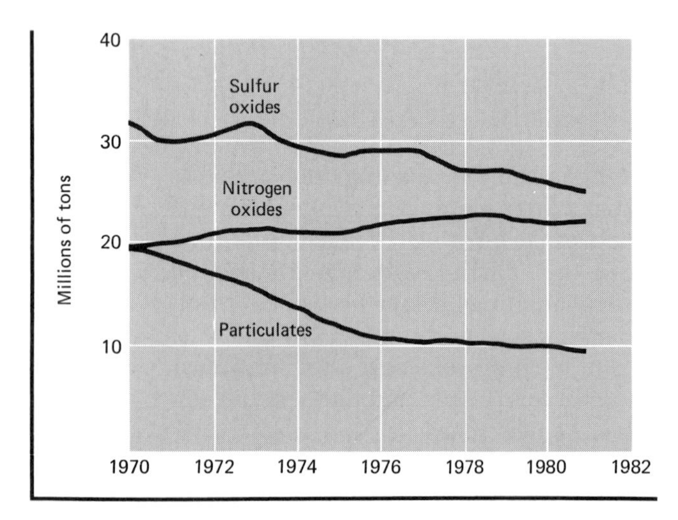

FIGURE 5.1 Annual emissions of sulfur oxides, nitrogen oxides, and particulates by all sources in the United States. Electric power plants burning coal produce the great bulk of the sulfur oxides and contribute significantly to the production of nitrogen oxides and particulates. Although coal use rose by 33% in the decade shown, particulate and sulfur oxide emissions declined as a result of controls placed on coal burning.

Undesirable by-products and risk are inescapable features of energy conversion. Our goal in this chapter is to examine coal-burning especially as it is related to coal-fired electric power plants that account for about two-thirds of the coal marketed.

5.2 MOLECULES AND CHEMICAL REACTIONS

Students generally include hydrogen, carbon, nitrogen, oxygen, and sulfur as words in their vocabulary. As noted in our study of electricity, these words identify atoms that have a specific number of electrons bound to nuclei by electric forces. Hydrogen has a single electron, carbon has six, and so on. The atoms are usually found in small associated groups called molecules. For example, the oxygen atoms we breathe enter our body in pairs, and the sulfur oxides alluded to in the previous section are configurations of sulfur and oxygen atoms. Chemical reactions usually involve molecular interactions, and often energy is liberated. Heat is liberated in gasoline burning through chemical reactions involving hydrocarbon and oxygen molecules. Similarly sulfur oxides are produced through chemical interactions of sulfur atoms and oxygen molecules. Understanding the rudiments of molecules and chemical reactions is essential for discussing the liberation of energy in burning coal and gasoline and the origin of undesirable by-products.

A normal atom is electrically neutral. Thus there is no electric force between separated atoms. But if the separation of atoms decreases sufficiently, the electronic charges of the atoms redistribute in such a way that attractive electric forces come into play tending to bind the atoms together. For example, hydrogen in the air around us is in the form of molecules each having two hydrogen atoms. The two atoms are close enough for the electron of one atom to be transferred to, or shared by, the other atom. The sharing of the two electrons results in a net attractive force holding the two atoms together. This type of binding is called covalent. There are various types of fairly complex binding (or bonding) having names like ionic, metallic, or hydrogen bonds. The details of these bonds are very important, but not necessary for our energy discussions. We need only know that the binding force for molecules is electric in nature and results from the redistribution of electronic charge.

It may seem strange that configurations of atoms are more likely to occur than single units. However, the determining factor is energy, and atoms like any physical system seek the lowest energy state. It is analogous to a marble rolling around on a desk top. Never do you see the marble spontaneously move to a higher level. But it may roll off the desk to the floor to be trapped at a lower energy level. To regain its original freedom on the desk top, it must acquire energy perhaps from a student picking it up. To say oxygen is more stable as a molecule having two atoms is saying the two-atom unit has achieved a lower energy level by being attracted together by electric forces. Energy must be added to the molecule to restore it to its condition of two separated atoms. Symbolizing the bound oxygen molecule as O_2 we symbolize the process as

$$O_2 + \text{energy} \rightarrow O + O.$$

The oxygen atoms are rearranged into different configurations but there is no change in the number of atoms in the rearrangement. This is a conservation-of-atoms law that applies to all reactions involving atoms and molecules. Energy is released when two oxygen atoms join to form an oxygen molecule. A symbolism for this process is

$$O + O \rightarrow O_2 + \text{energy}.$$

Energy is released in chemical reactions when atoms and molecules are rearranged to form more tightly bound lower energy configurations. This happens in the combustion of coal. Coal is primarily carbon (C). If the combustion is perfect, the carbon unites with oxygen (O_2) to form carbon dioxide (CO_2) and energy.

$$C + O_2 \rightarrow CO_2 + \text{energy}.$$

C + O O ⟶ C O O + energy

The combustion is not perfect and some carbon monoxide (CO) is formed.

$$2C + O_2 \rightarrow 2CO + \text{energy}.$$

C + C + O O ⟶ C O + C O + energy

Coal and oil contain sulfur left over from the sulfur in prehistoric plants and animals. When the coal is burned the sulfur (S) combines with oxygen to form sulfur dioxide (SO_2).

$$S + O_2 \rightarrow SO_2 + \text{energy}.$$

S + O O ⟶ S O O + energy

To a lesser extent other sulfur oxides such as SO_3 are formed. Air in the environment is the usual source of oxygen. About 21% of this air is oxygen in the two-atom molecular form. About 78% is nitrogen in a two-atom molecular form (N_2). In the presence of oxygen and the high temperature resulting from combustion, nitrogen and oxygen combine to form nitric oxide (NO) and nitrogen dioxide (NO_2).

$$N_2 + O_2 \rightarrow 2NO.$$

$$N_2 + 2O_2 \rightarrow 2NO_2.$$

Many other aspects of physics and chemistry can be built into these chemical reaction models. These energy ideas are sufficient to see the fundamental origin of undesirable by-products and heat energy. Let us proceed to examine the environmental and technological consequences of these by-products.

5.3 PARTICULATE MATTER

General Properties

Particulate matter emerging from the smokestack of a coal-burning electric power plant (Fig. 5.2) enters the atmosphere in the form of fine, solid particles. These particles intersperse with those from many other sources of particle-like pollutants. Some pollutants are in the liquid state and it is appropriate to term them particulates and consider them along with the solids.* Like a dust particle, particulates generally have jagged shapes. Nevertheless, particulates are classified according to size. The size refers

* The world *aerosol* is often used interchangeably with particulate. Some scientists, though, prefer to consider aerosols as particles with diameters less than some particular value, for example, 0.0001 meters.

FIGURE 5.2 The Killen Electric Generating Station near Manchester, Ohio, on the Ohio River is a modern coal-fired electric power plant that complies with strict emission standards. More than 99% of the mass of the particulates produced are captured before entering the atmosphere. (Photograph courtesy of Dayton Power and Light Company.)

to a length associated with an assumed shape of the particulate. If the particulate is assumed to be spherical, the size is determined by the diameter (or radius). Regardless of the assumed shape, the size is roughly the same unless the assumed shape is very odd. For simplicity, we presume the particulates are spherical and use the concept of mass density to deduce the diameter. Mass density is a measure of how much mass is contained in a prescribed volume. Formally,

$$\text{mass density} = \frac{\text{mass of an object}}{\text{volume occupied by the mass}}$$

or

$$\rho = \frac{M}{V}. \tag{5.1}$$

(The Greek letter rho [ρ] is the symbol used normally for mass density.) Mass density has units of mass divided by the units of volume. Metric units are kilograms per cubic meter (kg/m^3). Water has a mass density of 1000 kg/m^3. Regardless of geometric shape, one m^3 of water has a mass of 1000 kg. To avoid handling rather large numbers, density is often expressed in grams per cubic centimeter (g/cm^3). Because a kilogram = 1000 grams, and a cubic meter = 1,000,000 cubic centimeters, then water has a density of one g/cm^3. If you measure the mass of a certain amount of water to be 10 grams using a balance like that used to "weigh" a letter for mailing, you know its volume is 10 cm^3. Similarly, if the density of particulate matter is 2 g/cm^3, then, knowing its mass from some measurement, you can compute its volume from the expression

$$V = \frac{M}{\rho}. \tag{5.2}$$

For example, if a given particulate has a mass of one millionth of a gram (10^{-6} g) and a density of one g/cm^3, it has a volume of one millionth (10^{-6}) of a cubic centimeter. If this particulate were rolled into a tiny ball, it would have a diameter of 0.012 centimeter. A dime has a diameter of about one cm so the diameter of the particulate would be about one hundredth the diameter of a dime. This seems small but actually this is a fairly large particulate. The sizes of particulates range from about 0.00000002 cm to 0.05 cm. Rather than work with such very small numbers, it is common to express these sizes in terms of one millionth of a meter (or one ten-thousandth of a centimeter). One millionth of a meter is called a micrometer and is symbolized μm. On the micrometer scale, the particulate range 0.00000002 cm to 0.05 cm would be 0.0002 μm to 500 μm.

Particulate concentrations in the atmosphere are reported as the total mass contained in a cubic meter of air. Typical units are millionths of grams (micrograms) per cubic meter, symbolized μg/m^3. If a cube of air one meter on a side contained 100 millionths of a gram of particulates, we record the concentration as 100 μg/m^3. Mass concentrations range from about 10 μg/m^3 in remote nonurban areas to 2000 μg/m^3 in very heavily polluted areas. The annual average concentration in United States cities is about 100 μg/m^3. Even though the concentrations vary substantially throughout the country, the percentage having a particular size is essentially the same everywhere. Most of the particulates are very small—less than 0.1 μm in size. However, larger particulates with size greater than 0.1 μm account for about 95% of the total mass. Thus relatively few particulates account for the bulk of the mass. This fact is very important because smaller-sized particles may be more damaging to people than the larger-sized particles. It may be more important to reduce the number of smaller-sized particulates than to reduce the concentration of mass that tends to eliminate only the larger particulates.

Particulates evolve from several natural and industrial sources. Those smaller than one μm come principally from condensation and combustion processes. Those with sizes between one μm and ten μm come from such things as industrial combustion products and sea sprays. Sizes greater than ten μm result from mechanical processes such as wind erosion and grinding. The sizes and origin of atmospheric particulates are shown in Fig. 5.3.

Of the several environmental concerns over particulates in the air, the following are noteworthy.

1. The dirt particulates produce after settling to the ground is annoying and its removal requires energy and money.
2. Particulates produce effects detrimental to materials, plants, and animals, including human beings.
3. There is concern that accumulation of particulates in the atmosphere may alter the heat balance of the earth through reflection and absorption of solar radiation.

These effects are complicated by the fact that many of the particulates remain aloft for unusually long periods and in some cases become permanently air-

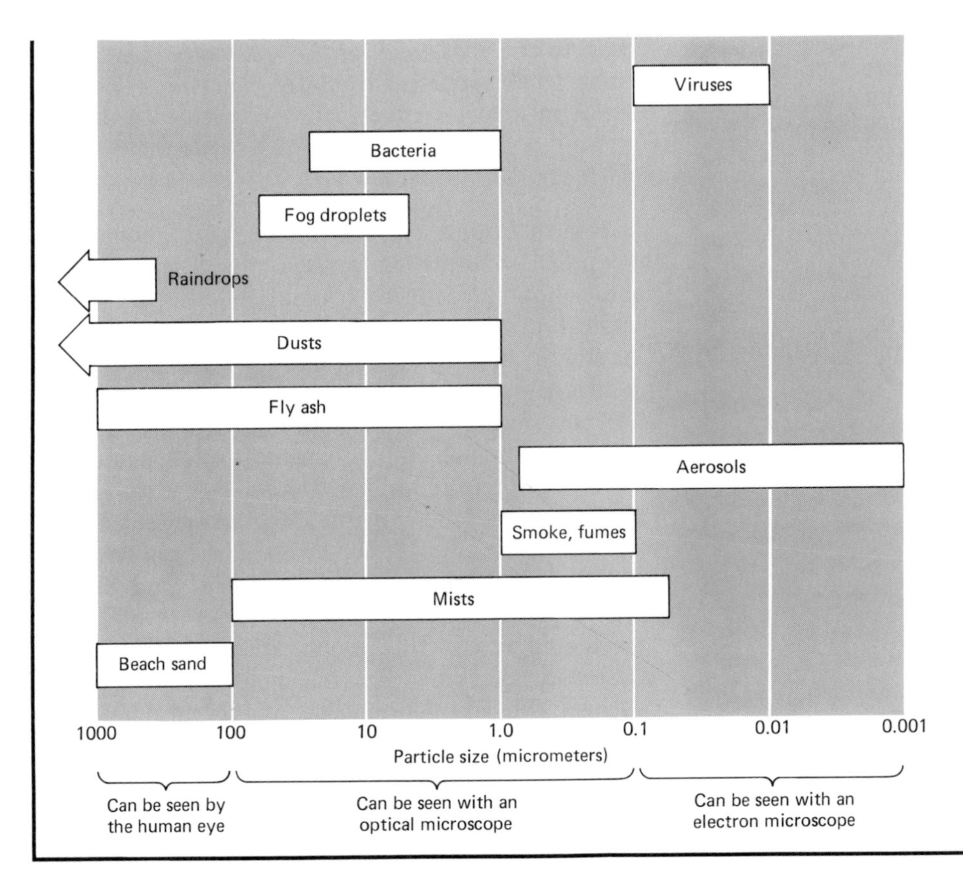

FIGURE 5.3 The sizes and origin of atmospheric particulates. Note that aerosols are labeled as particles with diameters of less than one micrometer. To obtain a feel for the size scale, think about raindrops and beach sand that you can see with your naked eyes. Baby powder is much finer, each particle ranging from about 0.5 to 50 micrometers. Molecules are much, much smaller and can be seen only through a powerful microscope.

borne. Let us examine the reasons for these phenomena and then elucidate the environmental effects.

Settling of Particulates in the Atmosphere

A solid plastic sphere dropped in a thick liquid like oil or shampoo is pulled downward by gravity and held back by (1) a viscous force exerted by the fluid on the sphere, and (2) a buoyant force tending to float the sphere. The viscous force is like that exerted on your hand when you hold it out the window of a moving car. The buoyant force causes a child's helium-filled balloon to rise.* As the sphere gains speed, the viscous force increases. Ultimately the force of gravity is balanced by the opposing viscous and buoyant forces and the sphere descends with constant speed. Any size sphere ultimately achieves constant speed, but the limiting speed achieved by spheres made of the same material increases as the diameter increases. Drop a handful of sand in a glass of water and watch the big grains touch bottom first. Particulates in the atmosphere are not plastic spheres and their motion is influenced greatly by winds, but the larger ones settle faster for exactly the same reason. Smaller particulates sometimes become permanently airborne and do not settle at all. A simple calculation using the rate concept shows why.

Table 5.1 shows the relation between settling velocity and particulate diameter. Let us determine the time it takes a 1-micrometer diameter particulate to settle 1 kilometer (0.62 mile). A 1-micrometer diameter particulate has a settling velocity of 0.004 centimeters/second. Thus

$$\text{settling time} = \frac{\text{distance traveled}}{\text{settling velocity}}$$
$$= \frac{100{,}000 \text{ centimeters}}{0.004 \text{ centimeters/second}}$$
$$= 25{,}000{,}000 \text{ seconds}$$
$$= \frac{25{,}000{,}000 \text{ seconds}}{3600\,(24) \text{ seconds/day}}$$
$$= 290 \text{ days.}$$

* Buoyant force is discussed in more detail in Section 8.5.

TABLE 5.1 Approximate settling velocities in still air for particles having a density of 1g/cm^3 (From *Air Quality Criteria for Particulate Matter,* National Air Pollution Control Administration publication No. AP-49).

Diameter (micrometers)	Settling velocity (centimeters/second)
0.1	0.00008
1	0.004
10	0.3
100	25
1000	390

Particulates with diameters less than one micrometer settle so slowly that they tend to migrate thousands of miles before settling to the ground. It is not uncommon to find that dust particles characteristic of the earth in Arizona have migrated as far east as New York. Or to find radioactive debris from nuclear bomb tests migrating across the Pacific Ocean to the east coast of the United States. Permanently airborne particulates cause concern for long-term weather effects.

Environmental Effects

Health Effects Several incidents of unusually high pollution levels were caused by peculiar weather conditions that produced stagnation of the air. (The physical origin of these conditions is discussed in Chapter 8.) In some areas, particularly London, the conditions were complicated by heavy fogs. Many of the incidences were accompanied by significant increases in health effects. The most-cited incidents of this type occurred in the Meuse Valley, Belgium (1930), Donora, Pennsylvania (1948), London (1952), and New York City (1953, 1966). The London episode produced some 4000 excess deaths. Some 5910 persons (42.7% of the total population) were affected to some degree in the Donora incident. Based on studies such as these, certain conclusions have been reached. They are not absolute but they are helpful as guidelines for setting standards. Two typical fairly reliable conclusions are as follows:

1. At concentrations of 750 $\mu g/m^3$ and higher for particulates on a 24-hour average, accompanied

FIGURE 5.4 Cincinnati City Hall photographed during cleaning in 1963 shows the accumulation of 34 years of dirt. (Photograph courtesy of the EPA.)

by sulfur dioxide concentrations of 715 μg/m^3 and higher, *excess deaths* and a considerable *increase in illness* may occur.

2. If concentrations above 300 μg/m^3 for particulates persist on a 24-hour average and are accompanied by sulfur dioxide concentration exceeding 600 μg/m^3 over the same period, *chronic bronchitis* patients will likely suffer *acute worsening of symptoms*.

Details leading to these conclusions are given in the literature.* The points to be made here about health effects are as follows:

1. There are documented correlations between high particulate concentrations and adverse human health effects.
2. Damage is done primarily to the respiratory system. In particular, those suffering from bronchitis and other respiratory ailments are adversely affected.
3. Sulfur oxides must also be considered when assessing effects of particulates because of synergistic effects.

More will be said about these effects in the discussion of sulfur oxides.

Effects on Materials The fallout of particulates produces an obvious unsightliness on buildings and streets (Fig. 5.4). Apart from this, particulates, especially when accompanied by sulfur oxides and moisture, can literally attack structural materials. For example, steel and zinc samples corroded three to six times faster in New York City than in State College, Pennsylvania, where the particulate concentrations were about 180 and 60 μg/m^3, respectively.

Effects on Vegetation There is little evidence of the direct effect on plant life of the mixture of

* See, for example, *Air Pollution: Threat and Response*, David A. Lynn, Addison-Wesley, Reading, Mass., 1976.

particulates typically found in the air. There is, however, evidence that specific types of particulates do produce damage. One of many examples is the case in which a marked reduction in the growth of poplar trees one mile from a cement plant was observed after cement production was doubled.

Federal Standards for Particulate Concentrations

Of several pieces of federal legislation designed to protect environmental quality, the 1970 Clean Air Act is the most forceful. The Act enables the Environmental Protection Agency (EPA) to set air quality standards for designated pollutants, and to promulgate and enforce the standards. Primary standards define levels of air quality judged necessary to protect the public health with an adequate margin of safety. Secondary standards define levels of air quality judged necessary to protect public welfare, for example, by protecting property, materials, and economic values. To effect the standards, the nation is divided into 247 regions and the states or parts thereof within a region are obligated to comply. A state may impose more stringent standards but cannot relax the standards. The federal standards for particulates are the following.

The maximum average 24-hour concentration that is not to be exceeded more than once per year is	
primary	260 micrograms per cubic meter of air,
secondary	150 micrograms per cubic meter of air.
The annual geometric mean* concentration cannot exceed	
primary	75 micrograms per cubic meter of air
secondary	60 micrograms per cubic meter of air.

* The most familiar mean (or average) value is formally called the arithmetic mean and is obtained by summing the quantities of interest and dividing by the number of quantities. For example, the average class grade for an examination is determined by adding all the grades and dividing by the number of students. The geometric mean is obtained by multiplying the quantities of interest and taking the *N*th root of the product where *N* is the number of quantities. For example, if 25 students took an examination, the geometric mean would be the 25th root of the product of all 25 grades.

To improve and maintain air quality, particulate emission limits are placed on coal-fired boilers. The 1970 Clean Air Act restricted particulate emissions to 0.1 pound for each million Btu of heat derived from burning. An Appalachian coal having 10% unburnable ash and liberating 12,000 Btu of heat when burned produces 8.3 pounds of particulates per million Btu. Thus the act requires capturing over 98% of the particulates generated. Still stricter limits were placed on units built after September 18, 1978. These systems can liberate to the environment no more than 0.03 pounds of particulates for each million Btu produced. Figures 5.1 and 5.5 graphically illustrate the decline of particulate emissions following imposition of these restrictions.

Particulate Collection Devices

No single particulate collection device can capture particulates of all sizes. Consequently, a utility employs combinations of collectors, each of which works well for a range of sizes. The most common collectors are called gravitational, cyclone, electrostatic precipitator, and fabric filtration. We have at hand the basic principles for understanding their operation.

Gravitational Collector The principle of the gravitational collector is shown in Fig. 5.6. The gases containing a suspension of particulates from the fuel combustion enter a large chamber. Like particulates in the atmosphere, they are influenced by both the gravitational force of the earth (pulling down) and retarding forces (drag and buoyant) due to the gaseous medium. Some of the particles will settle enough so that they will fall through the bottom of the container and be collected. As expected from our discussion of particulates settling in the atmosphere, the ones of larger diameter settle fastest and are the ones most likely to be collected. Gravitational collectors are extremely useful because of their simplicity and ease of maintenance, but they function efficiently only for particulate diameters greater than about 50 micrometers.

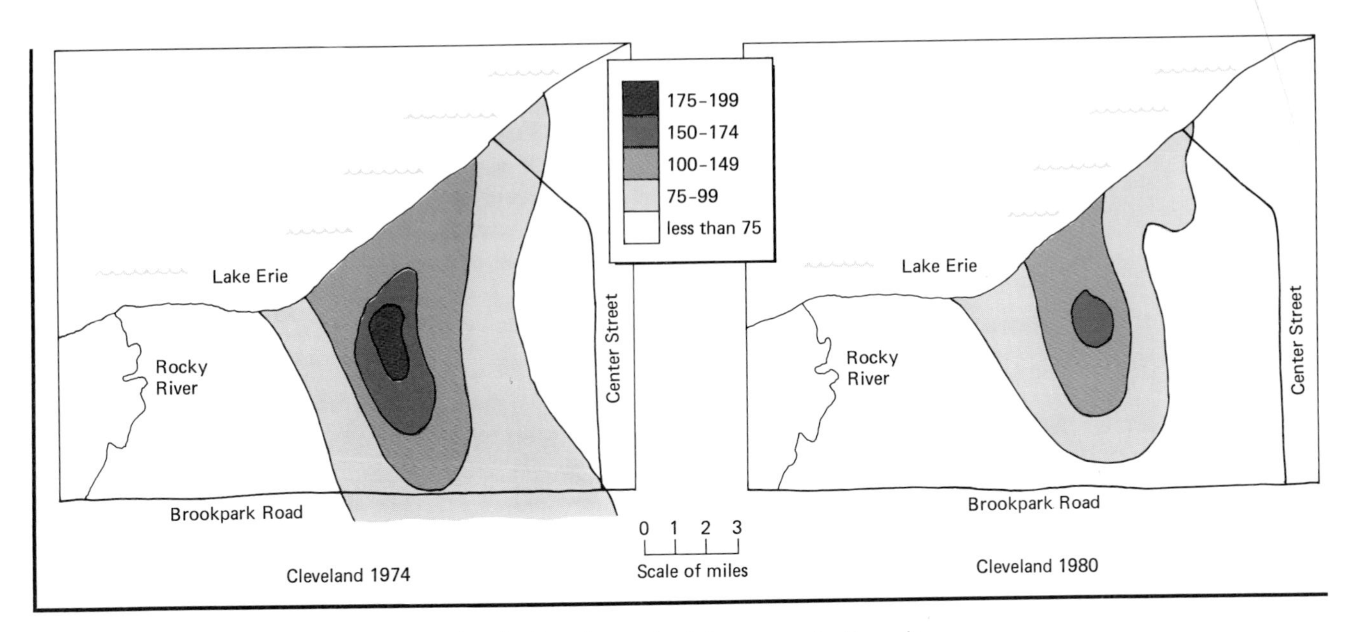

FIGURE 5.5 These two maps depict the decline in particulate concentrations in Cleveland, Ohio, between 1974 and 1980. Many other United States cities experienced a similar improvement in air quality. The shaded areas mark regions exceeding the federal primary standard. (Maps courtesy of the City of Cleveland.)

Cyclone Separator Any rapidly swirling air mass is loosely referred to as a cyclone. Such a condition is created with the combustion gases in a cyclone separator. As a result particulates are forced to the outer edge of the cyclone where they can be collected. In a cyclonic separator, the gas containing the particulates is forced down through a tapered cylinder to produce cyclonic action (Fig. 5.7). The

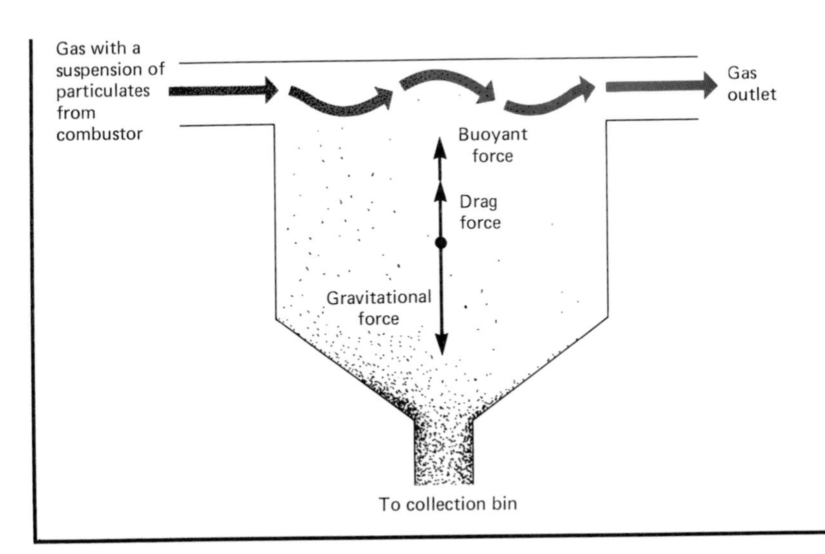

FIGURE 5.6 Schematic illustration of a gravitational collector. Particulates are influenced by forces exactly like particulates in the atmosphere. The gravitational force pulls them downward; a drag force and a buoyant force due to the air medium oppose the downward motion.

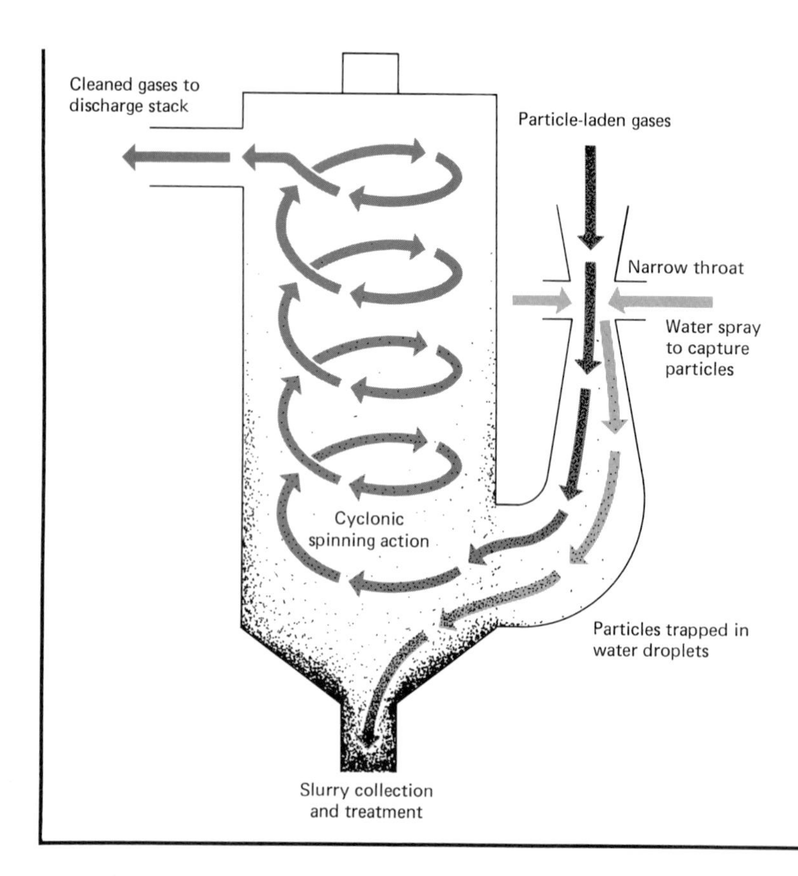

FIGURE 5.7 Schematic illustration of the cyclone particle collector. The cyclonic air motion forces particulates toward the walls of the collector. When the particulates strike the wall, they fall into a collection bin.

particulates move toward the walls of the cylinder and some strike the walls and fall into a collection bin. This type of device collects particulates as small as 5 micrometers.

Electrostatic Precipitator The electrostatic precipitator is the most efficient device for removing particulates with diameters less than 5 μm. It is based on the principle that unlike electric charges attract each other. In principle, the precipitator is a hollow container, say a cylinder, with wires located in it (Fig. 5.8). Provisions are made to make the wires strongly electrically negative and the cylinders strongly electrically positive. The electric force produced on electrons within the wires is large enough to literally pull electrons from the wires. The freed electrons then migrate toward the positively charged cylinder wall.

Some of the gas molecules, such as oxygen (O_2), will capture an electron and acquire a net negative charge (O_2^-). These molecules are accelerated toward the wall of the positively charged cylinder. During transit, the O_2^- ion may attach itself to a particle and the particle-O_2^- composite migrates toward the wall. Once the composite strikes the wall, it becomes electrically neutral and the electric force vanishes. The particulates are then removed from the walls of the container.

The electrostatic precipitator is the most efficient of the systems discussed and is capable of removing 99% of the total *mass* concentration. However, the efficiency for particulates less than 0.1 μm in diameter is poor. Even though the electrostatic precipitator may remove 99% of the mass, only about 5% of the *number* of particulates are accounted for because there are many more of the

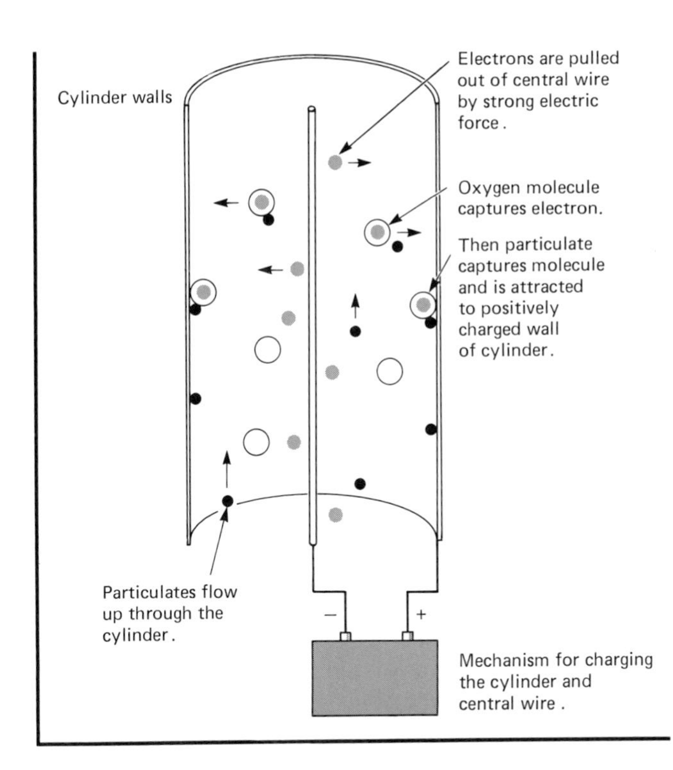

FIGURE 5.8 Schematic illustration of an electrostatic precipitator. The particulates having acquired a net negative charge are attracted toward the wall of the precipitator. When the particulates strike the wall, they become electrically neutral and fall into a collection bin.

smaller particulates in the gas effluent. The electrostatic precipitator has difficulty in extracting the small particulates. That electrostatic precipitators are effective is illustrated by Fig. 5.9 (a and b).

The Disposal Problem

In 1981 the electric utilities burned about 600 million tons of coal having about 10% unburnable ash. Because the collection efficiency is very good, about 60 million tons of coal ash were collected. For the most part, the ash is taken from the collectors and dumped into ponds where it settles to the bottom. Some commercial use is made of the by-product but it amounts to only about 13% of that collected. Some of the uses are as an aggregate for building bricks, improved traction material for car tires, fire-quenching material for coal mines, and fertility improver for soils.

(a)

(b)

FIGURE 5.9 The effectiveness of electrostatic precipitators for removing particulates escaping from a smokestack is illustrated in these pictures taken **(a)** without and **(b)** with the electrostatic precipitators in operation. The white clouds that appear in both photographs are composed of water vapor. (Photographs courtesy of Eastman Kodak Company.)

5.4 SULFUR OXIDES

General Properties

Sulfur dioxide (SO_2) and sulfur trioxide (SO_3) are the major sulfur oxides produced when coal is burned. However, the production of sulfur dioxide is some 40 to 80 times more probable than the production of sulfur trioxide. Sulfur dioxide is a nonflammable, nonexplosive, colorless gas. Most people can taste it in concentrations of about 2500 $\mu g/m^3$ in air. It produces a pungent, irritating odor for concentrations greater than about 8000 $\mu g/m^3$ in air.

It is easy to estimate how much SO_2 is produced from burning a ton of coal with a given sulfur content. However, because of the large variation in sulfur content, it is difficult to estimate how much is being produced in the entire country. Sulfur dioxide contributes about 16% of the weight of all atmospheric pollutants.

Environmental Effects

Health Effects The fate of sulfur dioxide in an atmosphere containing a variety of other pollutants is of significance because the sulfur dioxide reacts chemically and produces other more toxic species. For example, sulfur dioxide can be converted to highly corrosive sulfuric acid (H_2SO_4). One might think of this as happening by combining SO_2 and oxygen to form SO_3 according to

$$2SO_2 + O_2 \rightarrow 2SO_3.$$

SO_3 combines with water (H_2O) to form H_2SO_4 according to

$$SO_3 + H_2O \rightarrow H_2SO_4.$$

This reaction proceeds slowly because the formation of SO_3 by the first reaction above is slow. A more likely method of forming SO_3 employs a catalyst that modifies the reaction rate but is not consumed in the process. One possible catalyst is nitrogen dioxide (NO_2), a combustion product. Nitrogen and sulfur dioxide react according to

$$NO_2 + SO_2 \rightarrow NO + SO_3.$$

The NO_2 used to form SO_3 is then regenerated by the reaction

$$2NO + O_2 \rightarrow 2NO_2.$$

There are many catalysts in a polluted atmosphere contributing to the formation of sulfuric acid.

The effects of sulfur dioxide, like particulates, on human health are obtained from epidemiological studies. Although these effects are difficult to assess quantitatively, they have the distinct advantage of being derived from cases involving human subjects. Thus they are extremely valuable for deciding on allowable concentrations of various pollutants. Three fairly reliable conclusions from studies of this type are as follows:

1. At concentrations of about 500 $\mu g/m^3$ averaged over a 24-hour period, with low particulate levels, *increased mortality* rates may occur.
2. At concentrations of about 715 $\mu g/m^3$ averaged over a 24-hour period, accompanied by particulate matter, *a sharp rise in illness rates* for patients over age 54 with severe bronchitis may occur.
3. At concentrations of about 120 $\mu g/m^3$ averaged over a year, accompanied by smoke concentrations of about 100 $\mu g/m^3$, *increased frequency and severity of respiratory diseases* in school children may occur.*

It is important to note that (1) the elderly, the young, and persons suffering from respiratory ailments such as bronchitis and emphysema are particularly affected; and (2) there is an established synergistic effect between sulfur dioxide and particulates.

Effects on Vegetation Plants, in general, are easily damaged by sulfur dioxide (Fig. 5.10). There is, however, a wide variation in the amount of SO_2 needed to damage different types of plants. Acute injury occurs in alfalfa (a common food crop for livestock) at concentrations of about 3300 $\mu g/m^3$ for one

---

* More of the details leading to these and several other similar conclusions are given in *Air Quality Criteria for Sulfur Oxides*, EPA Publication AP-50.

FIGURE 5.10 Birch leaves. **(a)** This leaf shows injuries caused by sulfur dioxide. Compare with **(b)**, healthy leaf. (Photograph courtesy of the United States Department of Agriculture.)

hour. About 50,000 $\mu g/m^3$ for one hour is the concentration required to produce an equivalent damage in privet (a type of shrub commonly used for hedges). Damage has been reported for chronic exposures as low as 80 $\mu g/m^3$. The formation of sulfuric acid, as described earlier, that attacks plants is also a problem as are synergistic effects of sulfur dioxide with ozone and nitrogen dioxide.

Effects on Materials Combinations of particulates and sulfur dioxide do significant damage to materials. This is especially true if conditions are favorable for the formation of sulfuric acid. Building materials and statues often are discolored and disfigured from attacks by sulfuric acid. In addition, metals corrode and materials such as cotton, nylon, rayon, and leather are damaged by sulfur dioxide.

Federal Standards for Sulfur Dioxide Concentrations

The federal standards for sulfur dioxide concentrations are the following.

Primary
The maximum average 24-hour concentration that is not to be exceeded more than once per year is 365 micrograms per cubic meter.
The maximum average annual concentration is 80 micrograms per cubic meter.
Secondary
The maximum average 3-hour concentration that is not to be exceeded more than once per year is 1300 micrograms per cubic meter.

Substantial improvement has been made in the control of sulfur oxides since the 1970 Clean Air Act. The situation in Cleveland is typical (Fig. 5.11). Unlike the nationwide reduction of particulate concentrations, the reduction of sulfur dioxide concentrations has not evolved entirely from implementing proven technology. Rather it stems mostly from burning coal containing less than 1% sulfur. Because

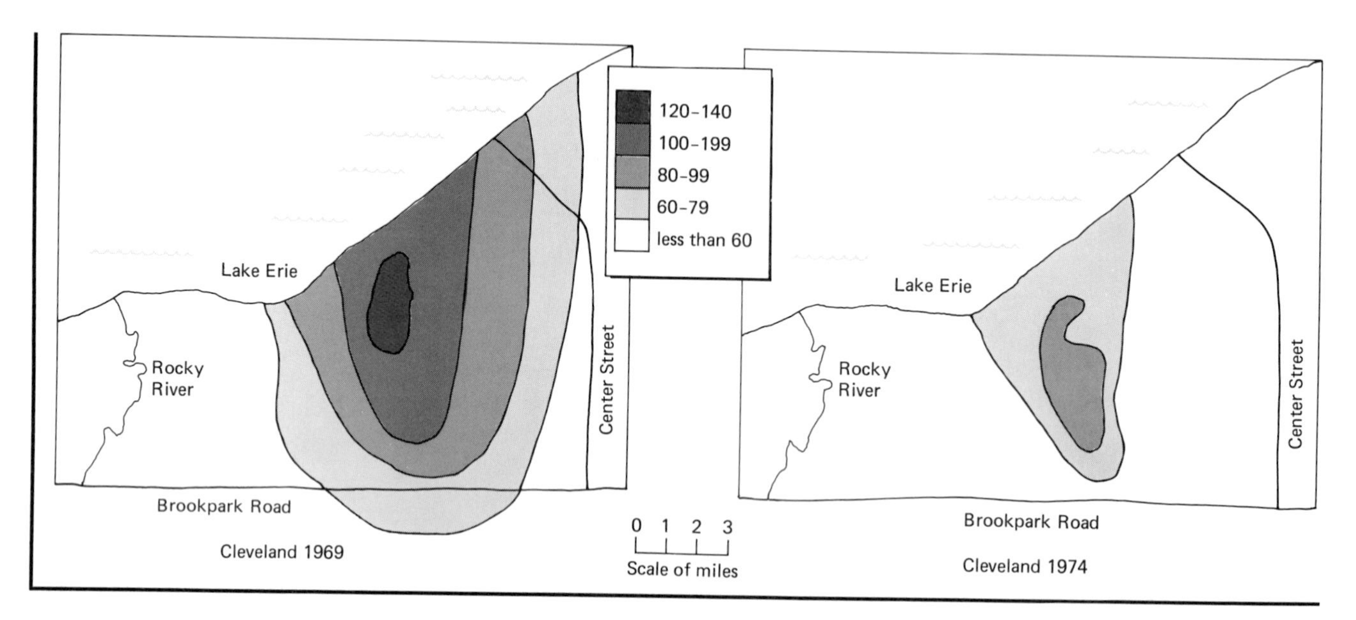

FIGURE 5.11 In Cleveland, Ohio, as in most cities, the sulfur dioxide concentration decreases as one progresses away from the center of the city. Significant reductions in the levels have occurred in Cleveland and many United States cities since 1968. For example, the yearly average concentration in the downtown area decreased from 134 $\mu g/m^3$ to 82 $\mu g/m^3$ between 1969 and 1974. By 1980 the yearly average dropped below 60 $\mu mg/m^3$ for the entire city. (Maps courtesy of the City of Cleveland.)

most of the low-sulfur coal is mined in the western United States, an economic penalty is being paid for the reduction of sulfur dioxide concentrations in the eastern United States. While there is promise for controlling sulfur dioxide emissions without resorting to low-sulfur coal, the technology is in a development and testing stage, and meeting the federal standards poses aggravating problems for individual states and their industries.

Methods for Controlling Sulfur Oxides

The 1970 Clean Air Act restricted emissions of sulfur oxides to 1.2 pounds per million Btu of energy derived from burning coal. To see what this means in terms of control consider an Appalachian coal having 3% sulfur and producing 12,000 Btu per pound of coal burned. A 3% sulfur content means 0.03 pounds of sulfur per pound of coal. Two oxygen atoms weigh about as much as one sulfur atom so that when sulfur is oxidized to produce SO_2, the weight essentially doubles. Thus if 0.03 pounds of sulfur are completely oxidized, one pound of this type of coal produces 0.06 pounds of sulfur dioxide implying 0.06 pounds per 12,000 Btu of heat, or five pounds per million Btu of heat. Thus reducing the emissions from five pounds per million Btu to 1.2 pounds per million Btu amounts to a 76% reduction. Utilities generally met this requirement by burning coal having less than 1% sulfur. Most of this coal is found in the western United States, more than a thousand miles from eastern markets. As for particulate emissions, stricter standards were placed on coal-fired boilers built after September 18, 1978. The 1.2 pounds of sulfur oxides per million Btu of heat was retained as a maximum, but an additional reduction was required that depended on the type of coal burned. A utility can no longer satisfy emission standards simply by burning low-sulfur coal. If low-sulfur coal is used, some control technology must be added to reduce the sulfur oxide emissions by as much as 70–90%. Generally, a utility employs some combination of

1. burning low-sulfur coal.
2. coal cleaning before burning.
3. removal of sulfur oxides following combustion.

In the future, a scheme called fluidized bed combustion may prove to be a worthwhile control technology.

Iron pyrite, chemical symbol FeS_2, is a yellowish metallic compound commonly called "fool's gold." Much of the sulfur in coal is pyritic. Other sulfur is bound in organic molecules. The pyritic and organic percentages depend on the coal deposit. Pyrites have a mass density some 3.5 times greater than that of the burnable product and it is possible to remove 70–80% of the pyritic sulfur by crushing the coal and employing mechanical separation methods (sifting). The removal of pyritic sulfur is a natural extension of coal preparation because coal is normally crushed and cleaned to remove unburnable products such as rocks. This sulfur removal technique is relatively simple and inexpensive but only about 14% of United States coal can be cleaned sufficiently to meet sulfur dioxide emission standards. Thus other sulfur removal technology must be used in addition to cleaning.

A flue is the opening in a chimney or a smokestack through which combustion gases such as sulfur oxides and nitrogen oxides rise. Unless captured, the gases escape to the atmosphere. Capturing sulfur oxides in the flue gas of a coal-fired boiler is called flue gas desulfurization. The most common method, called scrubbing, allows sulfur oxide gas to pass over a wet slurry of lime (calcium oxide, CaO) or limestone (calcium carbonate, $CaCO_3$). Chemical reactions remove 80–90% of the sulfur in the flue gas by producing solid calcium sulfite ($CaSO_3$) in a wet sludge requiring disposal. Using coal with 3.5% sulfur, the scrubber sludge is comparable in weight to the coal ash generated. For a large electric power plant, the sludge amounts to about 225,000 tons annually. Like coal ash, the sludge is usually deposited in a landfill or settling pond requiring between 400 and 700 acres to contain the sludge and ash generated in the 30-year lifetime of the plant. If disposal is not done carefully, it leads to water pollution and land degradation. Scrubbers are expensive and require about 5% of the energy output of a typical power plant. Their reliability is controversial but they are one of very few technologies available for meeting the stringent sulfur oxide emission standards.

A fluid flows easily into a container and assumes the container's shape. Thus a gas or a liquid qualifies as a fluid. However, a liquid or a gas could contain suspended solid particles and still be called a fluid. In fluidized bed combustion, pulverized coal is burned in a gaseous fluid containing powdered limestone. Sulfur oxides produced in the combustion environment react chemically with the limestone to produce calcium sulfite in a dry sludge. Coal ash and sludge are removed continuously. Fluidized bed combustion holds promise for efficient combustion of coal, efficient removal of sulfur oxides, and generation of a dry sludge that is relatively easy to handle. However, large-scale commercial feasibility has not been demonstrated.

Emission limits are also placed on nitrogen oxides produced from oxidation of nitrogen in air used to promote coal combustion. Depending on the type of coal burned in new coal-fired units built after September 18, 1978, nitrogen oxide emissions were restricted to 0.5–0.8 pounds per million Btu of heat produced. A utility approaches the nitrogen oxide requirement by controlling combustion rather than by an emission control technology.

REFERENCES

Fundamentals of Air Pollution, Samuel J. Williamson, Addison-Wesley, Reading, Mass., 1973.

Air Pollution: Threat and Response, David A. Lynn, Addison-Wesley, Reading, Mass., 1976.

Atmospheric Pollution: Its History, Origins and Preventions, A. R. Meetham, D. W. Bottom, S. Clayton, A. Henderson-Sellers, and D. Chambers, Pergamon Press, Oxford, 1981.

Health Effects of Fossil Fuel Burning: Assessment and Mitigation, Richard Wilson, Steven D. Colone, John D. Spengler, and David Gordon Wilson, Ballinger, Cambridge, Mass., 1980.

REVIEW

1. Describe the meaning of "pollutant" and "pollution."
2. What air pollutants are attributed to electric power plants?
3. How are particulates, sulfur oxides, and nitrogen oxides formed in coal combustion?
4. What are the environmental concerns for electric power plant emissions to the atmosphere?
5. If all lengths can be measured in meters, what is the rationale for introducing the micrometer unit of length for specifying particulate sizes?
6. Why is the settling time for particulates important?
7. Present a few arguments for the desirability of reducing pollution levels in urban areas.
8. How are atoms bound together to form molecules?
9. Explain the energy aspects of the chemical reaction
   $S + O_2 \rightarrow SO_2$.
10. Give a reason why a limit on the mass of particulates per cubic meter of air may not be completely appropriate for assessing health effects.
11. Name and describe briefly four devices used to collect particulates in a coal-burning electric power plant.
12. Describe the basics of a scrubber for removing sulfur oxide emissions.
13. Name the two ways that sulfur is generally tied up in coal.
14. What is meant by fluidized bed combustion?
15. During the last six years, the particulate concentrations in most American cities have
    a) risen sharply.
    b) declined.
    c) remained essentially constant.
    d) more than doubled.
    e) not even been monitored.
16. Among the several pollutants introduced into the atmosphere by a fossil-fueled power plant, the one that originates from a chemical impurity in the fuel is
    a) carbon monoxide.
    b) sulfur oxides.
    c) hydrocarbons.
    d) nitrogen oxides.
    e) particulate matter.
17. Pollution concentrations are generally measured in units of
    a) micrograms per cubic meter.
    b) micrograms.
    c) pounds.
    d) cubic meters.
    e) grams.
18. The pollutants of concern from an electric power plant are
    a) carbon dioxide, hydrocarbons, nitrogen oxides.
    b) sulfur oxides, particulates, nitrogen oxides.
    c) sulfur oxides, photochemical oxidants.
    d) nitrates, smog.
    e) carbon monoxide, hydrocarbons, nitrogen oxides.
19. The atoms in a molecule are tied together by
    a) nuclear forces.
    b) only gravitational forces.
    c) primarily electric forces.
    d) atomic strings.
    e) magic.
20. Pollution concentrations are usually measured in units of micrograms per cubic meter. A similar unit that could be used is
    a) micrograms.
    b) grams per cubic centimeter.
    c) pounds per square foot.
    d) newtons per cubic meter.
    e) micrograms per micrometer.
21. The word particulates when used in conjunction with coal-burning power plants means
    a) fine particles of sulfur produced from pulverizing coal.
    b) fine particles of unburnable products in the coal.
    c) the chemical formation of a product called particulates.
    d) fine particles produced from the use of scrubbers.
    e) the active ingredients of partichemical smog.
22. If a particulate concentration is measured to be 120 $\mu g/m^3$, then in 5 m^3 of air we would expect ______ μg of particulates.

    a) 24 b) 120 c) $\frac{5}{120}$ d) 6 e) 600

23. Electrostatic precipitators and gravity collectors make use of ____ and ____ forces, respectively. These forces are due to the properties of ____ and ____ possessed by the particulates.
    a) wind, settling, speed, inertia.
    b) electric, gravitational, charge, inertia.
    c) nuclear, chemical, mass, charge.
    d) electric, gravitational, charge, mass.
    e) kinetic, potential, speed, position.

24. The particulate collection devices used on coal-burning power plants
    a) remove almost all the mass.
    b) remove comparatively few very small particles.
    c) remove almost none of the sulfur oxides.
    d) are less effective than a scrubber for removing sulfur oxides.
    e) All of the above.

25. Studies show that the respiratory system is susceptible to damage from sulfur dioxide.
    a) true b) false

26. There is considerable environmental concern over the formation of sulfuric acid (H_2SO_4) starting from sulfur dioxide (SO_2) released from the burning of coal.

    $\_\_\_\_ + SO_2 \rightarrow NO + SO_3.$

    $SO_3 + \_\_\_\_ \rightarrow H_2SO_4.$

    a) NO_2, H_2O b) NO, H_2O c) NO_2, HO
    d) NO, HO_2

27. Two oxygen atoms weigh about as much as a single sulfur atom. This would mean that 1000 pounds of coal containing 5% sulfur would produce about ____ pounds of sulfur dioxide (SO_2) if all the sulfur were oxidized.
    a) 50 b) 150 c) 100 d) 10

28. Federal standards for particulate matter in the atmosphere are expressed in terms of micrograms of particulates per cubic meter of air. This measurement of particulate matter
    a) is especially useful because the massive particles make up the largest number of particulates.
    b) is especially useful because effective devices exist for removing *most* particles from exhaust gases.
    c) is not very meaningful because no effective way of reducing the mass of pollutants exists.
    d) may not be very useful because the vast majority of the particles make up only a small percentage of the mass of pollutants.
    e) is not currently in use.

29. Sulfur oxides produced in the generation of electric energy from burning coal
    a) are mainly responsible for the eye irritation experienced in most cities.
    b) are a natural product of burning a hydrocarbon fuel.
    c) result from oxidation of unwanted sulfur compounds in most coal.
    d) are also produced in copious amounts by automobiles.
    e) are a by-product of electrostatic precipitators.

30. One method for the removal of sulfur oxides produced from burning coal is called
    a) coal beneficiation. b) precipitation.
    c) flue gas desulfurization. d) cracking.
    e) sulfurization.

31. Sulfur oxide "scrubbers"
    a) are relatively inexpensive and are very efficient pollution control devices.
    b) represent the best means of reducing sulfur from coal prior to burning.
    c) have encountered virtually no resistance from the electric power industry.
    d) are very expensive and require substantial energy for operation.
    e) (a) and (b).

32. Which of the following is not a device for removing *particulate* emissions from coal-fired power plant emissions?
    a) electrostatic precipitator
    b) cyclonic particle collector
    c) gravitational collector
    d) flue gas desulfurizer
    e) bag filter.

33. If the federal and state allowable annual concentrations of some pollutant are labeled 100 $\mu g/m^3$ and 50 $\mu g/m^3$, then you know
    a) the 50 $\mu g/m^3$ is set by the state.
    b) the 50 $\mu g/m^3$ is a less stringent requirement.
    c) the 100 $\mu g/m^3$ is a more stringent requirement.
    d) the 100 $\mu g/m^3$ is set by the state.
    e) there is an error because the two standards cannot be different.

34. The main reason for the decline of sulfur oxide levels in some cities is
    a) the widespread use of electrostatic precipitators.
    b) the burning of low-sulfur western coal.

c) the conversion to nuclear energy.
d) a nationwide change in the prevailing winds.
e) more efficient boilers.

Answers

15. (b)	16. (b)	17. (a)	18. (b)
19. (c)	20. (b)	21. (b)	22. (e)
23. (d)	24. (e)	25. (a)	26. (a)
27. (c)	28. (d)	29. (c)	30. (c)
31. (d)	32. (d)	33. (a)	34. (b)

QUESTIONS AND PROBLEMS

5.1 MOTIVATION

1. A pollutant can be described as something affecting something we value. Why is sulfur dioxide considered to be a pollutant?
2. Figure 5.1 shows that ten million tons of particulates were emitted in 1980. To get a feel for the magnitude of such a quantity, consider the following. Suppose that a highway 50 ft wide extended for 3000 mi across the United States. If these ten million tons were spread uniformly over this highway, how deep would the layer be? Assume that one cubic foot of particulates weighs 100 lb.

5.2 MOLECULES AND CHEMICAL REACTIONS

3. Two persons tugging at opposite ends of a rope form a bound system analogous to two atoms bound together to form an oxygen molecule. How could you use three persons and three ropes to simulate a bound system similar to a CO_2 molecule?
4. An oxygen atom has about half the mass of a sulfur atom. Knowing the composition of the sulfur dioxide molecule (SO_2), explain why nearly 100 pounds of sulfur dioxide are produced from the oxidation of 50 pounds of sulfur.

5.3 PARTICULATE MATTER

5. Particles falling in the atmosphere are held back by a viscous force. The larger the particle, the larger the viscous force. How is this consistent with what you know about a parachutist floating gracefully to the ground?
6. Some time, take a handful of sand having many grain sizes and drop it into a clear bottle of water. Observe which particle sizes reach the bottom of the bottle first. Then draw an analogy between this effect and the settling of particulates in the atmosphere.
7. What are some of the geographical factors considered when choosing a site for an industrial city? How do some of these factors contribute to pollution problems?
8. A micrometer, symbolized μm, is a millionth of a meter. Express a particulate diameter of 0.001 centimeters in μm.
9. A sample of air is contained in a cubic container, each side having a length of 20 centimeters. Measurements show that the air contains 0.1 grams of particulate matter. Express the particulate concentration in micrograms per cubic meter of air. Remember, 1000 grams = one kilogram and 100 centimeters = 1 meter.
10. A classroom measures 10 meters $\times$ 15 meters $\times$ 4 meters. The suspended particulates in this room were determined to be 55 $\mu g/m^3$. What is the total mass (in grams) of suspended particulates in the room?
11. A high-quality coal from the eastern United States produces about 13,000 Btu of heat per pound. A low-sulfur western coal may produce 10,000 Btu per pound. If an electric power company can meet its daily demands with 5000 tons of eastern coal, how much western coal would be required?
12. For the sake of argument, suppose a sample of air contains 10,000 particles having a diameter of 0.1 μm and 100 particles having a diameter of 1 μm. All particles are made of the same material.
    a) Show that the total mass of 1 μm diameter particles is ten times greater than the total mass of the 0.1 μm diameter particles.
    b) If for health reasons, you wanted to remove 90% of the 0.1 μm diameter particles, would removing 90% of the total mass do the desired job?
13. The white chalk commonly used to write on classroom chalkboards is about one centimeter in diameter. A one centimeter length of the chalk has a mass of about $1\frac{1}{4}$ grams. If the chalk is pulverized to a fine powder and uniformly distributed throughout a classroom having dimensions 10 meters $\times$ 15 meters $\times$ 4 meters, what is the concentration of suspended chalk dust? How does this concentration compare with the primary standard for suspended particulates?
14. The concentration of suspended particulates in the atmosphere can be measured by forcing air through a semiporous paper that filters the particulates from the air. Determine the suspended particulate concentration from the measurements presented below:

    air flow rate = 0.71 cubic meters per minute,

air sampling time = 24 hours,

weight of filter paper
before sampling = 4.5722 grams,

weight of filter paper
after sampling = 4.7424 grams.

15. In a certain area, the average particulate concentrations for three consecutive days were 343, 216, and 343 micrograms per cubic meter.
    a) Was the federal primary standard violated during this period?
    b) Show that the arithmetic mean concentration is 301 $\mu g/m^3$ and the geometric mean concentration is 294 $\mu g/m^3$ (the geometric mean is described on page 88).

16. A certain particulate has a mass of 10 μg and a density of 1.25 $grams/cm^3$. What would a length of one of its sides be if it were a cube? Using Fig. 5.3 as a guide, what sort of particulate might this be? Remember, the graph uses micrometers for particulate size.

17. The differences in settling times for various sizes of particulates is striking. Show that this is true by calculating the time required for particles of diameters of 0.1, 1, 10, 100, and 1000 micrometers to settle a distance of one kilometer. Data for the settling velocities are given in Table 5.1.

5.4 SULFUR OXIDES

18. It is very hard to visualize the smallness of a millionth of a gram (μg). As a help, imagine you have one million sticks each having a length of one-half inch, and that these sticks are placed end to end in a straight line. If all the sticks are of one color except one which is placed at one end, how far would you have to walk from the opposite end to reach this odd-colored stick?

19. Sulfur trioxide (SO_3) can be formed by combining oxygen molecules (O_2) with sulfur atoms. Write a chemical reaction for this process.

20. There is considerable environmental concern over the formation of sulfuric acid (H_2SO_4) starting from sulfur dioxide (SO_2) released from the burning of coal. Fill in the steps in the equations below for a possible mechanism for producing sulfuric acid:

$$\_\_\_\_ + SO_2 \rightarrow NO + SO_3,$$

$$SO_3 + \_\_\_\_ \rightarrow H_2SO_4.$$

21. The total electric energy production in 1974 was 1.87 trillion kWh. Assume that this was produced by a coal-burning power plant with an overall efficiency of 40%. Estimate how much coal was burned. Assuming that the coal contained 2% sulfur, estimate the amount of sulfur dioxide produced. Compare with that quoted in Fig. 5.1.

22. A certain coal has 0.5% sulfur and produces 10,000 Btu of heat per pound. If all the sulfur is oxidized, show that the sulfur oxide emissions would be about one pound per million Btu of heat.

23. A certain coal has 5% sulfur and 10% unburnable ash. The sulfur is converted completely to sulfur dioxide. Explain why the weight of the sulfur oxides is essentially the same as the weight of the unburnable ash.

6

Automobiles and the Environment

6.1 MOTIVATION

It is an interesting exercise to list inventions having an impact on our lives. Television and computers are relatively new. Steam engines instigated the Industrial Revolution some 200 years ago. We can go on and on for the list is long but no entry is more important than the automobile. There are about half as many automobiles in the United States as there are citizens, and a driver has the freedom to make sundry short trips as well as extended trips over the interstate highway system built to accommodate the cars. The personal freedom to travel provided by the automobile has become an integral part of our life-styles. Prices are paid for massive use of automobiles, not the least of which is the price in human discomfort caused by the production of the undesirable

Our technological society is extremely dependent on private automobiles. Effluents from the multitude of automobiles have produced serious smog problems in nearly every American city. (Photograph by Chester Higgins. Courtesy of the EPA.)

by-products such as carbon monoxide, nitrogen oxides, and hydrocarbons, and the accelerated depletion of petroleum reserves. Automobile technology is changing in response to these societal problems. We delve into these problems and responses in this chapter.

6.2 FUNDAMENTAL ORIGIN OF EMISSIONS

A glance under the hood of a United States car quickly convinces you that the engine with all its auxiliaries for running an air conditioner, electric generator, windshield washer and wipers, hydraulic pumps, etc. is an intricate system. However, these auxiliaries do not function unless the engine converts the chemical energy of gasoline into kinetic energy of motion. It does this by igniting a mixture of gasoline vapor and air in the cylinders of the engine (Fig. 6.1). Because the burning takes place between the top of a cylinder and a piston, this type of engine is called an internal combustion engine. Expanding gases push the pistons downward. The linear motion of the pistons is converted to rotational motion by rods connected to a rotating shaft. This rotational motion is transferred to the wheels of the automobile by an appropriate linkage. A combustion cycle for one of the pistons in an engine is shown in Fig. 6.2. All of the carbon monoxide and nitrogen oxides and 60% of the hydrocarbons generated are combustion products. Of the remaining 40% of hydrocarbons, 20% of those released are a result of evaporation from the fuel tank and carburetor. The remaining 20% are from vapor "blowing by" the pistons and the cylinder walls of the engine. Let us look now at the formation of these pollutants.

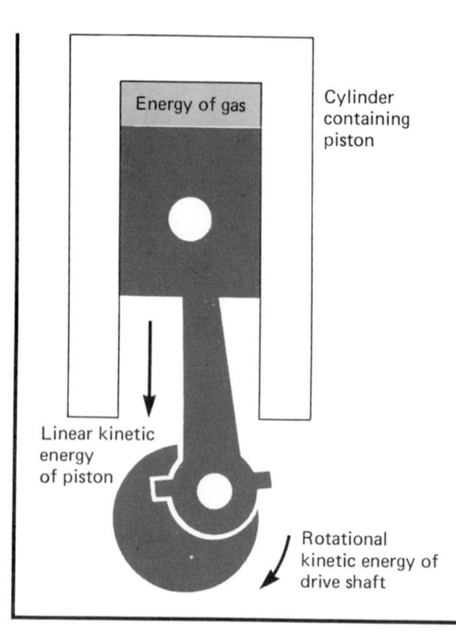

FIGURE 6.1 The energy conversion process in an internal combustion engine. Energy from expanding combustion gases is converted to linear kinetic energy in the piston and rotational kinetic energy in the drive shaft.

A class of hydrocarbons called alkanes has the chemical form C_nH_{2n+2}, where n is a number from one to ten. If $n = 5$, the molecule is C_5H_{12} and is named pentane. Gasoline with no special-purpose additives is a mixture of alkanes having five to ten carbon atoms. Isooctane is a special form of octane (C_8H_{18}) and is a particularly good automotive fuel. Isooctane has become the standard for comparison of other automotive fuels for the following reasons. With some gasolines, ignition occurs more as an explosion rather than as a smooth burning. This reduces the efficiency and produces an audible "knock" in the engine. A fuel of pure isooctane produces little knock while one of pure heptane (C_7H_{16}) knocks badly. The octane rating of a gasoline is a measure of the knock produced when the gasoline is used as an automotive fuel. Isooctane and heptane have been assigned octane ratings of 100 and 0, respectively. A mixture of 90% isooctane and 10% heptane has an octane rating of 90. When any other fuel, regardless of its chemical composition, is burned in a special engine and produces the same knock as 90% isooctane and 10% heptane, it is assigned an octane rating of 90. Since the inception of this octane rating procedure, fuels have been developed with antiknock properties superior to isooctane. These fuels are assigned octane ratings greater than 100.

Pollutants evolve in an automobile engine when the fuel is burned. Like the burning of coal in an electric power plant, the pollutants would be minimal if the combustion were complete, and if there

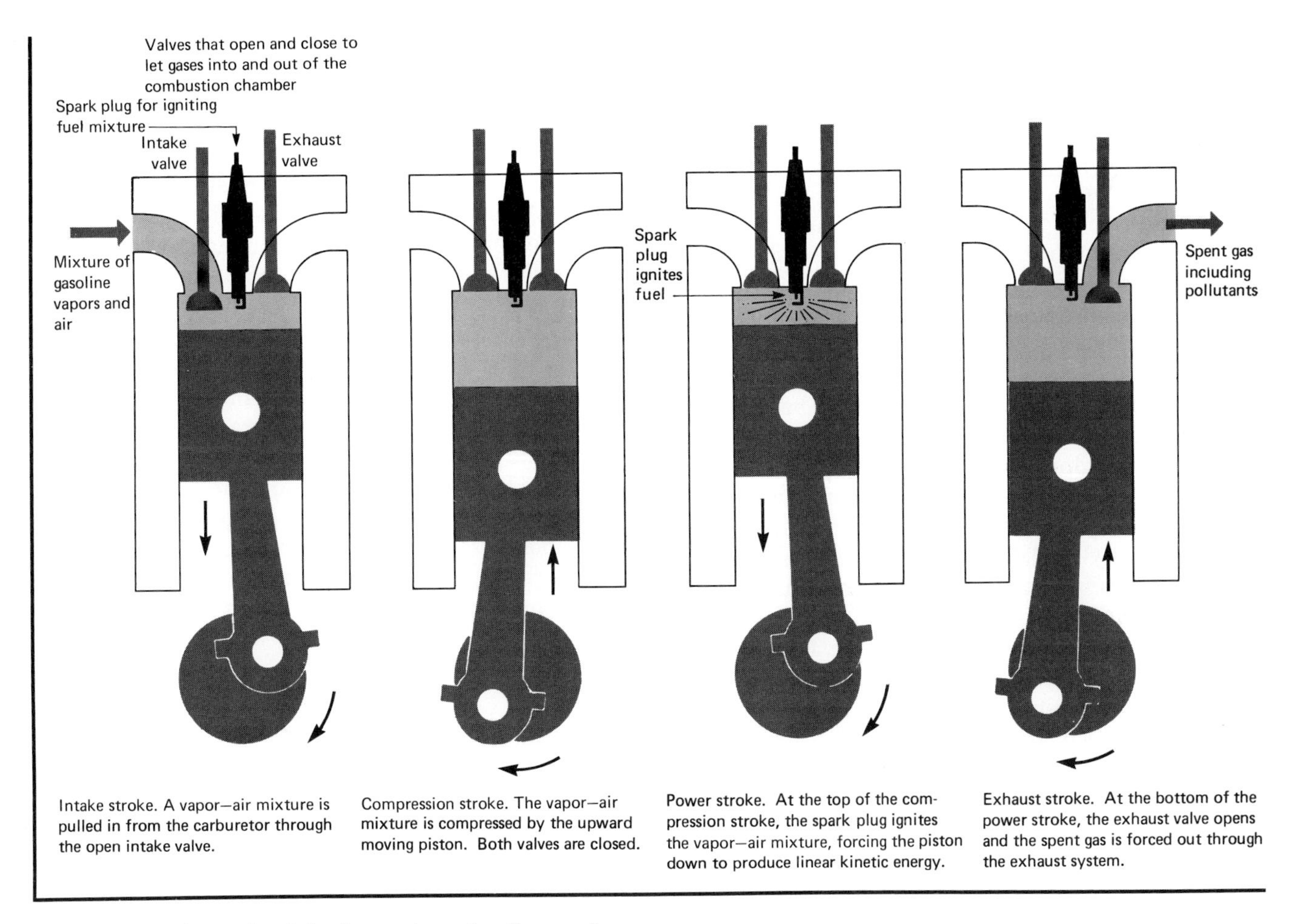

FIGURE 6.2 The cycle of the internal combustion engine.

were no side effects. To illustrate, consider the complete combustion of isooctane, which we might consider the standard fuel. Ideally, isooctane reacts with oxygen to form the products carbon dioxide, water, and energy.

$$\text{isooctane} + \text{oxygen} \rightarrow \text{carbon dioxide} + \text{water} + \text{energy.}$$

$$2C_8H_{18} + 25O_2 \rightarrow 16CO_2 + 18H_2O + \text{energy.}$$

Atoms are rearranged into lower energy configurations. The energy difference is liberated as heat. Except for long-term climatic effects (Chapter 8), the CO_2 is harmless. Similar reactions produce carbon monoxide (CO) as an undesirable by-product. For example,

$$2C_8H_{18} + 25O_2 \rightarrow 14CO_2 + 2CO + O_2 + 18H_2O + \text{energy.}$$

Oxygen for combustion is obtained from air normally constituted of 78% nitrogen in the form of N_2 and 21% oxygen in the form of O_2. N_2 is harmless. We inhale and exhale N_2 with each breath and do not find N_2 and O_2 interacting. But the high temperature and pressure environment in the combustion chambers of an automobile engine provoke the interaction of nitrogen and oxygen. The reactions producing most of the nitrogen oxide pollutants are

$N_2 + 2O_2 \rightarrow 2NO_2$ (nitrogen dioxide),

$N_2 + O_2 \rightarrow 2NO$ (nitric oxide).

These and other oxides of nitrogen are lumped together for environmental purposes and called nitrogen oxides, symbolized NO_x, and read as "NOX."

The efficiency of the internal combustion engine is directly related to the octane rating of the fuel. Throughout the years, the trend has been to increase the octane rating. One way of making a better fuel is to produce molecular structures of hydrocarbons with better burning characteristics—an involved and expensive process. The most common method of achieving a higher octane fuel is to add a compound called tetraethyl lead, $(C_2H_5)_4Pb$, in amounts of about two milliliters* per gallon. In addition to increasing the octane rating, the tetraethyl lead provides some lubrication and reduces valve burning. However, it produces material deposits such as metallic lead and lead oxides that remain in the engine and impair the functioning of the valves and spark plugs. To combat this, ethylene dibromide $(C_2H_4Br_2)$ is added to convert lead to lead bromide (PbBr) that escapes as a gas out the exhaust. Present day gasolines also contain antioxidants, metal deactivators, antirust and anti-icing compounds, detergents, and lubricants. Modern gasolines are far from the ideal isooctane suggested earlier.

Since the advent of 1975 models, United States automobiles have been equipped with a pollution control device called a catalytic converter (Sec. 6.6). Catalytic converters can be rendered ineffective by the lead compounds resulting from tetraethyl lead in gasoline. Thus cars with catalytic converters are designed to burn unleaded gasoline. To ensure that only unleaded gasoline is used in vehicles having catalytic converters, the opening in the filling port of the gasoline tank will not accept the nozzle from pumps issuing leaded gasoline.

* A milliliter is about the volume occupied by three pennies.

6.3 PHOTOCHEMICAL SMOG

Few metropolitan areas escape the eye-irritating effects of photochemical smog* produced from chemical reactions involving automobile effluents and sunlight. Many of the eye-irritating chemicals are vastly more complex than the relatively simple molecules produced in the combustion of gasoline. Although we will not present the complex chemistry, we can comprehend the basic mechanisms leading to formation of the irritants.

The most stable form of oxygen is the two-atom combination (O_2). Atomic oxygen (O) and a three-atom molecule called ozone (O_3) are extremely reactive in the presence of other atoms and molecules. Normal air contains very little atomic oxygen and ozone. When automobiles emit nitrogen oxides into the atmosphere, molecules of nitrogen dioxide (NO_2) absorb solar energy and separate into nitric oxide and atomic oxygen according to the reaction

nitrogen dioxide + solar energy →
nitric oxide + atomic oxygen,

NO_2 + solar energy → $NO + O$.

More than 99% of the atomic oxygen created by the breakdown of NO_2 combines with molecular oxygen to form ozone.

atomic oxygen + molecular oxygen→ozone,

$O + O_2 \rightarrow O_3$

However, some atomic oxygen reacts with hydrocarbons from automobile exhausts in a series of complex chemical reactions. Nitrogen dioxide and ozone are regenerated and some very irritating chemicals such as formaldehydes, peroxyacyl nitrates (PAN), and acroleins are produced. The original high NO concentration from automobile emis-

* The word smog is a combination of parts of the words **smo**ke and **fog** and was used to designate the type of polluted conditions that often occurred in London. Smog now describes a chemical condition which is much more complex than this relatively simple mixture.

sions is diminished by reactions such as

$$NO + O \rightarrow NO_2.$$

$$NO + O_3 \rightarrow NO_2 + O_2.$$

Figure 6.3 records graphically the photochemical smog process in a large city. These processes are responsible for the characteristic haze over many cities (Fig. 6.4) and it is the chemicals that produce the eye-irritating effects.

Table 6.1 summarizes the characteristics of photochemical smog and gives the corresponding London smog data for comparison.

Smog tends to be problematical in cities where industry and large concentrations of motor vehicles exist, and every sizeable city in the United States has smog problems. The Los Angeles area is particularly troubled because of characteristic weather conditions that produce an atmospheric lid on the air near the ground for about 100 days of the year (see Chapter 8). The situation is complicated by mountains and hills to the north, east, and south preventing flushing of the trapped smog. Four hundred years ago Sir Francis Drake referred to this area in his ship's journal as the "bay of smokes." Denver, Colorado, with ever increasing industrial and urban expansion, has similar problems.

6.4 EFFECTS AND EXTENT OF AIR POLLUTANTS

Carbon monoxide is a colorless, odorless, and tasteless gas. Yet exposure to an atmosphere of air in which only one part out of each 1,000 parts (by volume) is carbon monoxide can produce unconsciousness in one hour and death in four hours. Essen-

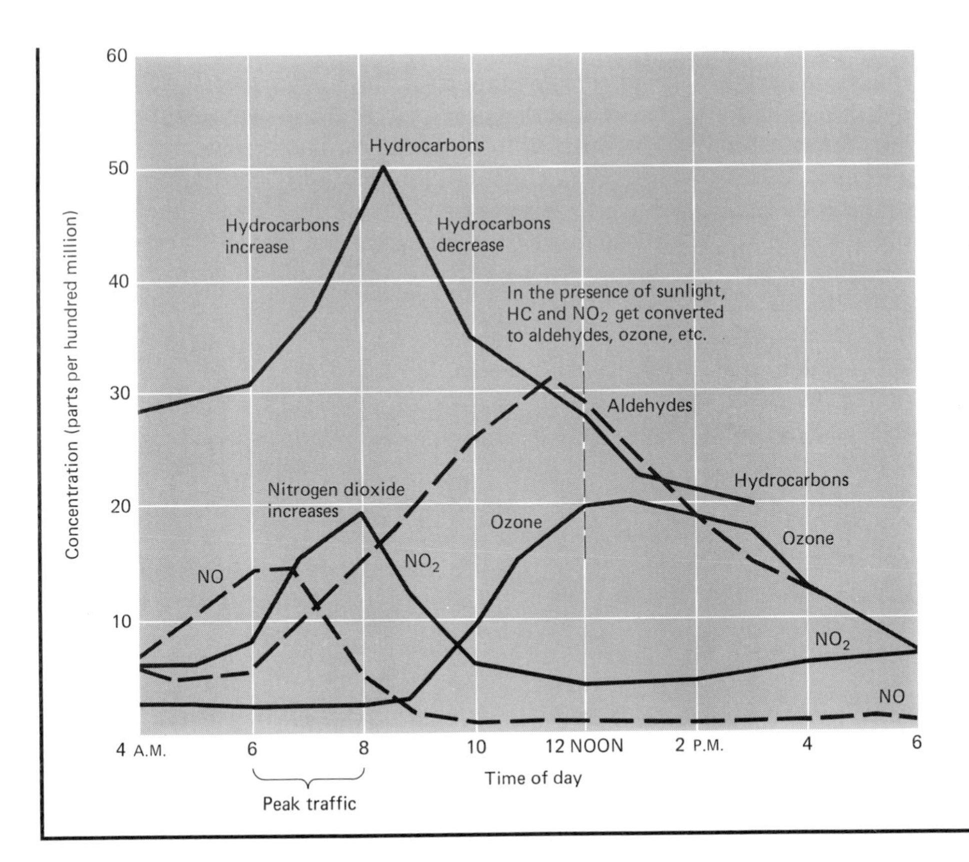

FIGURE 6.3 Hourly variations of the constituents of an intense photochemical smog. When commuter traffic materializes between 6 and 8 A.M., hydrocarbons and nitrogen oxide concentrations build up. The nitric oxide (NO) levels exceed the nitrogen dioxide (NO_2) levels because the combustion process strongly favors the formation of NO. In the presence of the morning sunlight, NO_2 begins to break down into NO and O and the photochemical smog process begins. The hydrocarbon levels decrease because they get converted to the eye-irritating chemicals such as aldehydes. The peak smog concentration occurs in early afternoon when the sun's rays are most intense. (From *Photochemistry of Air Pollution,* Philip A. Leighton, Academic Press, Inc., 1961.)

FIGURE 6.4 A visual illustration of clear conditions and an intense smog formation in Los Angeles. (United Press International photograph. Reproduced by permission.)

tially, a person suffocates from lack of oxygen in the body. The bloodstream carrying oxygen throughout the body has a much greater affinity for carbon monoxide than for the life-supporting oxygen. When one breathes air containing carbon monoxide and oxygen, the bloodstream preferentially absorbs carbon monoxide. When the carbon monoxide is carried to components of the body expecting oxygen, the chemistry is wrong, and the person suffocates. The burning of any carbon-based material—gasoline, coal, charcoal, tobacco—produces carbon monoxide to some extent. Therefore, one should ensure adequate ventilation for the effluents. It is extremely dangerous to operate an automobile in a closed

TABLE 6.1 Some distinguishing characteristics of photochemical and London smogs.

Characteristic	Photochemical	London
Major fuels involved	petroleum	coal and petroleum products
Principal constituents	O_3, NO, NO_2, CO, organic products	particulates, CO, sulfur compounds
Types of reactions	photochemical and thermal	thermal
Time of maximum occurrence	midday	early morning
Principal effects	eye irritation	bronchial irritation, coughing
Visibility	about 0.5 to 1 mile	less than 100 yards

garage or to tolerate a faulty exhaust system on an automobile or to use a charcoal grill in closed quarters. Many charcoal bags have the following words printed on them in prominent type—**Warning:** Do not use for indoor heating or cooking unless ventilation is provided for exhausting fumes outside. Toxic fumes may accumulate and cause death.

The direct effect of nitrogen oxides and hydrocarbons on plant life (Fig. 6.5) and animal life is not of great concern. Rather it is the consequences of photochemical smog which is the major concern. The effects generally attributed to smog are

1. eye irritation,
2. characteristic damage to vegetation,
3. respiratory distress and even death,
4. reduction in visibility,
5. objectionable odors,
6. excessive cracking of rubber products and damage to other materials.

Some specific effects of air pollutants of concern are briefly summarized in Table 6.2. Using the specific effects as criteria, the Environmental Protection

(a)

(b)

FIGURE 6.5 (a) Ozone damage to a White Cascade petunia. **(b)** Leaf of an ozone-sensitive variety of tobacco shows white spots characteristic of air-pollutant damage called weather fleck. (Photographs courtesy of the United States Department of Agriculture.)

TABLE 6.2 Brief summary of air pollutant effects. Note that all exposures are given in micrograms per cubic meter.

Effects associated with oxidant concentrations in photochemical smog			
Effect	*Exposure (micrograms/m³)*	*Duration*	*Comment*
Vegetation damage	100	4 hours	leaf injury to sensitive species
Eye irritation	Exceeding 200	peak values	such a peak value would be expected to be associated with a maximum hourly average concentration of 50 to 100 $\mu g/m^3$
Impaired performance of athletes	60–590	1 hour	exposure for one hour prior to athletic event
Effects associated with carbon monoxide			
Effect	*Exposure (micrograms/m³)*	*Duration*	*Comment*
Bodily damage	30,000	8 hours or more	impairment of visual and mental acuity
	200,000	2–4 hours	tightness across the forehead, possible slight headache
	500,000	2–4 hours	severe headache, weakness, nausea, dimness of vision, possibility of collapse
	1,000,000	2–3 hours	rapid pulse rate, coma with intermittent convulsions
	2,000,000	1–2 hours	death

Source: The data were taken from the Air Quality Criteria Handbooks by the U.S. Department of Health, Education and Welfare and the Environmental Protection Agency.

Agency has arrived at the standards shown in Table 6.3. For these pollutants, the primary and secondary standards are identical. While the standards are being questioned by reliable authorities, the general consensus is that they are very stringent.

Although Los Angeles has come to be identified with automotive air pollution in this country, it is by no means the only city with problems. Most cities will have trouble meeting the standards set by the Environmental Protection Agency.

6.5 THE FEDERAL EMISSION STANDARDS

Emissions from motor vehicles must be regulated if stringent federal standards are to be met. A program to gradually reduce emissions began in California in 1961 and nationwide in 1968. Amendments to the 1970 Clean Air Act required an ultimate reduction of carbon monoxide and hydrocarbon emissions from light-duty vehicles to 90% of the measured 1970 values and an ultimate reduction of nitrogen oxide emissions to 90% of the measured 1971 values. The ultimate requirements for carbon monoxide and hydrocarbons in the first column of Table 6.4 were met by 1981 model vehicles. Whether or not the ultimate emission standards are firm remains to be seen. Automobile manufacturers see them as overly restrictive and argue that the 1980 values are more reasonable. Even if these maximum emission requirements are met, the federal air quality criteria could be exceeded if there were sufficient automobiles.

6.6 APPROACHES TO MEETING FEDERAL STANDARDS

Several options for meeting the maximum emission requirements come to mind. These approaches

TABLE 6.3 Federal standards for carbon monoxide, hydrocarbons, photochemical oxidants, and nitrogen oxides.

Pollutant	Maximum average concentration not to be exceeded more than once a year (micrograms per cubic meter of air)	Averaging period
Carbon monoxide	10,000 40,000	8 hours 1 hour
Hydrocarbons	160	3 hours (6–9 A.M.)
Photochemical oxidants*	160	1 hour

The annual arithmetic mean concentration of nitrogen oxides cannot exceed 100 micrograms per cubic meter of air.

* Photochemical oxidants are primarily ozone but they also include other chemical compounds created by the smog conditions.

include

1. modifying the present internal combustion engine,
2. developing a new fuel for the present internal combustion engine,
3. removing the pollutants from internal combustion engine effluents, and
4. developing alternates to the internal combustion engine.

All of these options are being studied. The major focus, however, is on removal of the pollutants from the effluents and, to a lesser extent, modification of the internal combustion engine and some of its auxiliary components.

Removal of Pollutants from the Internal Combustion Engine Effluents

Figure 6.6 shows some of the devices used on automobiles to curb pollutant emissions. All those devices excepting the catalytic converters are engine modifications. The bulk of the pollutant control, however, is done by the catalytic converters. Let us now look a little more deeply into these control devices.

Proper operation of an internal combustion engine requires a tight seal between the piston and the cylinder walls. This seal is never perfect and some gaseous products leak out of the combustion region into a chamber called the crankcase. It was standard practice until 1961 to shuttle these emissions through a vent in the crankcase to the atmo-

TABLE 6.4 Federal timetable for maximum allowable light-duty vehicle emissions as established by amendments in 1977 to the 1970 Clean Air Act. The data are expressed in grams emitted per mile of travel.

	Ultimate	1977	1978	1979	1980	1981
Carbon monoxide	3.4	15.0	15.0	15.0	7.0	3.4
Hydrocarbons	0.41	1.5	1.5	1.5	0.41	—
Nitrogen oxides	0.4	2.0	2.0	2.0	2.0	1.0

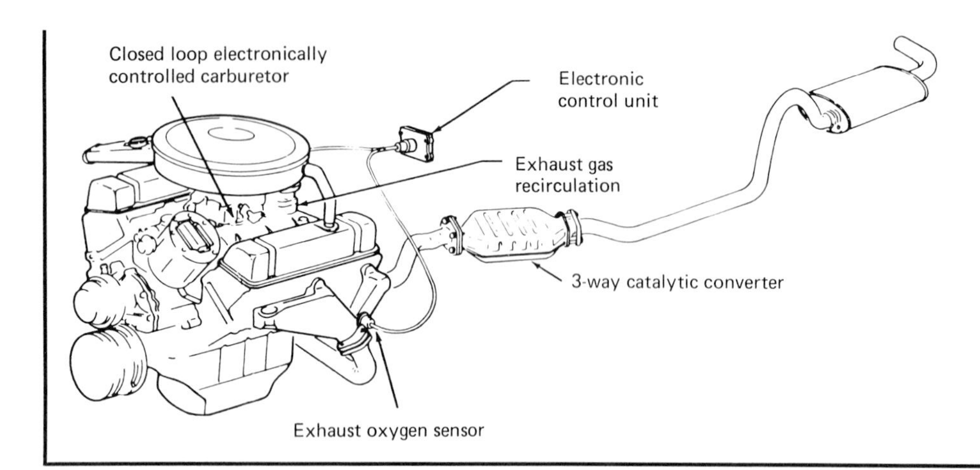

FIGURE 6.6 A typical advanced emission control system for limiting pollutant emissions from an automobile. (Photograph courtesy of General Motors Corporation.)

sphere. However, the emissions can be recirculated into the air-intake system of the engine (Fig. 6.6). All United States cars include them as standard equipment. The connection between the crankcase and carburetor includes a positive crankcase ventilation (PCV) valve to ensure unidirectional flow. Conscientious motorists see to it that the PCV valve works properly.

Smelling gasoline fumes is evidence of the evaporative ability of automobile fuel. The smell is more noticeable on a hot day. This indicates that the rate of evaporation is related to temperature. As long as the fuel tank is vented to the atmosphere, hydrocarbons in the gasoline vapor escape to the environment. Significant evaporative losses also occur at the carburetor that converts the liquid gasoline to a mixture of air and gasoline vapor. During operation of the engine, this mixture is injected into the combustion chambers and ignited. When the engine is shut down and the cooling system deactivated, the temperature of the carburetor rises from the operating temperature of about 150° F to about 200° F. As a result, the liquid fuel bowl of the carburetor warms and the fuel may even boil. These vapors escape to the environment if the carburetor is vented to the atmosphere.

There are two methods for controlling these evaporative losses. One utilizes the crankcase as a storage volume for the vapors. The vapor is pumped out of the storage into a condenser that returns liquid fuel to the gasoline tank. The second method utilizes an activated carbon filter to trap the vapor and hold it until it can be fed back into the carburetion system.

Combustion gases produced in the internal combustion engine include carbon monoxide, nitrogen oxides, and hydrocarbons that must be removed to make ready for the next cycle. These gases are forced into an exhaust pipe that leads to the atmosphere surrounding the car. To control emission of these undesirable gases, the catalytic converter intercepts their flow toward the environment. A catalytic converter uses an appropriate chemical catalyst to convert carbon monoxide (CO) and hydrocarbons to harmless carbon dioxide (CO_2) and water vapor (H_2O) and the nitrogen oxides to molecular nitrogen (N_2) and oxygen (O_2). All these products from the catalytic converter are common components of the air we breathe. Note carefully that the operations on carbon monoxide and hydrocarbons are distinct from the operation on the nitrogen oxides. The former are converted to different molecular forms containing more oxygen; the undesirable gases are said to be oxidized. The nitrogen oxides are broken down from a molecular form containing oxygen to forms that do not contain oxygen. They are said to be reduced. Thus there are two distinct types of catalytic processes—oxidizing and reducing. Modern converters are carefully designed to accomplish oxidation of carbon monoxide and hydrocarbons and reduction of nitrogen oxides. A variety of oxidation catalysts are possible but the ones mentioned most often are platinum and palladium metals. Platinum and palladium are rare, expensive metals that must be imported.

The chemistry involving the oxidizing catalyst is complex but the basic idea is straightforward. Molecular oxygen (O_2) reacts with the catalyst and separates into two oxygen atoms each bound to a constituent of the catalyst:

molecular oxygen + catalyst →
atomic oxygen bound to constituents of the catalyst.

Symbolically,

$$O_2 + \text{catalyst} \rightarrow (O \leftrightarrow \text{catalyst}) + (O \leftrightarrow \text{catalyst}).$$

This symbol means that an oxygen atom is bound to a constituent of the catalyst.

The bound oxygen–catalyst system reacts with carbon monoxide to produce carbon dioxide. The catalyst is freed in the process and the carbon dioxide escapes out the exhaust system:

$$CO + (O \leftrightarrow \text{catalyst}) \rightarrow CO_2 + \text{catalyst}.$$

Note that the catalyst used to instigate the process is regenerated in the oxidation of the carbon monoxide molecule. This process requires a comparatively low temperature (about 800° F). The platinum and palladium metals are deposited on a porous inert substrate. The exhaust gases circulate over the substrate and the catalytic reactions take place (Fig. 6.7). Catalytic converters first appeared on the 1975 model cars. Interim experience indicates that the oxidizing catalytic converter is very effective for reducing carbon monoxide and hydrocarbon emissions.

6.7 POSSIBLE ALTERNATIVES TO THE CONVENTIONAL INTERNAL COMBUSTION ENGINE

No one questions the convenience and reliability of the modern automobile with its internal combustion engine. But if the engines are fostering photochemical smog and contributing to the rapid depletion of petroleum resources, why is it so difficult to replace this engine with a less polluting and more energy-efficient counterpart? Economics is a major factor. An engine with low emissions is useless if no one can afford to buy it. There are, however, some unique features of an automobile, and the internal combustion engine is one of the very few energy sources that can do the job. These features are easy to assess. Let us start by considering what is asked of a contemporary automobile. For the most part it takes us to work, to shop, to visit friends and relatives, and to sundry other places. In fact, about 60% of the mileage is accumulated in trips of less than three miles. However, a contemporary car powered by an internal combustion engine

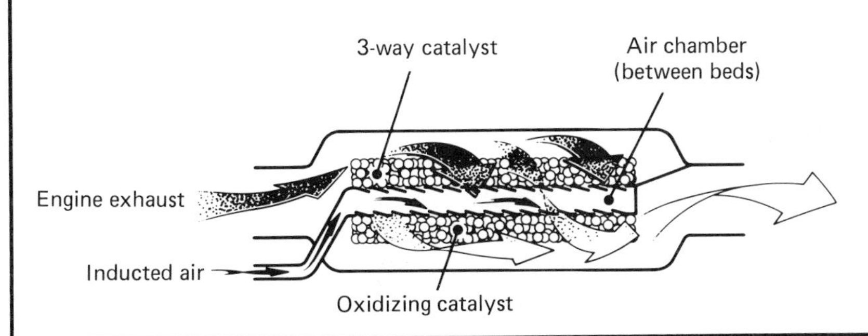

FIGURE 6.7 Cutaway view of a catalytic converter that can be placed in the exhaust line of an automobile to limit the emission of carbon monoxide, hydrocarbons, and nitrogen oxides. The exhaust gases flow through appropriate catalytic materials and are converted to harmless gases. (Drawing courtesy of General Motors Corporation.)

has the energy and power to travel nonstop for 200–300 miles on an interstate highway at speeds exceeding the existing speed limits. Any other engine providing the versatility of the internal combustion engine must have comparable energy and power characteristics. The prospects for building an alternative kind of engine are extremely limited and like the internal combustion engine, the alternatives would not be pollution-free. Steam engines and gas turbines are sometimes looked upon as possibilities. Steam engines have a number of disadvantages. They are bulky, massive, expensive, require a start-up time and, because of the necessary boilers, might be dangerous. Gas turbines (Fig. 6.8) similar to those used in jet aircraft are extremely reliable but they are efficient only at high speeds. To take advantage of the efficiency, a complex linkage between the turbine and wheels is required. The future of the turbine in an automobile is questionable but possibilities exist for using turbines in buses and trucks for long-distance trips with few stops. Alternatives to the gasoline-powered internal combustion engine come and go. The diesel engine, which is also an internal combustion engine, is making an impact. Diesels are more efficient than gasoline engines. They generate the same type of pollutants as the conventional internal combustion engine but the engines have some unique characteristics that make the pollutants somewhat easier to control. Electric vehicles do not have energy characteristics comparable to conventional cars. For example, they cannot travel long distances. Nevertheless, they could be extremely useful for much of the short-trip driving. Let us examine the features of these two possibilities for replacing the conventional internal combustion engine.

The Diesel Engine

Probably you have seen a large tractor-trailer truck with smoky, smelly exhausts cruising on a highway. Probably this vehicle was powered by a diesel engine. The diesel engine has many features of a gasoline-powered internal combustion engine. However,

FIGURE 6.8 The characteristics of a gas turbine engine are best suited for large trucks and buses. The truck shown here is a prototype powered by a gas turbine engine. (Photograph courtesy of General Motors Corporation.)

rather than using an electric spark to ignite an air–fuel mixture in the combustion chamber, the diesel engine takes advantage of heating a gas by compression. When you push the piston of a bicycle pump to force air into a tire, the connecting hose warms from compression of the air. The cycle of a diesel engine starts by compressing air in the combustion chamber. When the temperature of the air rises sufficiently, fuel is injected into the combustion chamber and is ignited by the high-temperature environment. The remainder of the cycle is very similar to a gasoline-powered internal combustion engine. The diesel engine has several attractive features. It operates on a lower-grade fuel; you may have noticed that diesel fuel is slightly less expensive than gasoline at truck stops. The absence of spark plugs simplifies the ignition system. Finally, the diesel engines are generally more efficient than gasoline engines. While diesel engines are extremely popular in United States trucks, buses, and trains, they have not been popular in United States cars. The relatively long "warm up" time of the diesel engine and its relatively high noise level have probably contributed to its unpopularity. Because the government has mandated automobile fuel economy measures, the United States automobile industry is reevaluating the diesel engine for family car use. The pollution and fuel economy characteristics are attractive, and both United States and foreign models are becoming increasingly popular.

Electric Vehicles

Golf provides a livelihood for a few and a diversion for many. Many golf spectators tool effortlessly and silently between strokes in vehicles powered by electric motors energized by batteries. After a few trips over the course, the battery's energy is exhausted and it is recharged with energy from an electric power plant. A few golfers motor to the course in electric vehicles only slightly more sophisticated than the carts carrying them around the course. The batteries in these vehicles, like those in the simpler golf cart counterparts, face frequent chargings. There could well be a future for electric cars but as long as the energy-storing capacity of batteries remains low, electric cars will be confined to short trips.

An electric vehicle does not emit pollutants in its operation but it does contribute to pollution indirectly. Electric energy for the batteries would probably come from a conventional electric power plant. A coal-burning plant produces sulfur oxides and particulates. An automobile engine produces about 30 pounds of carbon monoxide for each 10 gallons of gasoline burned. Because of the differences in efficiency, an electric vehicle has to convert the equivalent of only about 2.5 gallons of gasoline to deliver the same energy as a conventional car converting 10 gallons of gasoline. Generating the electrical energy equivalent of 2.5 gallons of gasoline by burning coal produces about two pounds of sulfur oxides. Thus the weight of sulfur oxides produced is somewhat less than the weight of carbon monoxide produced from burning 10 gallons of gasoline in an internal combustion engine. This example does not mean that the pollution aspects of electric and internal combustion vehicles are the same. Rather it shows that electric vehicles contribute to pollution and to depletion of fossil fuels. Pollution control is transferred from automobiles to an electric power plant, a concept having considerable merit.

Conventional lead-acid batteries like those used for electrical services in present-day automobiles deliver about 10 watt-hours per pound to the energy of an electric car. Gasoline delivers about 1000 watt-hours per pound to the energy of a conventional car, or about 100 times more energy per pound than a battery. For the same weight, batteries deliver 100 times less energy than gasoline. If a car can go 300 miles at 60 miles per hour on a tank of gasoline, it could go only three miles at the same speed using batteries of the same weight as the gasoline. Certainly, additional energy can be obtained by utilizing more batteries, but this presents problems. For example, a 15-gallon tank of gasoline weighs about 100 pounds. Thus the weight of batteries required to deliver the same energy would be 10,000 pounds. Easing the speed requirements reduces the energy requirements. There is little passenger room in an electric vehicle because the batteries occupy a sizable portion of the vehicle volume. In the event of

FIGURE 6.9 To many people, an electric vehicle conjures an image of a modified horse-drawn carriage. But as illustrated by this photograph, contemporary electric vehicles are attractive as well as functional. (Photograph courtesy of General Motors Corporation.)

an accident there is some concern for the ramifications of acid spilled from the batteries. Until batteries that are more energy efficient are developed, the electric vehicle with proven batteries is relegated to the class of special-purpose vehicles for short trips (Fig. 6.9).

Experimental batteries with energies of 100 watt-hours per pound are being built. Like any new product they are expensive, but if their technical worth proves out, their price will surely decrease. These batteries increase the range of an electric car to 150 miles in urban traffic on a single charge. A range of 150 miles would handle all but major interstate trips. Batteries with 300 watt-hours per pound are feasible but the timetable for development is around the year 2000.

REFERENCES

AIR POLLUTION

The references given at the end of Chapter 5 are appropriate for this chapter.

ELECTRIC AUTOMOBILES

Electric Automobiles, William Hamilton, McGraw-Hill, New York, 1980.

The Complete Book of Electric Vehicles, Domus Books, Chicago, 1979.

The Electric Car Book, Love Street Books, Louisville, Ky., 1981.

DIESEL AUTOMOBILES

Modern Diesel Cars, Jan P. Norbye, Tab Books, Blue Ridge, Pa., 1978.

REVIEW

1. Outline the cycle of an internal combustion engine.
2. What are the products of complete combustion and incomplete combustion of gasoline?
3. Describe the meaning of the octane rating of gasoline.
4. What is the purpose of tetraethyl lead in gasoline?
5. How is photochemical smog formed? Why is smog particularly bad in the Los Angeles area?
6. What is the concern for excessive amounts of carbon monoxide in the atmosphere?
7. Name four environmental effects generally attributed to photochemical smog.
8. What options exist for controlling undesirable emissions from automobiles?
9. What is a catalyst? What is a catalytic converter?
10. Distinguish between oxidizing and reducing catalytic converters.
11. What alternatives are there for the conventional internal combustion engine?

12. What are the attractive features of diesel-powered cars?
13. Discuss both the positive and negative features of electric vehicles.
14. Tetraethyl lead is used as an additive to gasoline in order to
    a) make it less odorous.
    b) make it burn more completely.
    c) increase the octane rating.
    d) reduce the amount of nitrogen oxides in the exhaust.
    e) convert carbon monoxide to a less toxic gas.
15. Perfect combustion of pure isooctane (gasoline) gives rise only to _____, _____, and _____ while a deficiency of oxygen will generally also produce _____ and _____.
    a) carbon dioxide, water, energy, carbon monoxide, unburned hydrocarbons.
    b) sulfur dioxide, water, energy, smog, eye irritation.
    c) nitric oxide, carbon dioxide, water, sulfur oxides, smog.
    d) aldehydes, acroleins, kerogens, particulates, sulfur oxides.
    e) any of the above will form a correct completion.
16. The production of nitrogen oxides in an automobile is a result of
    a) incomplete combustion of the gasoline.
    b) releasing nitrogen oxides that are a component of gasoline.
    c) catalytic converters.
    d) oxidation of nitrogen taken into the engine.
    e) unburned gasoline.
17. About five pounds of nitrogen oxides are produced for every ten gallons of gasoline burned in an internal combustion engine. An average car uses 750 gallons of gasoline each year and there are about 100 million cars in the United States. Hence, the amount of nitrogen oxides produced annually in the United States is approximately
    a) 375 pounds.
    b) 18.8 million tons (one ton = 2000 pounds).
    c) 375 thousand pounds.
    d) 750 million pounds.
    e) none of the above.
18. The presence of hydrocarbons in an automobile exhaust is a result of
    a) incomplete combustion of the gasoline.
    b) releasing nitrogen oxides that are a component of gasoline.
    c) catalytic converters.
    d) oxidation of nitrogen taken into the engine.
    e) hydrogenation of gasoline.
19. The poisonous gas contained in emissions from the exhaust of an automobile is
    a) carbon dioxide.
    b) arsenic.
    c) carbon tetrachloride.
    d) nitrogen oxide.
    e) carbon monoxide.
20. Nearly all cities in the United States are plagued by photochemical smog in the summer months. Of those constituents of photochemical smog listed below, the one of most concern is
    a) sulfur dioxide.
    b) carbon monoxide.
    c) ozone.
    d) hydrocarbons.
    e) sulfates.
21. Pick the reaction that "kicks off" the formation of photochemical smog.
    a) $2H_2 + O_2 \rightarrow 2H_2O$.
    b) $NO + O \rightarrow NO_2$ + energy.
    c) $SO_3 + H_2O \rightarrow H_2SO_4$.
    d) NO_2 + solar energy $\rightarrow NO + O$.
    e) None of the above.
22. The main environmental concern about nitrogen oxides and hydrocarbons involves their
    a) direct effect on the human respiratory system.
    b) chemical explosive characteristics.
    c) role in the formation of photochemical smog.
    d) effect on agricultural products.
    e) role in the formation of acid rain.
23. One should expect that carbon monoxide (CO) could be burned to produce energy because
    a) carbon monoxide is a hydrocarbon.
    b) carbon monoxide is a result of incomplete combustion of carbon and therefore should undergo further combustion.
    c) carbon monoxide is closely related to methane gas that burns easily.
    d) carbon monoxide is toxic.
    e) carbon monoxide can be hydrogenated.
24. Manufacturers of charcoal for outdoor grills warn consumers not to burn the charcoal in a closed area

like an indoors room. The warning refers to the hazards associated with

a) carbon dioxide.
b) nitrogen oxide.
c) photochemical smog.
d) carbon monoxide.
e) particulates.

25. Carbon monoxide is recognized as an air pollutant. This is because

a) of its role in the formation of photochemical smog.
b) of its biological effect on human beings.
c) of its effect on plants.
d) it tends to deactivate catalytic converters.
e) of its role in the greenhouse effect.

26. Pollutant control in a United States car is done primarily with

a) a catalytic converter.
b) an electrostatic precipitator.
c) tetraethyl lead.
d) higher octane gasoline.
e) fluidized bed combustion.

27. Catalytic as used in the phrase "catalytic converter" means that

a) industry has coined a clever word for a gas-saving device.
b) a chemical catalyst is used in the conversion of pollutants.
c) a chemical named catalyst is employed.
d) the manufacturer of the device is named catalyst.
e) the pollutants are called catalysts.

28. Gasoline containing lead compounds cannot be used in automobiles equipped with a catalytic converter because

a) the converter will not remove the lead from the exhaust gases.
b) the lead renders the catalyst ineffective.
c) the converter permits the use of high compression ratios and makes the use of lead unnecessary.
d) lead compounds lead to the production of sulfuric acid in the converter.
e) the lead combines chemically with the catalyst to introduce a very toxic lead-platinum compound into the exhaust gases.

29. Catalytic converters removing carbon monoxide are called oxidizing because

a) oxygen is united with carbon monoxide.
b) oxidizing is a trade name for the converter.
c) oxygen is removed from the carbon monoxide.
d) the carbon monoxide is reduced to carbon and oxygen.
e) oxygen is used as a catalyst.

30. Batteries for an electric automobile are likely to be recharged by electricity produced by a coal-burning electric power plant. Therefore, it is appropriate to say that electric automobiles have no part in air pollution problems.

a) true b) false

31. The use of electric vehicles is presently limited because the vehicles have

a) poor acceleration.
b) inadequate energy for long trips.
c) excessive pollution characteristics.
d) poor starting characteristics.
e) complicated mechanical linkages.

Answers

14. (c)	15. (a)	16. (d)	17. (b)
18. (a)	19. (e)	20. (c)	21. (d)
22. (c)	23. (b)	24. (d)	25. (b)
26. (a)	27. (b)	28. (b)	29. (a)
30. (b)	31. (b)		

QUESTIONS AND PROBLEMS

6.2 FUNDAMENTAL ORIGIN OF EMISSIONS

1. Two gasolines have octane ratings of 50 and 100. What is the octane rating of a gasoline having equal portions of these two gasolines?
2. About five pounds of nitrogen oxides are produced for every ten gallons of gasoline burned in an internal combustion engine. An average car uses 750 gallons of gasoline each year and there are about 100 million cars in the United States. Estimate the amount of nitrogen oxides produced in a year.
3. Knowing that the complete oxidation of heptane (C_7H_{16}) produces only carbon dioxide (CO_2) and water (H_2O), complete the following reaction.

$$C_7H_{16} + 11O_2 \rightarrow \_\_\_\_ + \_\_\_\_ .$$

6.5 THE FEDERAL EMMISION STANDARDS

4. To what extent do nationwide economic conditions influence the public's desire for pollution controls on automobiles?

6.6 APPROACHES TO MEETING FEDERAL STANDARDS

5. Nitric oxide (NO) can be rendered harmless if it is converted to molecular nitrogen (N_2) and water vapor (H_2O) through thermal reactions with ammonia (NH_3). Fill in the blanks in the chemical reaction below for this process.

   _____ NO + 4_____ → 5_____ + _____ H_2O.

6. The two reactions below can be used to convert SO_2 to SO_3. The end product of the first reaction (NO_2) is used as part of the input for the second reaction.

   $2NO + O_2 \rightarrow 2NO_2$
   $2NO_2 + 2SO_2 \rightarrow 2NO + 2SO_3$.

   How many molecules of NO and NO_2 are consumed in the execution of these two reactions? What is the role of NO and NO_2 in the reactions?

6.7 POSSIBLE ALTERNATIVES TO THE CONVENTIONAL INTERNAL COMBUSTION ENGINE

7. It is common to use electric-powered service vehicles inside merchandise warehouses and manufacturing facilities. What advantages do electric vehicles have over gasoline-powered vehicles for this function?
8. Why isn't a battery-powered motorcycle practical?
9. The fuel gauge on a gasoline-powered car registers the amount of gasoline remaining in the tank. What would the fuel gauge on an electric car register?
10. Electric vehicles once outnumbered gasoline-powered automobiles and the future for electric cars looked promising. What do you think led to the demise of the electric car?
11. What characteristics of electric golf carts make them attractive?
12. A certain car has a 15-gallon gasoline tank. With a full tank, the car weighs 2500 pounds. Assuming a gallon of gasoline weighs eight pounds, what percentage of the total weight is due to the fuel? If the car were electric and the batteries had an energy capacity (watt-hours per pound) ten times less than gasoline, what percentage of the 2500 pounds of weight would represent the weight of the batteries?
13. The transformation of energy from fossil fuels to mechanical energy in an automobile via electric generation and battery-powered motors involves several steps with characteristic efficiencies.
    a) Power plant efficiency: 40%
    b) Ten % energy loss through transformers and transmission to the city
    c) Ten % energy loss through transformers and distribution within the city
    d) Battery-charging efficiency: 80%
    e) Motor efficiency: 90%

    Show that the overall efficiency is 23%. How does this compare with the efficiency of a typical automobile?

7

Thermodynamic Principles

7.1 MOTIVATION

Most drivers view an automobile engine as a mysterious device that derives energy from gasoline and converts some of it to energy of motion. Many drivers know how many miles the car travels for each gallon of gasoline purchased, but few realize that only about 25% of the energy is converted into energy of motion. Most drivers are also aware that oil is added periodically to an automobile engine and that the oil serves as a lubricant to reduce friction between moving parts. Friction might be suspected as the cause for the low efficiency of the engine; we might suppose that if we were sufficiently ingenious or if we had a better lubricant, we could make an automobile engine with an efficiency close to 100%. We might draw the same conclusion if we were to read that

Nature provides natural thermodynamic systems in the form of underground reservoirs of hot water and steam. Near The Geysers in northern California, the Pacific Gas and Electric Company generates electricity from energy derived from undergound steam reservoirs. (Photograph courtesy of the Pacific Gas and Electric Company.)

the efficiency of a steam turbine is about 50%. Friction is an impotant consideration, but a more fundamental cause of low efficiency is found in the thermodynamics of the energy conversion process. The steam turbine, like all heat engines, draws heat energy from a source (a container of steam, for example), converts some of the heat energy into mechanical energy (rotating shaft), and releases heat energy to a lake, for example. Typically, as much heat energy is released as is converted to mechanical energy. Because there are so many large steam turbines in operation, especially in the electric power industry, disposal of the unused heat energy is a major engineering consideration in an electric power plant.

Our goals in this chapter are several. First, we want to learn enough thermodynamic principles to appreciate the reasons for the low efficiency of a commercial steam turbine. Second, we want to use these principles to understand the methods being used to dispose of heat in environmentally acceptable ways.

7.2 BASIC THERMODYNAMICS

Pressure and Temperature

Both the modern steam turbine and its older counterpart, the steam engine, are devices that convert the energy of an expanding gas to mechanical energy (Fig. 7.1). While the turbine has largely replaced the engine as a practical device, we shall discuss the principle of a steam engine because it is easier to analyze. Our conclusions will also apply to the steam turbine.

We start with the observation that an expanding gas pushing against a piston does mechanical work on the piston. The expanding gas does work just as you would do work by pushing on the piston with your hand (Fig. 7.2). If we knew the force exerted on the piston by the gas, then (as in Section 2.6) we could compute the work by multiplying the force and the distance the piston moves ($W = Fd$). We shall relate the force to pressure, which along with temperature are basic physical quantities in thermodynamics (Fig. 7.3). For convenience, we symbolize these quantities as P and T.

Both pressure and temperature are common words. Pressure implies a force. Being pressured to study for an exam suggests that we are forced to do so. Automobile drivers are conscious of the air pressure in the tires. When a balloon bursts, it is because the air pressure within the balloon exceeds the breaking strength of the balloon. Anything that supports the weight of some object or some material has a pressure exerted on it. The surface of the earth supports the weight of the air atmosphere above it; therefore, the atmosphere exerts a pressure on the earth's surface. These examples suggest that the concepts of force and pressure are the same. However, there is an important difference. The concept of pressure incorporates the area over which the

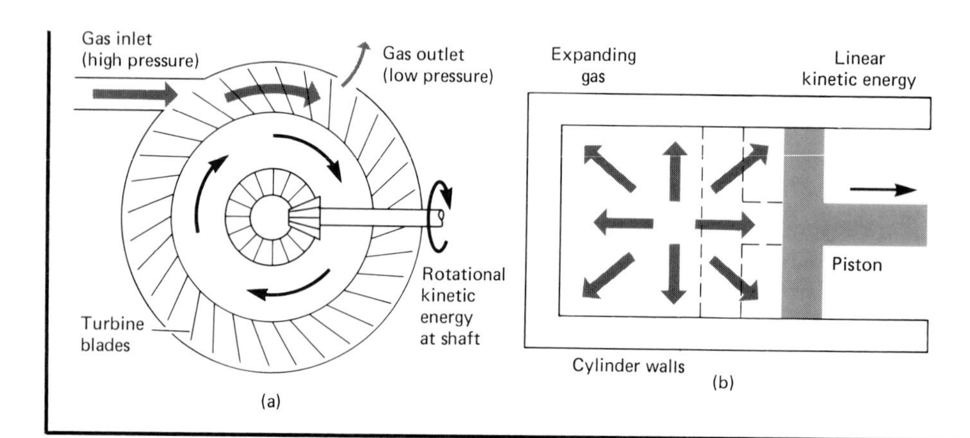

FIGURE 7.1 (a) Schematic illustration of the operation of a turbine. A gas or vapor exerting a force on the blades of the turbine produces rotational kinetic energy. (b) Schematic illustration of the operation of a steam engine. A gas or vapor exerting a force on a piston produces linear kinetic energy.

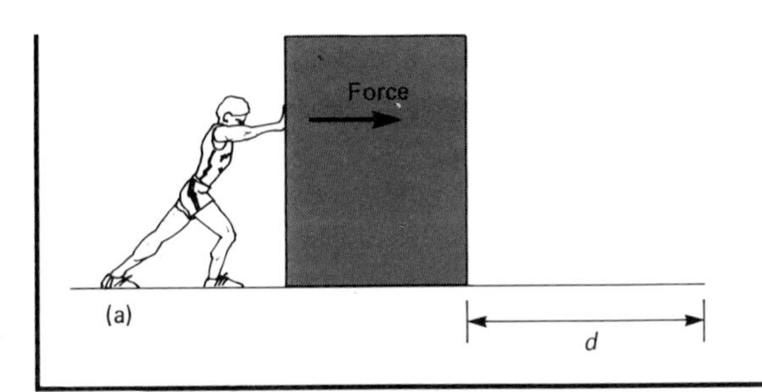

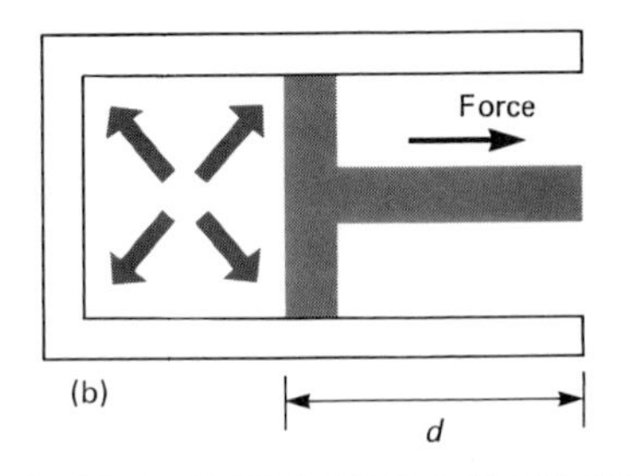

FIGURE 7.2 (a) A person does work when moving a box. (b) A gas in a cylinder does work when moving a piston.

force acts.

$$\text{pressure} = \frac{\text{force on a surface}}{\text{area over which the force acts,}}$$

$$P = \frac{F}{A}. \tag{7.1}$$

Pressure has units of force divided by units of area. Typical units are newtons per square meter (N/m^2) and pounds per square inch (lb/in^2). Atmospheric pressure at sea level is about 100,000 N/m^2 (15 lb/in^2). This means that each square meter on the surface of the earth at sea level supports a weight of 100,000 newtons (22,500 pounds) of air. Note that pressure can be very large for a moderate force if the contact area is very small. A very large pressure can be exerted with a thumbtack with minimal force because the contact area of the point of the tack is very small. This is why it is easy to press a thumbtack into a bulletin board.

There is probably no scientific word more familiar than temperature. Nearly all houses have a thermostat to regulate the indoor temperature. Outdoor temperature is a regular feature of TV newscasts and newspapers. However, temperature is a much subtler concept than pressure. Pressure measures a mechanical quantity, but temperature is a measure of the sensation of hot and cold. A thermometer assigns a number related to this sensation. On the Celsius temperature scale, an equilibrium mixture of ice and water at atmospheric pressure at sea level is assigned 0° C. Water boiling at atmospheric pressure at sea level is assigned 100° C. Other temperatures are measured relative to these two temperatures. The ice and steam points are assigned numbers 32 and 212, respectively, on the familiar Fahrenheit scale, and 273 and 373 on the Kelvin scale (Fig. 7.4). The Celsius, Fahrenheit, and Kelvin scales are related by

$$\mathrm{F} = \frac{9}{5}\mathrm{C} + 32. \tag{7.2}$$

$$\mathrm{K} = \mathrm{C} + 273. \tag{7.3}$$

Note that if you let C = 0 and C = 100 in Eq. (7.2), you get F = 32 and F = 212, as expected, for the

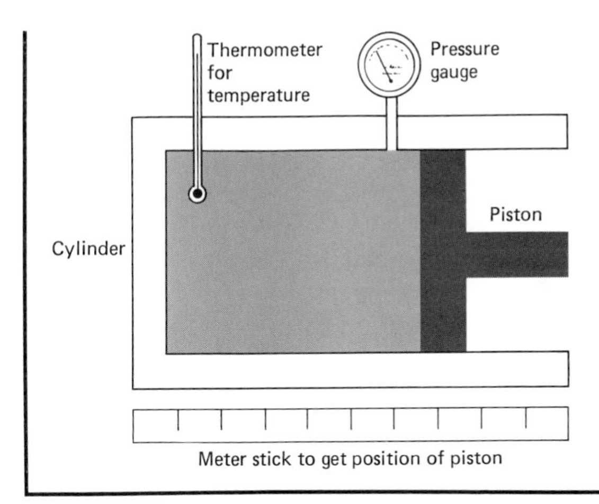

FIGURE 7.3 Illustration of an operational way of measuring the thermodynamic variables of a gas in a cylinder. The thermometer is similar to a glass thermometer often used to measure body temperature. The pressure gauge functions like a gauge used to record the air pressure in an automobile tire. The cylinder is shaped like a soup can so that the volume equals the area of the bottom multiplied by the distance from the piston to the bottom of the cylinder.

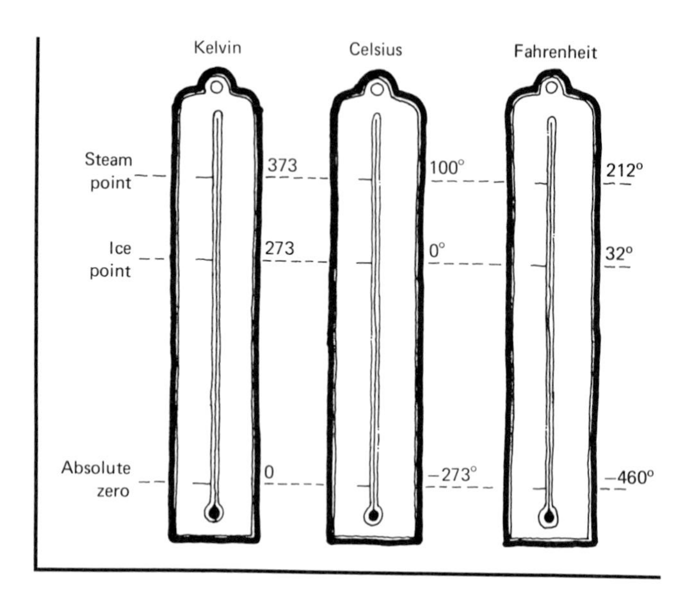

FIGURE 7.4 Reference temperatures for the three common temperature scales.

corresponding Fahrenheit temperatures. And for the same Celsius temperatures, you get 273 and 373, as expected, for the corresponding Kelvin temperatures. If a room temperature is reported as 20° C, the Fahrenheit temperature is $F = \left(\frac{9}{5}\right) 20 + 32 = 68°$ F and the Kelvin temperature is $K = 20 + 273 = 293$ K. We state the temperatures as 20 degrees Celsius, 68 degrees Fahrenheit, and 293 kelvins.

Internal Energy

The Astrodome in Houston is a completely enclosed sports complex where fans cheer (and boo) and players compete. TV cameras scanning the arena display pictures of people in all sorts of activities involving kinetic energy—because of their motions—and gravitational potential energy—because of their elevation above the playing field. If we were sufficiently ingenious, we could add up all these energies and call the sum internal energy because the energy is contained within the boundaries of the arena. When you inflate a balloon, you fill it with molecules. There is no TV camera to scan inside the balloon, but the molecules are present in incessant motion. Physicists have learned how to compute and measure the total energy of the molecules, but the computation is beyond the scope of this study. However, some of the salient results are of interest here.

Let us begin by considering ordinary air in a closed container such as an inflated football. The ball contains many trillions of molecules bouncing off its walls like balls rattling around in a vibrating box (Fig. 7.5). Pressure is exerted on the walls of the football because of collisions of the molecules with the walls. It is easy to feel this pressure by pressing your fingers against the ball. There are several ways to change the pressure. If some of the air is let out of the football, the pressure decreases and the ball deflates. Squeeze the football and the pressure increases. Squeezing reduces the volume of the ball and reduces the distance a molecule must travel before hitting another wall. Heating the football causes the pressure to increase. This effect is commonly observed with automobile tires on hot days or when tires warm after being run on a highway for an extended time. Although the pressure changes resulting from letting gas out and changing the volume can be understood fairly easily, the connection between temperature and pressure is rather subtle. A pressure increase, regardless of how it is made, means that the molecules are hitting the walls with greater frequency. This increase in frequency could come from reducing the volume as suggested earlier or by increasing the speed of the molecules. If the volume does not change, then the increased collision frequency must come from an increase in the speed of molecules. Greater molecular speed, like greater automobile speed, means greater kinetic energy. So when the gas warms and the pressure increases, the kinetic energy of the molecules increases. Knowing that other forms of energy exist, we might suspect that some energy goes into forms other than kinetic energy. This is indeed the case, but it is not necessary to include these other potential energy forms in these considerations because the behavior of many gases is explained by assuming that all the molecular energy is kinetic.

Some thought should convince you that one direction of molecular motion in the football is just as likely as any other—the motion of molecules is completely random. There is also evidence that all molecules do not have the same speed. Perhaps this is not obvious, but it is not hard to accept. The molecules of a gas are somewhat like bees buzzing around in a hive. Surely we would not expect every bee to be moving with the same speed as all the other bees. So we cannot talk meaningfully about the kinetic energy of a particular molecule. But we can refer to the average kinetic energy of a molecule because this concept does not distinguish a particular molecule. There is considerable experimental and theoretical evidence to support the hypothesis that the average random kinetic energy of a molecule in a gas is directly proportional to the kelvin temperature of the gas. If the kelvin temperature doubles, the average kinetic energy of a molecule doubles. Summarized as an equation,*

$$E = \frac{3kT}{2}. \tag{7.4}$$

This random kinetic energy is called thermal energy. The energy derived from a gas to run a steam turbine comes from the random energy of the gas. If energy is extracted from the gas, the temperature decreases.

We have used a gas as the basis of our discussions because it is of particular interest at this time. However, liquids and solids are also composed of atoms, or molecules as the case may be, and they possess thermal energy. The higher the temperature of an object, the greater its thermal energy.

Heat

While work is a concept having broad everyday usage, the principles of physics allow us to define it precisely and quantitatively. Heat falls into the same category. When someone says, "The house is cold, we need heat," the request is clear. Similarly, we know that exercise produces both warmth and an increase in body temperature, and we know energy

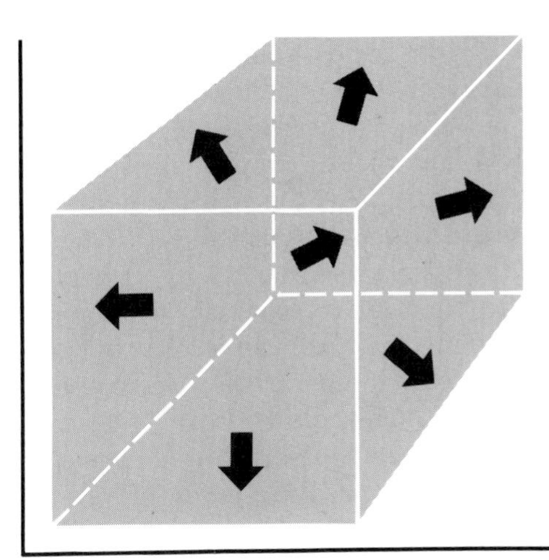

FIGURE 7.5 Molecules in a container push on the walls. The pushing gives rise to a pressure. The greater the speed and frequency of the collisions, the greater the pressure.

is derived from food. So we make some intuitive connection between heat, energy, and temperature. Sticking your hand in water at 90° C produces a greater sensation of being hot than if you stuck your hand in water at 40° C. Accordingly we say the 90° C water contains more energy and energy was transferred to your hand by the water. We distinguish carefully the energy content of the water and the energy transferred. The energy *content* is the *internal* energy. It is energy within the system—water in this example. It is energy arising from thermal motion, that is, the kinetic energy of the atoms and molecules and whatever forms of potential energy that might be involved. Generally, only thermal energy resulting from thermal motion is important. As the temperature increases, the internal energy of a system increases because the thermal energy increases. Energy in *transit* is called *heat*. Heat energy is transferred between two objects (hand and water, for example) only if there is a temperature difference. The energy flows from the object having the higher temperature. If your hand is at 30° C and you stick it in ice water, heat flows from your hand to the water—your hand cools. The internal energy of your hand decreases; the internal energy of the water increases. If your hand is at 30° C and you stick it in water at 90° C, heat flows

* The letter k in the formula for average random kinetic energy is called Boltzmann's constant. It is extremely small and has a numerical value of 1.38×10^{-23} joules/(K × molecule).

from the water to your hand—your hand warms. The internal energy of your hand increases; the internal energy of the water decreases.

The calorie (cal) and British thermal unit (Btu) are units used to quantify this concept. If the temperature of one gram of water is raised by 1° C, the water is said to have absorbed one calorie of heat. The kilocalorie (kcal) or ordinary food calorie is equal to 1000 calories defined in this way. A Btu is the heat required to raise the temperature of one pound of water by 1° F. If in a household hot-water system we want to know how much heat is required to raise the temperature of 20 gallons (165 pounds) of water from 80° F to 140° F, it suffices to say

$$\begin{aligned} \text{heat} &= \text{weight} \times \text{temperature change} \times 1\ \text{Btu per (pound} \times {}^\circ\text{F)}, \\ &= (165)\,(60)\,(1) \\ &= 9{,}900\ \text{Btu}. \end{aligned}$$

When we discuss the storage of thermal energy in solar heated houses (Chapter 10), we will want to know how much heat is required to raise the temperature of a mass of rocks a certain number of degrees. Thus it is also meaningful to know the heat required to raise the temperature of one gram (or one pound) of any substance by one degree Celsius (or by one degree Fahrenheit). This is called the specific heat. Generally, the specific heat depends on temperature, but for most calculations it is assumed to be constant. The specific heats of some common substances are shown in Table 7.1. If in a solar energy application we want to know the heat required to raise the temperature of a 1000-kilogram block of concrete from 25° C to 30° C, we look up the specific heat of concrete (Table 7.1) and write

$$\begin{aligned} \text{heat} &= \text{mass} \times \text{temperature change} \times \text{specific heat}, \\ &= (1{,}000{,}000\ \text{grams})\,(5^\circ\ \text{C})\,(0.22\ \text{cal/g} \times {}^\circ\text{C}), \\ &= 1{,}100{,}000\ \text{cal}, \\ &= 1{,}100\ \text{kcal}. \end{aligned}$$

First Law of Thermodynamics

An energy converter changes energy forms but the sum total of the energy at any time remains constant. A ball rising above the ground loses kinetic energy, but its gravitational potential energy increases accordingly so that the total energy is the same at any height. With the recognition of internal energy and heat as energy in transit, we now have to account for new energy forms. Nevertheless, the energy conservation principle still prevails. This principle is called the first law of thermodynamics. If 1000 joules of heat are absorbed by the gas in a cylinder and the piston does no work by moving, we say the heat transfer increases the internal energy of the gas (Fig. 7.6). Similarly, if we see a piston move in a cylinder removed from any external source of energy, we say the energy for the work came from the internal energy of the gas. If 1000 joules of work are done, the internal energy decreases 1000 joules. As a general rule, the change in internal energy equals the heat added to the system (for example, the gas in the cylinder) less the amount of work done by the system.

TABLE 7.1 Specific heat of some common substances. The data were taken from *Handbook of Chemistry and Physics*, CRC Press, Inc., Boca Raton, Fla., and *Applied Solar Energy*, Aden B. Meinel and Marjorie P. Meinel, Addison-Wesley, Reading, Mass., 1976.

Substance	Specific heat $\left(\frac{\text{calories}}{\text{gram} \cdot {}^\circ\text{C}}\right)$
Water	1.00
Aluminum	0.22
Iron	0.11
Copper	0.093
Concrete	0.22
Brick	0.20
Dry earth	0.19
Wet earth	0.50

Steam Turbine

A steam turbine in an electric power plant is a massive, expensive unit requiring impressive engineering and technology to build (Fig. 7.7). The energy principle is straightforward (Fig. 7.8): the turbine

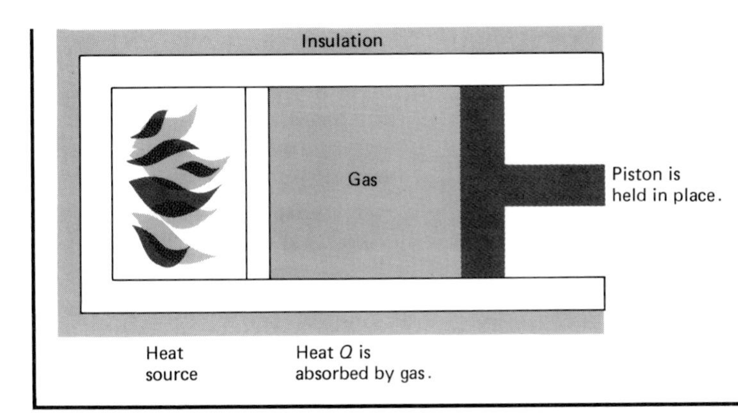

FIGURE 7.6 The gas in a cylinder with a fixed piston absorbs heat Q from a heat source. No work is done by the gas because the piston doesn't move, but the internal energy and temperature of the gas increase.

converts thermal energy to mechanical motion, and performs its task in cycles. Understanding the cyclic aspect is very important. Before delving into the cycle of a steam turbine, let us look at an analogous situation. Imagine (but do not experiment!) filling a container with water, tightly closing it with a cap, and placing the container onto a burner. Turn the burner on and the water absorbs energy and changes to steam. The sealed container is moved from the burner and some of the energy from the steam is removed and converted to mechanical motion. There is no change in the amount of steam in the container but its temperature has dropped because the steam has less energy. The container is now placed in contact with a cooler surface, energy is removed, the steam condenses to water, and the temperature drops to its initial value. The container is placed back on the burner to begin another cycle. No water was lost in the cycle and there is no difference between the thermal energy content of the water between the beginning of one cycle and the start of another.

The energy-producing cycle in a steam turbine begins at the boiler with water at a temperature roughly that of water from a tap in a house. Heat, labeled Q_{hot} in Fig. 7.8, is added to the water, converting it to steam having a temperature several hundred degrees higher. The thermal energy of the molecules in the vaporized water has increased significantly. The steam molecules are directed via pipes toward the blades of the turbine where they do work on the turbine blades, giving some of their energy to the shaft of the turbine. Having lost energy, the temperature of the steam decreases after impinging on the turbine blades. In order to return the water

FIGURE 7.7 The people in this photograph are standing alongside a modern steam turbine that powers an electric generator in an electric power plant. The viewers are facing the turbine part of the combination. Some appreciation for the size of the operation is conveyed by the relative size of the people and the machinery. (Photograph courtesy of Dayton Power and Light Company.)

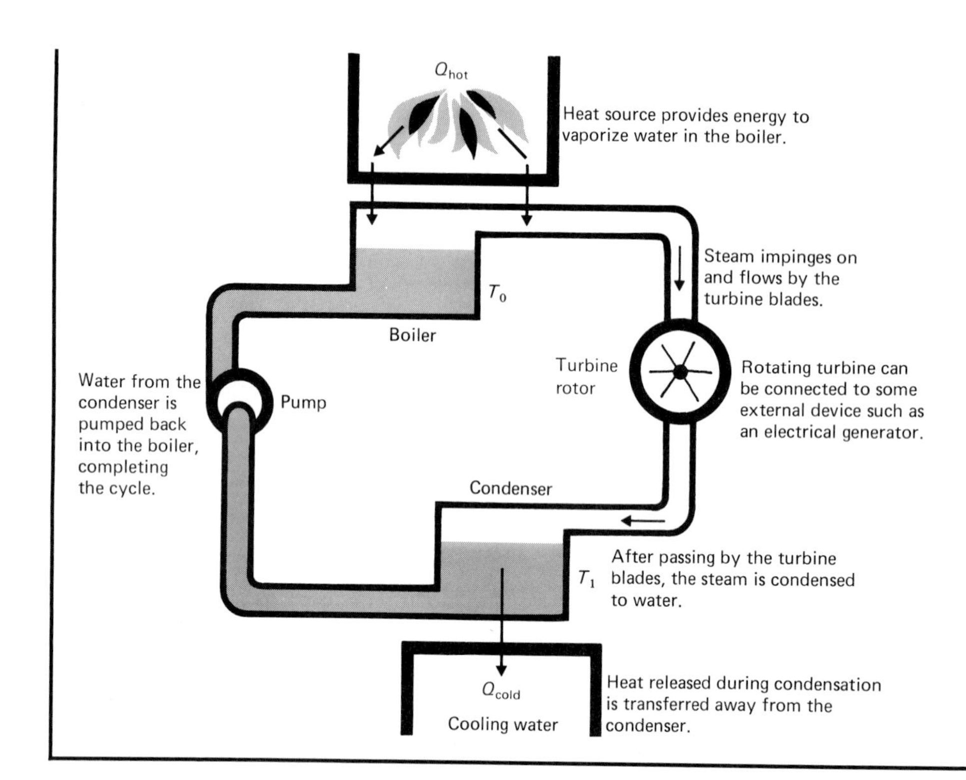

FIGURE 7.8 Schematic diagram of energy processes in a steam turbine. The boiler, turbine, condenser, and pump are separated to illustrate the cyclic flow of steam and water. The combination of all four components constitutes the steam turbine heat engine.

molecules to their initial liquid condition in the boiler, the steam at the lower temperature must be condensed. This operation is done at the condenser. The heat labeled Q_{cold} is liberated at the condenser. This heat must be removed. Otherwise the condenser warms. Following condensation, the water is pumped back to the boiler completing the cycle. Two features are especially important.

1. Unless there are leaks in the pipes of the turbine system, the same water is circulated continuously. The water changes from liquid to steam and from steam to liquid, and exchanges energy with the turbine blades and condenser, but the same water molecules are involved in the cyclic process.
2. The energy state of the water at the end of the cycle is unchanged. Thermodynamically, we would say there is no change in its internal energy. In the cycle, heat (Q_{hot}) was added to the water to turn it into steam, some energy from the steam (W) was given to the turbine, and the remainder (Q_{cold}) was absorbed by the condenser.

Accounting for these energies in accordance with the first law of thermodynamics, we may write

energy acquired by the rotating turbine equals energy of the steam entering the turbine minus energy of the steam leaving the turbine.

Summarizing as an equation

$$W = Q_{hot} - Q_{cold}.$$

Because the function of the turbine is to convert thermal energy to mechanical motion, we can write its efficiency as

efficiency = energy acquired by the turbine ÷ energy of the steam entering the turbine

$$\epsilon = \frac{W}{Q_{\text{hot}}} = \frac{Q_{\text{hot}} - Q_{\text{cold}}}{Q_{\text{hot}}}. \tag{7.5}$$

This efficiency equation applies to any heat engine, large or small. If, for example, a steam turbine in each cycle of its operation extracts 100 Btu of heat to vaporize water and in passing through the turbine the steam molecules deliver 40 Btu of energy to the turbine, we record its efficiency as

$$\epsilon = \frac{40}{100} = 0.4 \text{ or } 40\%.$$

Forty percent of the energy extracted from the energy source is converted to useful work.

Be very sure that you understand what the efficiency equation (Eq. 7.5) is saying. Q_{hot} is the energy from which the mechanical motion is obtained. Clearly, we would like to convert as much of this energy as possible into mechanical motion. That is the reason for the heat engine. The efficiency improves as the heat rejected to the condenser decreases. In the previous numerical example, the heat rejected was 60 Btu. If we could reduce this to 30 Btu, the efficiency would increase to 0.7 or 70%.

An automobile engine is a form of heat engine. The energy for an expanding gas comes from igniting a gasoline vapor and air mixture. As citizens interested in automobiles, we see that some engines are constructed better than others and we see that some engines perform better than others. We expect differences in the construction and performance of steam turbines. Regardless of its complexity, a heat engine must have a working fluid such as a gas at a temperature we label T_{hot} and a lower temperature region for the fluid to move into. In a steam turbine, this lower temperature region is provided by the condenser. If there is no temperature difference, the working fluid—steam, for example—is not going to flow anywhere and the turbine will deliver no mechanical energy. As the temperature difference increases, the energy delivered to the turbine increases and its efficiency increases accordingly. If for some reason there are limitations on the two temperatures, is it conceivable that there is a limit on the efficiency of a heat engine that must operate between these two temperatures? Is it conceivable that, no matter how clever or skillful a designer and builder may be, there is a limit to the performance of a heat engine that must operate between two given temperatures? Probably these questions were posed at the inception of heat engines but it was not until 1824 that a French engineer Sadi Carnot provided an answer that has borne the test of time. Carnot imagined a perfectly constructed engine—no friction, no heat losses to the outside world, nothing involving construction that would detract from its operation. His theoretical engine, subsequently called a Carnot engine, was reversible so that it could function as a refrigerator as well as an engine. Carnot argued that the least amount of heat that could be rejected in a cyclic heat engine using Q_{hot} units of energy in each cycle involving temperatures T_{hot} and T_{cold} was

$$Q_{\text{cold}} = Q_{\text{hot}} \cdot \frac{T_{\text{hot}}}{T_{\text{cold}}}.$$

Carnot's conclusion requires that the temperatures be expressed on the kelvin scale. Remember, for a given amount of heat (Q_{hot}) the efficiency improves as the heat rejected (Q_{cold}) decreases. If we substitute Carnot's result into the general efficiency equation (Eq. 7.5) we arrive at

$$\epsilon = 1 - \frac{T_{\text{cold}}}{T_{\text{hot}}}. \tag{7.6}$$

No result has more practical significance than Carnot's efficiency equation. Knowing only two temperatures, you can determine the largest efficiency for any heat engine utilizing these two temperatures. For example, if the steam temperature entering a turbine is 400 K and the condenser is maintained at 300 K, then no engine using these temperatures can have an efficiency exceeding

$$\epsilon = 1 - \frac{300}{400} = 0.25 \quad \text{or} \quad 25\%.$$

Any heat engine will have imperfections that prevent attainment of the theoretical limit.

Steam enters a steam turbine in an electric power plant at a temperature of about 800 K. The condenser is maintained at about 300 K. Thus the largest possible efficiency is about 63%. Imperfections in the steam turbine reduce the efficiency to about 50% which seems low. (In the energy model of an electric power plant in Section 4.12 we used an efficiency of 47% for the steam turbine.) When you realize that with constraints on the temperatures, 50% compared with a limit of 63% is really quite good.

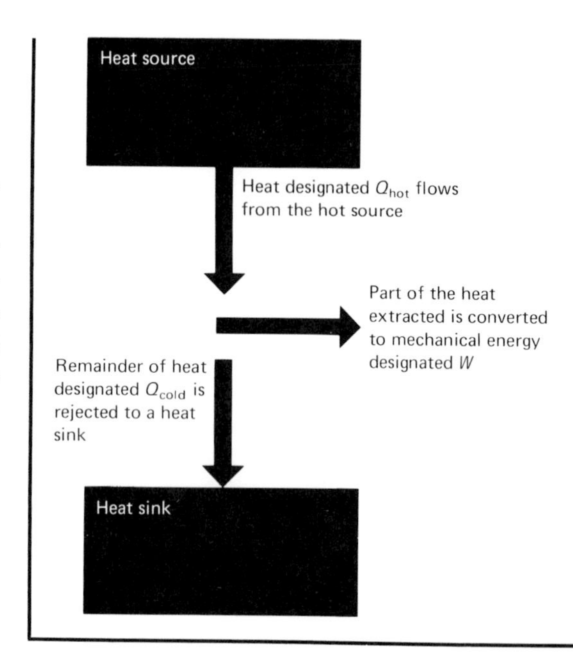

FIGURE 7.9 In each cycle of its operation, a steam turbine draws heat (Q_{hot}) from a hot source, converts part of this heat to mechanical energy (W), and rejects the unused heat (Q_{cold}) to a sink at a temperature lower than the source.

Second Law of Thermodynamics

Place a marble on a slanted floor and it may move spontaneously to a lower level but you never see the marble move spontaneously to a higher level. The spontaneous movement of energy is always from higher energy to lower energy. Similarly, if you place your hand in ice water heat never moves spontaneously from the cold water to your warm hand. The direction of heat flow is from the warmer to the cooler. These observations are the essence of the second law of thermodynamics. In simplest form the law requires a temperature difference for the spontaneous movement of heat that must move from warmer to cooler. There are many ramifications of the second law of thermodynamics. Two that play a part in our energy discussions involve heat engines and refrigerators. Any heat engine, regardless of complexity, operates in cycles, extracts heat from a reservoir, and converts a portion of the heat to useful energy (Fig. 7.9). The second law of thermodynamics says that the engine cannot convert all the heat into useful energy. Some of the heat extracted must end up in some area at a lower temperature. Succinctly, the second law of thermodynamics precludes the building of a heat engine with 100% energy conversion efficiency, and establishes a maximum efficiency governed by the Carnot equation (Eq. 7.6). Nothing prevents moving a marble from the floor to a table top but something has to do work in order to do it. When you pick the marble up and place it on the table, you do work because gravity works against you. Similarly, nothing prevents forcing heat from a cool region to a warmer region. But forcing this heat transfer requires work. Freezing compartments of refrigerators are cooled by removing thermal energy. The energy movement does not occur spontaneously because it is warmer outside the refrigerator. Work is required to transfer the energy to the outside. The role of electric motors on refrigerators, air conditioners, and heat pumps (Chapter 13) is to do the necessary work. Consumers pay electric power companies for energy to run the motors. It would be nice to freeze water in a refrigerator freezer and not have to pay for an energy expenditure. However, the second law of thermodynamics dashes this wishful thinking by requiring that work be done to force heat to flow from a cool region to a warm region. A marble on top of a staircase has gravitational potential energy. Each steplike movement toward the floor reduces its energy. When it reaches the floor and is trapped in the confines of the room, its ability to do work is halted. When steam flows out of a boiler, each steplike movement toward a lower temperature reduces its energy. When the steam reaches the temperature of the environment, there is no lower temperature to accept it and its ability to do work is halted. The first law of thermodynamics ensures that no energy is lost in the traversal from boiler to environment. However, once the temperature reaches that of the environment, the ability to do further work is lost because there is no connecting region at a lower temperature. When heat flows from a reservoir such as a boiler, no engine can convert more of the heat to useful work than a Carnot engine. The work is called the available energy and the heat needed to produce it is called the thermal energy input. The available energy is theoretically available from the thermal energy input. Even though the thermal energy input is never lost because it ends up somewhere at a lower temperature, the available energy is lost because there will never be another opportunity to convert the heat to mechanical energy or electricity or any other form of energy. Energy available for work is lost; it is gone forever!

7.3 DISPOSAL OF WASTE HEAT

A steam turbine efficiency of 50% means two units of thermal energy produce one unit of useful mechanical energy and one unit of thermal energy that is rejected to the environment. It seems odd that thermal energy is thrown away to the environment when water is returned to the boiler only to be reheated. To understand this apparent contradiction, you must realize it is desirable to extract as much energy as possible from the high-temperature steam. Thus the temperature difference between the steam and the low-temperature region has to be made as large as possible to make the turbine efficiency as large as possible. After relinquishing energy to the turbine, the steam temperature drops but it is still steam, and it condenses when entering the low-temperature region. Water absorbs energy when it vaporizes and liberates energy when it condenses. (Steam can produce more severe burns than water at the same temperature because of the energy released on condensation.) Unless the thermal energy from condensation is removed from the low-temperature part of the turbine, that is, from the condenser, the condenser temperature rises reducing the turbine efficiency (a reduction which is highly undesirable). A large turbine may handle two billion watts of heat at its input which means it is liberating one billion watts of heat at the condenser. Hence large amounts of heat must be removed and disposed of in an environmentally acceptable manner. The heat is removed by a continuous flow of water that warms when it contacts the condenser. This cooling system is completely isolated from the water and steam flow in the turbine. The amount of warming depends on how long the cooling water is in contact with the condenser. The shorter the time of contact, the smaller the increase in temperature. Thus the larger the rate of flow of water into the condenser, the smaller the temperature rise. Generally, the rate is such that the water warms 5 to 10 C° (10 to 20 F°).

You should also realize that although there is a lot of energy in the water that is cooling the condenser, its fairly low temperature limits its use. The total energy of any system that has energy distributed among its parts is a combination of the energy of each part and the number of parts. Each student in a large class may have a small amount of money but the monetary worth of the class may be large because the total worth is the product of the money per student and the number of students. If the temperature of the cooling water is only a few degrees above its normal temperature, it means the average thermal energy per water molecule is fairly small. However, the number of molecules in the cooling water is very large making the total energy very large because the total energy is the product of energy per molecule and the number of molecules.

An adequate supply of cooling water is essential for locating an electric power plant. The shores of the Great Lakes and the major rivers are, therefore, favorite locations. Large volumes of water taken from a natural source and returned 5 to 10° C warmer can alter the ecostructure. Increasing the temperature of a body of water above its normal value produces a variety of changes on the quality of the water. Those changes of concern to plant and animal life in the water are the following:

1. As the temperature of the water increases, the capacity to hold oxygen dissolved in the water decreases. At the same time, the need for oxygen by aquatic life increases.
2. Addition of heat to a body of water can cause stratification with each stratum having a different temperature.
3. Chemical reactions are accelerated at elevated temperatures.
4. Concentrations of undesirable chemicals increase because of accelerated evaporation of water.

The anticipated effects of elevated temperatures on aquatic life include the following:

1. Decreased spawning success of fish.
2. Decreased survival probability of young fish.
3. Limitation of migration patterns.
4. Changes in processes that depend on biological rhythms.
5. Increased susceptibility to disease.
6. Alteration of the balance between plant and animal life.

Water quality standards promulgated and enforced by the Environmental Protection Agency generally prohibit taking water directly from a source such as a lake or river to cool a steam turbine condenser, and then returning the warmed water from the condenser to the same source. The alternative is to transfer the thermal energy deposited in condenser cooling water directly to the atmosphere via a cooling tower. Whether heat from the condenser is transferred to a river or lake or to a cooling tower, it eventually ends up in the atmosphere. A cooling tower is an intermediary between the condenser and the atmosphere. To understand how cooling towers work, we need some basic knowledge of heat transfer.

Conduction and Convection

When some object is placed in contact with another object at a lower temperature, the hotter one cools down and the cooler one warms up until an equilibrium temperature is achieved. We say *heat* has flowed (or been conducted) from the hotter to the cooler object. Thinking in terms of molecules we would say the average kinetic energy of the molecules in the hotter object has decreased and the average kinetic energy of the molecules in the cooler object has increased. At equilibrium, the average kinetic energy is the same (Fig. 7.10). The transfer of energy has taken place by molecular collisions in much the same way energy is transferred by colliding billiard balls.

There is an enormous difference in the heat-conducting properties of materials. With your bare hand you can hold a wooden-handled spoon in boiling water without eliciting any hot sensation. But you quickly drop the spoon if it is made of metal. This is because heat is conducted from the water

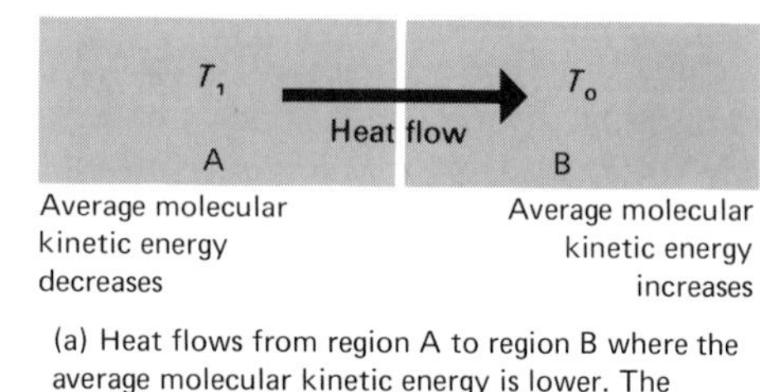

(a) Heat flows from region A to region B where the average molecular kinetic energy is lower. The energy transfer produces a decrease in the average molecular kinetic energy of the warmer region (A) and an increase in the average molecular kinetic energy of the cooler region (B).

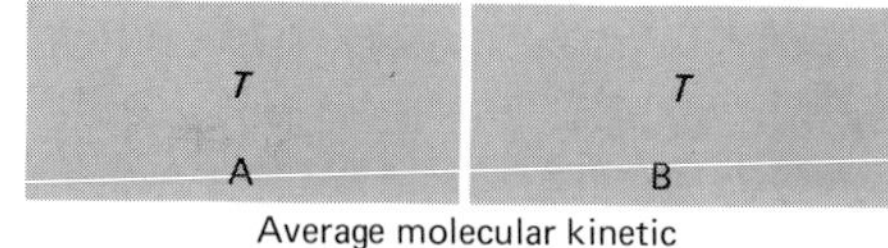

(b) An equilibrium temperature is established when the average molecular kinetic energy is the same in both regions.

FIGURE 7.10 Atomic interpretation of heat transfer. Initially region A has a higher temperature than region B.

much more readily by metal than by wood. Metals tend to be very good heat conductors. The free electrons that make metals good electrical conductors are instrumental in heat conduction. Liquids and gases are generally poor conductors of heat and electricity. Air conducts heat very poorly. A skier wears clothing with many air pockets to prevent body heat losses to the cold outdoors.

Heat conduction depends on several factors. In wintertime, we don several layers of clothing to prevent bodily heat loss. Therefore, heat conduction depends on the thickness of the conductor—the thicker the conductor, the smaller the heat loss. If you want to warm your cold hand by touching another object, you touch your entire hand rather than the tip of a single finger. The conduction of heat to your hand depends on the contact area between the conductor and source—the greater the area, the greater the heat transfer. Finally, experience tells you body heat loss in wintertime depends on the outdoor temperature. As the temperature difference between your body and the outdoors increases, the heat loss increases. We can relate all these factors to the rate of heat conduction with a single equation:

$$\text{rate of heat loss} = \frac{(\text{heat conductive property of the material}) \times (\text{contact area of the conductor}) \times (\text{temperature difference})}{\text{thickness of the conductor}},$$

$$\frac{H}{t} = \frac{kA\,(T_{\text{warm}} - T_{\text{cool}})}{d}. \tag{7.7}$$

The heat conductive property symbolized by k is called the thermal conductivity. Thermal conductivity is expressed in a variety of units. If heat flow is measured in watts, lengths in meters, and temperature in °C, then thermal conductivity is measured in watts per meter per degree Celsius. In short,

$$\frac{\text{watts}}{\text{m}\cdot{}^{\circ}\text{C}}.$$

Table 7.2 lists some representative thermal conductivities. The wide variation in values represents the enormous difference in the heat conduction properties of materials. Air is one of the poorest heat conductors of all the materials listed. That is why in winter one wears loose-fitting coats and jackets that trap air spaces to diminish conductive heat losses from the body. You might also wonder why the walls of a home are often filled with rock wool, which is not cheap, rather than with air which is a poorer heat conductor. The reason is that rock wool forms a physical barrier which restricts heat losses from movement of the air. This movement of air is an example of heat transfer via convection which we discuss later.

TABLE 7.2 Some selected thermal conductivities. A more complete list is found in *Handbook of Chemistry and Physics,* CRC Press Inc., Boca Raton, Fla.

Material	Thermal conductivity $\frac{\text{watts}}{\text{m}\cdot{}^{\circ}\text{C}}$
Silver	410
Copper	390
Concrete	0.8
Building brick	0.6
Glass (window)	0.8
Rock wool	0.04
Wood (typical)	0.04
Air	0.024
White pine	0.11
Oak	0.15
Water	0.6
Styrofoam	0.01

Thermal conductivity values are of enormous practical value. A designer of a house needs to know how much heat will be conducted through the walls. Given the dimensions of the wall, the type of material, and the difference in temperature, the calculation is straightforward. Probably, you are not a house designer and such calculations are of little interest. As an energy-conscious citizen, however, you should be interested in how different materials compare in heat conducting properties. For example, there is a significant difference between the amount of heat transferred through a single-pane glass window and through a four-inch thick rock wool insulation in the walls. For the same contact area and temperature difference, the heat loss through the glass is over 200 times that lost through the rock wool insulation. Heat loss through windows can be reduced by using two panes with an enclosed layer of air and by using storm windows. But even the best-designed window systems conduct heat about five times more rapidly than an insulated wall. We will say more about these heat losses in the chapter on energy conservation.

When heat is conducted from one object to another, there is no transfer of material. The energy transfer is a result of molecules touching other molecules but not relinquishing their positions. It is also possible to transfer energy by literally moving material. If a mass of air moves into a region of lower temperature, then energy will be transferred. How might this be accomplished in practice? Suppose a room is warmed with a heater on the floor (Fig. 7.11). The air near the source warms, the energy increases, and the density of the air decreases. The less dense air rises in the room like a helium-filled balloon and is replaced by cooler (higher density) air falling from the top of the room. As a result, a circulation of air is set up. This circulation of air is called convection. Heat is circulated in houses primarily by convection. Convection currents also give rise to wind. Because convection also occurs in liquids, it is partially responsible for ocean currents.

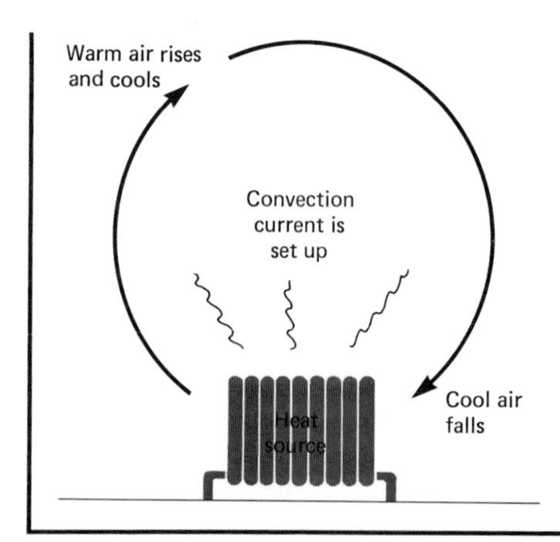

FIGURE 7.11 Manner in which a circulating air current is produced in a room.

There is another heat transfer process called radiation (Chapter 8). It is the transfer of energy by electromagnetic waves. This mechanism is an important one but it is not primarily involved in any manufactured system for disposing of heat from a steam turbine.

Evaporation

To conclude this discussion of heat-transfer, let us consider evaporation. Evaporation is literally the escape of molecules from a liquid. As an analogy, imagine bees in a box having one end covered by a thin membrane that some energetic bees can penetrate (Fig. 7.12). Their motion is haphazard, and there is a mixture of energies and speeds. Bees with sufficient energy escape, but those energetic ones near the surface are more likely to escape. The same is true of molecules in a liquid. On the surface of any liquid there is, effectively, a membrane because of a surface tension (Fig. 7.13). Surface tension supports small bugs skating around on a water surface. To escape from the liquid, molecules must penetrate the surface. Only the more energetic molecules, especially those near the surface, are able to escape.

The origin of the molecular force causing surface tension is understandable. Molecules experience attractive molecular forces from their neighbors. A molecule in the interior of a liquid (Fig. 7.13) is closely surrounded on all sides by similar molecules. Therefore, it experiences forces from all directions. However, a molecule at the surface has few or no neighbors beyond the surface. As a result, it is subject only to forces tending to pull it back into the liquid. A molecule escaping does work against the attractive force of the neighbor molecules. This results in an increase in the energy of the escaping molecule and a decrease in the thermal energy of the liquid.

If these forces did not exist, every molecule would readily escape. Since only the most energetic molecules escape, the average kinetic energy of the liquid decreases and the temperature decreases. Evaporation is therefore a cooling process. This is why

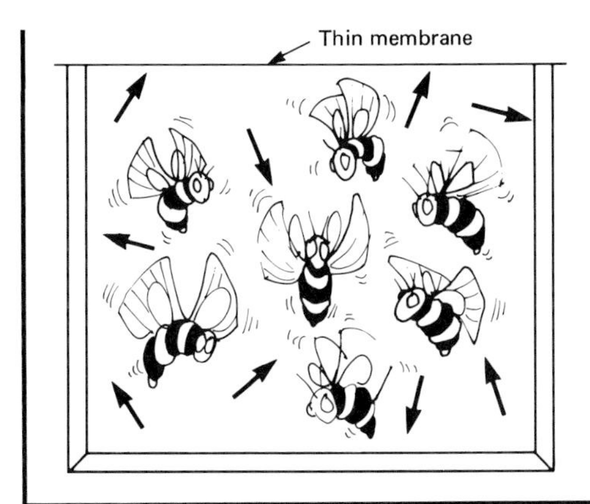

FIGURE 7.12 A box covered with a thin membrane and containing bees is analogous to a container of liquid.

your skin feels cool when perfume evaporates from it. Because evaporation happens on a surface, it follows that the larger the surface area, the greater the cooling effect.

Condensation is the reverse of evaporation. In condensation a molecule in the vapor above a liquid is captured by the liquid. In any situation, both evaporation and condensation occur simultaneously. If evaporation predominates, there is a loss in the volume of the liquid and a cooling of it. The rate of evaporation depends strongly on the amount of vapor above the liquid. The smaller the amount of vapor, the greater the evaporation rate. There is a limited amount of vapor in air before the vapor begins to condense. This amount, as we would expect from everyday experience, depends on the temperature. When air at a given temperature contains the maximum amount of vapor, it is said to be saturated. The word *humidity* refers to the amount of water vapor in the air. The ratio of the amount of vapor actually in the air to the amount that would saturate the air at that same temperature is called the *relative humidity.* Expressed as a percent,

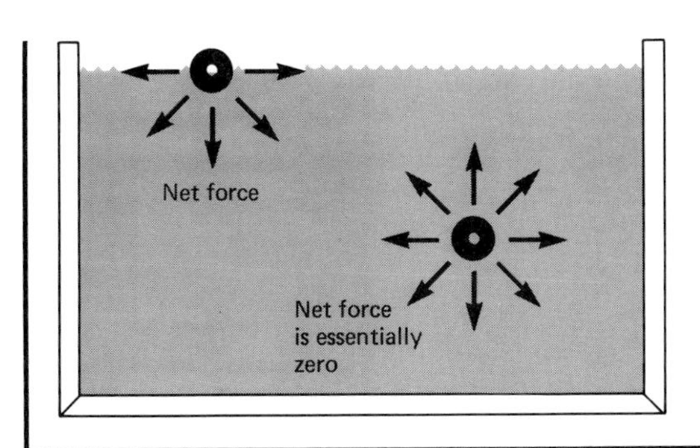

FIGURE 7.13 Schematic illustration of forces on a molecule in the interior and on the surface of a liquid. The net force is zero on an interior molecule but there is a net force on a surface molecule that pulls it toward the interior.

$$\text{relative humidity} = \frac{\text{actual vapor content}}{\text{saturated content}} \times 100\%. \tag{7.8}$$

The evaporation rate depends on the relative humidity. Relative humidity is measured by determining the difference in the dry- and wet-bulb temperatures. The dry-bulb temperature is the reading of an ordinary thermometer in the environment in question. If the bulb of a similar thermometer is surrounded by a moist cloth, it is referred to as a wet-bulb thermometer (Fig. 7.14). The wet-bulb thermometer has a lower reading because the evaporation of water molecules lowers the temperature. The amount of lowering is a measure of the relative humidity. Normally, a table* is used to determine the relative humidity from the temperature readings. The wet-bulb temperature is the lower limit for cooling processes utilizing an evaporative process. The dry-bulb temperature is the lower limit for a cooling process based on a conduction process. This turns out to be quite important for two cooling schemes for the condenser of a steam turbine.

Methods of Disposing of Heat

The common methods of disposing of heat from the condensers of steam turbines are shown in Table 7.3. We will now proceed to discuss these methods.

---

* See, for example, *Handbook of Chemistry and Physics,* CRC Press Inc., Boca Raton, Fla.

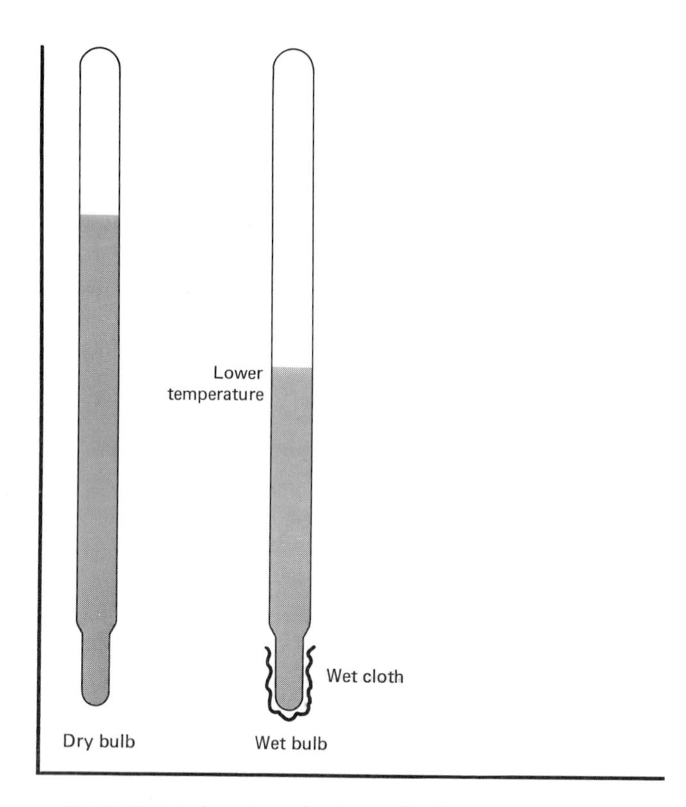

FIGURE 7.14 A wet- and a dry-bulb thermometer. The relative humidity is obtained from the difference in the readings of the two thermometers.

Once-through Cooling

In once-through cooling, water is drawn from a source, passed through the condenser, and returned to the source at a different location. Rivers, lakes, reservoirs, ponds, estuaries, and oceans are possible sources. The heated water is dispersed in a variety of ways. In a river, the natural flow is utilized. In all cases, the heat is dissipated from the body of the water to the atmosphere primarily by evaporation. Because evaporation occurs on the surface of a liquid, heated water is dispersed to as large a surface area as possible. The longer the cooling water stays in the condenser, the more heat it absorbs and the more its temperature increases. To keep the temperature of the cooling water from rising more than 15° F, water must flow through the condenser at a rate of about 0.01 gallons per second for each kilowatt of electric power generated. Hence a large plant (1,000,000 kilowatts) requires cooling water at a rate of about 10,000 gallons/second. If the temperature increase were to be lowered to 5° F, the water flow would increase to about 30,000 gallons/second. Water pumped out of the river must be replenished by the natural river flow at a rate at least equal to the rate of flow through the condenser. Few rivers can meet these requirements and, as a result, rivers are not being considered so frequently as sources for dissipating heat. The sheer size of these flow rates is difficult to imagine until we realize that such rates approach those supplying

TABLE 7.3 Common methods of heat disposal.

Method	Physical method	Comments
Once-through cooling	Evaporation, radiation, conduction	Utilizes a natural (lake, river, ocean) or artificial (cooling pond) water source
Wet-type cooling tower	Evaporation, convection	Recirculates water to condenser after evaporative process
Dry-type cooling tower	Conduction, convection	Heat dissipated to air through heat exchanger (radiator)

the water needs of a city like Los Angeles.

Cooling ponds are artificial bodies of water used to dissipate heat by evaporation. They are used when a natural supply is not available and land is inexpensive. As we might expect, large areas are required. A large plant requires a pond with a surface area of 1000 to 2000 acres (the playing area of a baseball field is about three acres). Most of the cooling ponds in this country are located in the southwestern states.

In areas in the United States where they are abundant, large and deep lakes seem to be candidates as sources of cooling water. As far as predicting the thermal effects, they might seem at first to present an ideal situation. But unlike rivers, for which the water flows are pretty well defined, lakes are subject to a variety of flow patterns and conditions that are strongly dependent on the weather. One feature, known as thermal stratification, makes lakes highly attractive sources of cooling water. In summer, sunlight penetrates and warms the upper portion of the lake. Because water is a poor heat conductor (about like window glass), little heat is conducted from the warmed upper layers to the cooler underlying water. Additionally, the density of the warm upper layer is less than the cooler underparts so the upper layer tends to float somewhat like wood. Hence, in summer the lake stratifies into a cool lower region called the hypolimnion and a warm surface region called the epilimnion. Connecting the cool hypolimnion and the warm epilimnion is the thermocline having varying temperature (Fig. 7.15). In winter, lakes tend to achieve a nearly uniform temperature.

From the standpoint of the cooling condenser of a steam turbine, it is desirable to extract water from the cool hypolimnion and disperse the heated water from the condenser over the surface of the warm epilimnion. Done properly, little or no increase in temperature of the surface water results. Although this is an ideal engineering feature, there is an undesirable environmental feature. Oxygen dissolved in water is vital for plant and animal life in the water. The dissolved oxygen in water in the hypolimnion is lowered by biological activity. If this water is used to cool a condenser and then returned to the epilimnion, the water returned is low in oxygen. Thus plant and animal life in the epilimnion is deprived of oxygen.

The oceans constitute a nearly inexhaustible source of cooling water and surely they will be tapped. There are unique problems, such as corrosion, that result from the salt in the water. Solving these problems are expensive propositions.

Cooling Towers

Wet- or dry-type cooling towers are normally used when using a natural or artificial watercourse is unfeasible. They are called wet or dry depending on whether water vapor is or is not released to the atmosphere in the cooling process. Because these

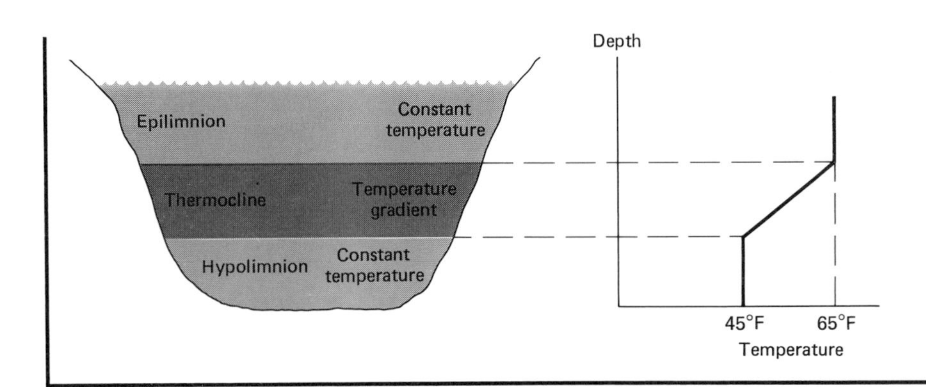

FIGURE 7.15 Schematic diagram of the thermal stratification of a lake. The temperature is constant in the epilimnion (upper region). The temperature gradually decreases in the thermocline and levels again in the hypolimnion (lowest region).

towers must duplicate the cooling task usually assigned to a large body of water like a river or a lake, they must necessarily be enormous. Typical cooling towers (Fig. 7.16) are some 400 feet high and as much as 400 feet in diameter at the base. Two to three percent of the cost of a nuclear power plant is for the cooling tower. Constructing these towers presents some challenging engineering and environmental problems.

The wet-type tower utilizes evaporative cooling (Fig. 7.17). The lowest cooling temperature is therefore the wet-bulb temperature (Fig. 7.14). To promote evaporation water is dispersed over a large surface area. This is achieved by either producing water droplets or forming thin films. To remove the evaporated water molecules either a natural or forced-air draft is created in the tower. The natural draft works by the ordinary chimney effect; the forced-air draft uses a fan to pull the air through

FIGURE 7.16 Cooling towers for the Rancho Seco nuclear-electric generating station located near Sacramento, California. Note the relative size of the towers, the generating facility, and the automobiles. (Photograph courtesy of the Sacramento Municipal Utility District.)

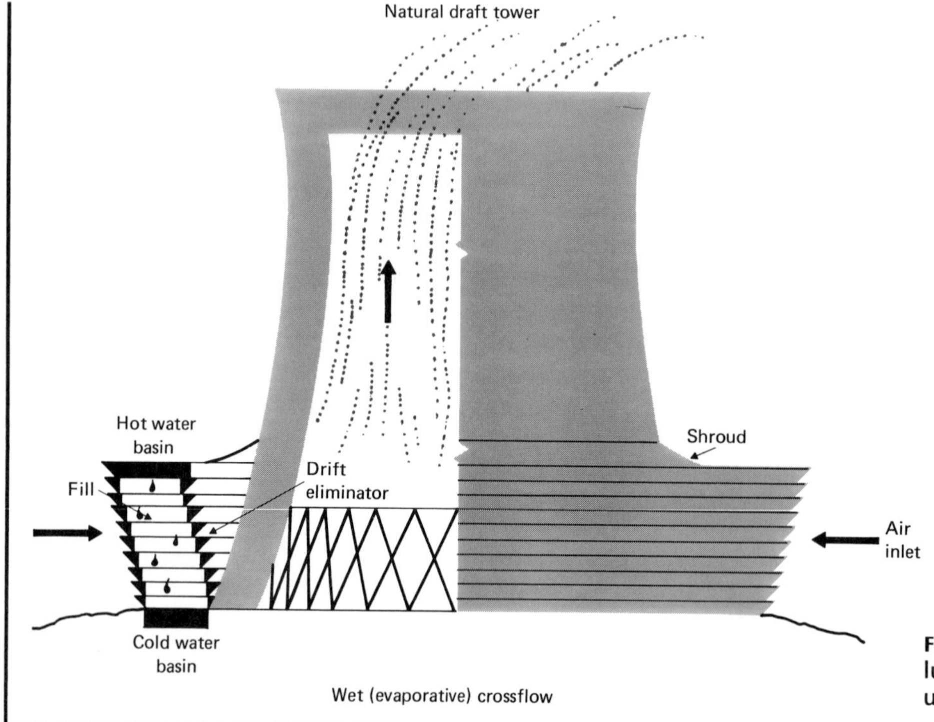

FIGURE 7.17 Schematic drawing illustrating the principles of the natural draft wet-cooling tower.

the top or force it up from the bottom. About 2% of the warm water returned from the condenser is lost through evaporation—a figure which is deceptively small. A large facility requires evaporative losses of about 225 gallons of water per second. This is equivalent to a daily rainfall of about one inch over an area of one square mile. Hence, severe fogging and icing may occur in the vicinity of the plant on a cold day.

In principle, the dry-type tower is a huge radiator of the type found in most automobiles. The water to be cooled is pumped through a heat exchanger (Fig. 7.18) having a huge surface in contact with air. Heat is transferred from the exchanger to the air by conduction and is carried away by either forced or natural convection currents.

The dry-type system seems ideal because it is closed and there is essentially no water loss. However, it is the most expensive of the systems discussed and it produces an immense heat plume because the energy is transferred directly to the atmosphere. The heat plume is so pronounced that it could produce local alteration of the weather. Systems of this type have never been used in the United States in conjunction with electric power plants, but they are used on a limited scale in Europe. As the availability of cooling water declines, dry-type cooling towers are becoming increasingly attractive.

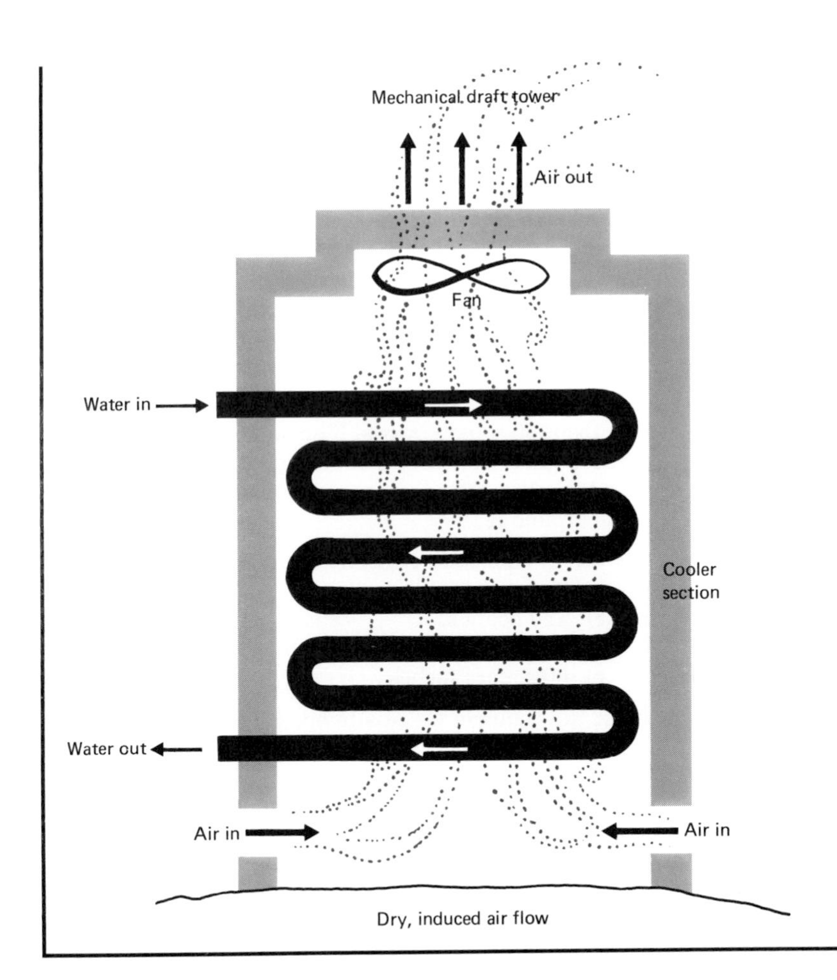

FIGURE 7.18 Schematic drawing illustrating the principle of the mechanical draft dry-cooling tower.

REFERENCES

BASIC THERMODYNAMICS

Most physics and physical science texts will devote a number of pages to the study of thermodynamics. Two such texts that would be useful for expansion of the material presented in Chapter 7 are

Conceptual Physics, Jae R. Ballif and William E. Dibble, Wiley, New York, 1971.

Physics in Perspective, Eugene Hecht, Addison-Wesley, Reading, Mass., 1980.

Some interesting insight into thermodynamics and the role of the steam engine in the Industrial Revolution are presented in Chapters 10 and 11 of

Physics and Its Fifth Dimension: Society, Dietrich Schroeer, Addison-Wesley, Reading, Mass., 1972.

DISPOSAL OF WASTE HEAT

"Thermal Pollution and Aquatic Life," John R. Clark, *Scientific American* **220,** 19 (March 1969).

"Cooling Towers," Riley D. Woodson, *Scientific American* **224,** 70 (May 1971).

Introduction to Environmental Science, Joseph M. Moran, Michael D. Morgan, and James H. Wiersma, Freeman, San Francisco, 1980.

Environmental Pollution, 2nd ed., Laurent Hodges, Holt, Rinehart and Winston, New York, 1977.

Energy and Environment, Edward H. Thorndike, Addison-Wesley, Reading, Mass., 1976.

REVIEW

1. Describe the operation of a heat engine in energy terms.
2. How are pressure and force alike and how do they differ?
3. Distinguish between the ideas of temperature and heat.
4. Distinguish between thermal energy and heat.
5. What is the meaning of a calorie and a British thermal unit?
6. What is specific heat and what is the practical significance of specific heat?
7. How does the first law of thermodynamics incorporate the principle of conservation of energy?
8. Explain why the efficiency of a heat engine depends on the temperature difference of the "hot" source and "cold" connecting region.
9. Name four physical effects from increased temperatures of a body of water.
10. What are some possible effects of elevated temperatures on aquatic life?
11. What is the ultimate disposition of heat rejected by a steam turbine?
12. What happens to the energy of molecules in a gas when the temperature of the gas decreases?
13. Identify the source of the mechanical energy produced by a steam turbine.
14. What is the atomic explanation for the conduction of heat from an object to a cooler object?
15. Describe conduction, convection, and evaporation.
16. Why is evaporation a cooling process?
17. Describe the concept of relative humidity.
18. Name three ways of disposing heat to the atmosphere without first rejecting heat to a lake, river, or stream.
19. What is a cooling tower?
20. Distinguish between wet-type and dry-type cooling towers.
21. The surface area of the soles of an adult male's shoes is about 60 square inches. If an adult male weighing 180 pounds distributes his weight over 60 square inches, then the pressure exerted by the person on the ground is
    a) $\frac{1}{3}$ pound/square inch.
    b) 3 pounds/square inch.
    c) 180 pounds.
    d) 10,800 pounds/square inch.
    e) 60 square inches.
22. On the common Celsius temperature scale, the ice and steam points of water are assigned values of 0° C and 100° C, respectively. On the less common, but more specific, Kelvin scale the ice and steam points of water are assigned values of 273 K and 373 K, respectively. Using these data or recalling from memory, pick the correct relationship between Celsius and Kelvin temperatures.

a) $K = \frac{5}{9}C + 32$ b) $K = C + 273$
c) $K = C - 273$ d) $C = K + 273$

23. Two ordinary mercury-type thermometers are identical in construction. One is calibrated in degrees Celsius, the other in degrees Fahrenheit. When both are inserted into a bottle of warm water, we would expect
    a) the Celsius thermometer to display a smaller number.
    b) the Fahrenheit thermometer to display a larger number.
    c) the levels of mercury to be the same in both thermometers.
    d) (a) and (b) are both correct.
    e) (a), (b), and (c) are all correct.

24. It would be appropriate to measure temperature in ______ and heat in ______.
    a) degrees and calories
    b) Btu and Btu
    c) calories and calories
    d) calories and degrees
    e) both (b) and (c) are correct answers

25. Units for describing the flow of heat through a window could be
    a) calories per second.
    b) Btu per day.
    c) joules per second or watts.
    d) Btu per month.
    e) all of the units above could be used.

26. Heat is
    a) the transfer of energy by virtue of a temperature difference.
    b) measured with thermometers.
    c) completely absent from the environment on a very cold winter day.
    d) another word for temperature.
    e) a conceptual idea that is quantified by kelvin degrees.

27. Water has a higher specific heat than soil. Therefore if you add equal amounts of thermal energy to equal masses of water and soil, then
    a) the soil will undergo the greater temperature change.
    b) the water will undergo the greater temperature change.
    c) the temperature change for water and soil will be the same. Specific heat has nothing to do with temperature changes.
    d) it is impossible to tell which undergoes the greater temperature change. Specific heat has nothing to do with temperature changes.
    e) the water conducts the heat away faster than the soil.

28. Specific heat is defined as the heat required to raise 1 gram of a substance, 1° C. Concrete has a specific heat of 0.22 calories/gram × °C. That is, 0.22 calories of heat are required to raise 1 gram of concrete, 1° C. If 22,000 calories of heat are added to 1000 grams of concrete the temperature increases
    a) 100° C. b) 1° C. c) 44° C.
    d) 10° C. e) 22° C.

29. The thermal energy of a substance
    a) is independent of its temperature.
    b) decreases as its temperature increases.
    c) depends only on pressure.
    d) is in direct proportion to its kelvin temperature.
    e) is independent of the mass of the substance.

30. One version of the second law of thermodynamics says
    a) refrigerators must reject the same amount of heat that they remove from the cold interior.
    b) refrigerators must reject less heat than they remove from the cold interior.
    c) energy is conserved in any energy converting process.
    d) an isolated system always tends toward order.
    e) heat flows naturally from hot to cold.

31. The first law of thermodynamics states
    a) an engine must reject more heat than it extracts.
    b) an engine cannot be 100% efficient.
    c) the total energy does not change in any energy transformation.
    d) a refrigerator must reject the same amount of heat that it extracts.
    e) energy is conserved during cyclic processes *only*.

32. The maximum efficiency of a heat engine is given by

$$\frac{T_H - T_C}{T_H}$$

where T_H is the kelvin temperature of the hot reservoir and T_C is the temperature of the cold region.

One plausible reason for accepting this relation is

a) the efficiency is zero if $T_H = T_C$.
b) the efficiency has units of kelvins.
c) the efficiency increases as the temperature difference $T_H - T_C$ decreases.
d) it clearly does not matter whether you substitute temperature values in kelvins or degrees Celsius.
e) it involves thermal energies as represented by the notation T_H and T_C.

33. A certain steam turbine customarily operates with steam (hot source) at 800 K and a condenser (cold region) at 300 K. Given a choice of increasing the efficiency by *either* increasing the steam temperature by 100 K or decreasing the condenser temperature by 100 K, which option would you take?

a) Decrease the condenser temperature by 100 K.
b) It makes no difference. The efficiency is independent of the temperature.
c) Increase the steam temperature by 100 K.
d) It makes no difference. A 100 K change produces the same increase in efficiency.

34. A heat engine extracting one Btu of heat in each cycle and converting the one Btu to one Btu of mechanical energy does not violate the first law of thermodynamics.

a) true b) false

35. A heat engine operating between 1000 K and 450 K could not have an efficiency greater than

a) 55%. b) 50%. c) 30%. d) 20%.
e) 10%.

36. Some 1,500,000 Btu of thermal energy are generated each second at the condenser of a commercial electric power plant. This energy is carried away by water flowing through the condenser. Although the amount of energy carried away each second is quite large, the temperature of the water does not increase much because

a) the thermal conductivity of the water is very small.
b) the rate of flow of the water is very small.
c) the amount of water absorbing the thermal energy is comparatively large.
d) the poor thermal conduction between the condenser and the water.
e) thermal energy is inversely proportional to the kelvin temperature.

37. In the heat conduction process between two objects,

a) molecules are conducted from one object to the other.
b) electrons are conducted from one object to the other.
c) heat particles are conducted from one object to the other.
d) energy is transferred from one object to another by molecular collisions.
e) all of the above.

38. If you place your hand in water at a temperature of 140° F, then heat flows from ______ to ______ by the heat transfer method of ______.

a) the water; your hand; conduction
b) your hand; the water; conduction
c) your hand; the water; convection
d) the water; your hand; convection
e) your hand; the water; evaporation

39. The thermal conductivity of glass is 25 times larger than the thermal conductivity of rock wool. For the identical area and temperature difference the rate of flow of heat through one inch of glass is ______ times ______ than the flow of heat through two inches of rock wool.

a) 50; less
b) 12.5; greater
c) $\frac{2}{25}$; less
d) 50; greater
e) 25; greater

40. In the summer months, the surface of a pond or lake is warmed by solar radiation. However, the lower portion of the pond or lake does not warm appreciably because of the

a) low thermal conductivity of water.
b) high specific heat of water.
c) mineral content in the lower portion.
d) high density of fish in the lower portion.
e) reflectivity of the surface of the water.

41. If a certain volume of air can hold one gram of water and it actually holds 0.5 grams, we would say that the ______ is ______.

a) surface tension; 50%
b) relative humidity; 50%
c) thermal conductivity; 25%
d) convection; 75%
e) conduction; 100%

42. The purpose of a cooling tower at an electric power plant is to
    a) transfer directly to the atmosphere the heat of condensation produced at the condenser of the turbine.
    b) vent radioactivity to the environment.
    c) vent particulates and sulfur oxides to the environment.
    d) transfer to the environment the heat produced by the large step-up transformers at the power plant.
    e) cool water prior to entering a nuclear reactor.
43. An ordinary mercury thermometer reads 70° F in a certain room. If the bulb-end of the thermometer is wrapped in a wet cloth, we would expect the thermometer to read ____ because ____.
    a) more than 70° F; evaporation warms the thermometer
    b) 70° F; the wet cloth has no effect
    c) less than 70° F; evaporation cools the thermometer
    d) 32° F; the wet cloth will freeze
    e) 212° F; the wet cloth will cause the temperature to rise

Answers

21. (b)	22. (b)	23. (e)	24. (a)
25. (e)	26. (a)	27. (a)	28. (a)
29. (d)	30. (e)	31. (c)	32. (a)
33. (a)	34. (a)	35. (a)	36. (c)
37. (d)	38. (a)	39. (d)	40. (a)
41. (b)	42. (a)	43. (c)	

QUESTIONS AND PROBLEMS

7.2 BASIC THERMODYNAMICS

1. The electrical resistance of a wire changes with temperature. How could you conceive a thermometer based on this effect?
2. Why does a large tractor exert considerably less pressure on the ground than a hypodermic needle on the skin of a patient?
3. When a gas is compressed by pushing on a piston in a cylinder, why does the internal energy of the gas increase?
4. What is the direction of heat flow when
    a) you place your hand in water at 40° C?
    b) you place your hand in ice water?
5. The rubber hose of a bicycle tire pump noticeably warms during the inflation of a tire. What is the origin of the energy producing the warming of the hose?
6. Why does a child see her birthday party balloon become smaller in the cold winter air?
7. Work done on a tennis ball by a racket increases the kinetic energy (or speed) of the ball. How does this work–energy principle apply when work is done on a gas in a cylinder by compressing it with a piston?
8. Use an energy argument to explain why heat does not flow spontaneously from an object to another object at a higher temperature.
9. The average random kinetic energy of a single molecule is directly proportional to the Kelvin temperature ($E = \frac{3}{2}kT$). Why is it that even if the temperature outside a house is fairly low, say 253 K (−20° C), there is still an enormous amount of thermal energy in the ground and in the surrounding atmosphere?
10. After an athletic event one often sees people milling around the area. Yet at an exit gate there is an orderly flow of people. How is this analogous to the conversion of random molecular motion into ordered motion in a steam turbine?
11. Liquid nitrogen boils at 77 K. Determine its boiling temperature on the Celsius and Fahrenheit scales.
12. The temperature in a room is usually about 68° F. Show that this corresponds to 20°C and 293 K.
13. Suppose you were not satisfied with any of the three temperature scales mentioned in the text and you chose to assign numbers of 100 and 200 to the ice and steam points. If you labeled your scale °M, show how you could convert °M to °C.
14. How does a temperature *change* of 1° C compare with a temperature *change* of 1° F? How does a temperature *change* of 1° C compare with a temperature *change* of 1 K?
15. Atmospheric pressure at sea level is about 15 pounds per square inch. How big a square on the earth's surface is needed so that the total force exerted on this square equals the weight of a 135-pound person?
16. A ten-ton tractor distributes its weight over two tracks each having an area of 20 square feet. What is the pressure in pounds per square foot? If the diameter of a hypodermic needle is $\frac{1}{64}$ inch, how much pressure is exerted on the skin by a one-lb force supplied by a nurse? (You will need to determine the cross-sec-

tional area of the needle. The formula for computing this is $A = 0.785d^2$ where d is the diameter.)

17. Very few substances have a higher specific heat than water. Consequently, water is especially useful for cooling the condenser of a steam turbine. To show why, determine the temperature rise of one kilogram of water and one kilogram of copper if each absorbs 1000 joules of heat. (Specific heats are given in Table 7.1.)

18. An average home uses about 20 gallons of hot water per day for each occupant. Complete the table below to determine the daily cost of hot water in a house using an electric hot water heater.

    Number of occupants = 5.

    Initial water temperature = 25° C.

    Hot water temperature = 70° C.

    20 gallons of water = 75 kilograms.

    Specific heat of water = 1 calorie/gram × °C.

    One kWh = 860,000 calories.

    Cost = 9¢ per kWh.

    $\_\_\_\_ \dfrac{\text{gallons}}{\text{day person}} \times \_\_\_\_ \text{ persons} = \_\_\_\_$ gallons

    $\_\_\_\_ \text{ gallons} \times \_\_\_\_ \dfrac{\text{grams}}{\text{gallon}} = \_\_\_\_ \text{ grams}$

    $\_\_\_\_ \text{ grams} \times \_\_\_\_ \dfrac{\text{calories}}{\text{gram} \times {}^\circ\text{C}} \times \_\_\_\_ \ {}^\circ\text{C}$

    $= \_\_\_\_ \text{ calories}$

    $\_\_\_\_ \text{ calories} \times \_\_\_\_ \dfrac{\text{kWh}}{\text{calorie}} = \_\_\_\_ \text{ kWh}$

    $\_\_\_\_ \text{ kWh} \times \_\_\_\_ \dfrac{¢}{\text{kWh}} = \_\_\_\_ ¢$

    Therefore, the daily cost is ____ ¢.

19. A household hot water heater generally has a reservoir of about 80 gallons. Assuming water from the local supply is at a temperature of 65° F, how many Btu of heat are required to raise the temperature of 80 gallons of water to 135° F? (The weight of 20 gallons of water is 165 pounds.)

20. A 1000-kilogram concrete pillar in a house stores thermal energy it acquires from sunlight entering through windows in the house. The temperature of the pillar rises to 30° C on a bright, sunny day.
    a) How many calories of thermal energy are available from this pillar when it cools to 24° C? (See Table 7.1 for the specific heat of concrete.)
    b) If the nightly heat requirement to maintain the house at 24° C is 7 million calories, what fraction of the heat load could be provided by the concrete?

21. At a Kelvin temperature T a container of N molecules has a certain amount of thermal energy. How many molecules are required in a container at one-half the Kelvin temperature to have the same amount of thermal energy?

22. Ten trillion (10^{13}) molecules from a container at 300 K are transferred to another container having 5 trillion (5×10^{12}) molecules at 200 K. What is the equilibrium temperature in the combined system?

23. The average random kinetic energy of a molecule is given by

    $$E = \tfrac{3}{2} kT$$

    where

    $$k = \frac{1.38 \times 10^{-23} \text{ joules}}{K \cdot \text{molecule}}$$

    and T is the Kelvin temperature.
    a) Calculate the average random kinetic energy of a molecule at room temperature (300 K).
    b) Burning a pound of coal produces about 13,000 Btu of heat. How many molecules at 300 K are required to produce a total amount of average random kinetic energy equivalent to burning a pound of coal? (A Btu = 1055 joules.)

24. As you hold an ice cube, is there anything in the first law of thermodynamics preventing a spontaneous transfer of energy from ice to your hand? If this energy movement did happen, what changes would result in the temperatures of the ice and your hand?

25. Heat does not flow spontaneously from one object to another object at a higher temperature. But is it possible to contrive a set of conditions where heat can be transferred from one object to another at a higher temperature? Explain.

26. If a friend asks you to *prove* the second law of thermodynamics, why would you tell her it can't be proven?

27. Advertisements for air conditioners label the cooling capacity in units of Btu, for example, 10,000 Btu. Although time is not mentioned explicitly in the evaluation, why is it very important?

28. The oceans are an enormous source of energy. Why doesn't someone construct an engine that extracts heat from the surface of the ocean, converts some of this energy into useful work, and rejects the remaining energy back into the surface of the ocean?

29. Since a motor is basically a generator running backwards, why couldn't the electric energy output of a generator be fed to a motor and the mechanical energy output of the motor fed to the generator so that a perpetual motion machine results? Try to identify all the forms of energy involved and use the first law of thermodynamics to explain your answer.

30. An engine in each cycle takes in 50,000 Btu of heat and releases 20,000 Btu of heat.
    a) How much thermal energy is converted to work?
    b) What is the efficiency of the engine?
    c) What physical principle did you use for part (a)?

31. As a way of seeing how temperature changes affect power plant performance, assume that a steam turbine operates at maximum efficiency. The temperature of the heat source is 900 K and heat is rejected at 300 K. The efficiency improves if either the temperature of the heat source increases or the temperature of the cooling water decreases. Given an option of either a 25 K increase in heat source or a 25 K decrease in cooling water, what is the best choice? Cooling water for a steam turbine is often drawn from a lake. Do you think it would be worth the trouble to use water from the lower part of a lake rather than from the upper part?

32. A heat engine having an efficiency of 40% is designed to reject heat to a cool region having a temperature of 300 K. What is the lowest possible temperature for the hot source of this system?

33. As a result of inflating a bicycle tire with a hand pump, 3100 joules of work are done on the air and 500 joules of heat escape to the atmosphere. What is the change in internal energy of the gas?

34. The addition of 500 calories of heat to the gas in a cylinder containing a piston causes 250 calories of work to be done by the piston. What is the change in internal energy of the gas?

35. A participant in a laboratory exercise on heat engines reports his measurements as

    heat taken in during each cycle = 4320 Btu,
    work produced in each cycle = 5,275,000 joules.

    Comment on the results of the experiment using the laws of thermodynamics as a basis for discussion.

36. A steam turbine operating at 50% efficiency has a thermal power input of 2000 megawatts and delivers 1000 megawatts of heat to the condenser. This means the condenser must absorb 1,000,000,000 joules of heat energy each second.
    a) If each second this energy is deposited to a tank of water and you allowed the temperature to rise 10° C, how many kilograms of water would be required? Remember one calorie produces a 1° C temperature increase in 1 gram (0.001 kilogram) of water.
    b) How many gallons of water would be required? (One gallon of water has a mass of 3.8 kilograms.)

    If a 10° C temperature difference is maintained between the inlet and outlet water lines to the condenser, then water must flow through the condenser at the rate calculated in part (b).

37. Efficiency is a measure of engine performance.

$$\text{efficiency} = \frac{\text{work produced}}{\text{energy input}}.$$

    Using the laws of thermodynamics, explain why the efficiency is always a number less than 1.

38. If in the operation of a refrigerator, X Btu are taken from its cold interior, Y Btu are delivered to the surroundings, and Z Btu of work are required to run the refrigerator, determine from the data given below which of the laws of thermodynamics are violated.
    a) $X = 1000$ $Y = 1000$ $Z = 0$
    b) $X = 1000$ $Y = 2000$ $Z = 1000$
    c) $X = 2000$ $Y = 1000$ $Z = 1000$

39. The total energy consumption in the United States in 1981 amounted to 7.39×10^{16} Btu whereas the total electric energy production amounted to 2.32×10^{12} kWh (1 kWh = 3413 Btu).

a) What percentage of the total energy is electric?
b) About 76% of this *electric energy* was generated by fossil fuel plants. What percentage of the *total energy* was generated by fossil fuel plants?
c) If the average energy conversion efficiency for fossil fuel plants was 35%, what percentage of the total energy was used at the input of the power plants?
d) If 45% of the energy computed in part (c) is rejected as heat, what fraction of the total energy is rejected into the environment in the form of heat?

7.3 DISPOSAL OF WASTE HEAT

40. A glass window is installed in a brick wall 4 inches thick. How thick would the window have to be so that the heat loss through the window and an equal area of the brick are the same?
41. The inside of a house is maintained at 20° C when the outdoor temperature is 0° C. What is the rate of heat loss through a 0.5 centimeter thick pane of glass 60 centimeters wide and 120 centimeters long?
42. Why does a pan of water on an electric stove heat faster if it is in firm contact with the heating coil rather than being as little as the thickness of a dime above the heating coil?
43. Some materials are better heat conductors than others. From your experience of touching the handles of spoons immersed in a hot liquid like coffee, judge which is a better heat conductor—metal or plastic.
44. The greater the diameter of a wire the smaller its resistance to the flow of charge. Why would you expect this same relationship for the flow of heat in a pipe? Where do you see this idea put into practice in a home?
45. Aluminum, a metal, is a significantly better heat conductor than wood. Yet aluminum-frame windows are very common. What makes aluminum windows so attractive?
46. From the standpoint of heat conduction, what is the advantage of wearing loose-fitting clothes in wintertime?
47. Compare the heat flow through a brick wall and a concrete wall of the same thickness and dimensions when the temperature difference across the two walls is the same.
48. Some windows are constructed of two thicknesses of glass with an air space in between. From a heat-loss standpoint, why is this arrangement better than a solid glass window of the same total thickness as the two layers and the air space?
49. Why is it common practice to wrap hot water pipes in a building with a glass wool material similar to the rock wool mentioned in Table 7.2?
50. What is the main heat transfer mechanism in the draft created in a fireplace?
51. Why is it that most of the heat generated in a fireplace usually escapes up the chimney?
52. From the standpoint of energy conservation, why is it important to keep a chimney blocked off when an open fireplace is not being used?
53. Water often collects in puddles on a tennis court following a rain. To promote drying, the water is spread over the surface of the court. Why does this accelerate drying?
54. Sweating is a cooling process for the human body. What heat transfer process is involved in this cooling effect?
55. Why does an area of skin feel cool when touched with perfume or shaving lotion?
56. What is the meaning of a relative humidity reading of 50%?
57. Describe the cooling system on an automobile and compare it with that used in a dry-type cooling tower.
58. A 1000-megawatt electric power plant using a wet-tower cooling method disperses about 30 cubic feet of water to the atmosphere during each second of operation. If the water is distributed uniformly over a square mile of land, show that it corresponds to a daily rainfall of about one inch.
59. In a day's time the amount of water flowing through the cooling condenser of a large electric power plant is comparable to the amount of water handled by the entire water system of a large city. This may seem incredible but it is easy to see why. Water for all uses amounts to about 150 gallons per day per person in a city. How much water would be used in a day in a city having five million people? Water flows through the cooling condensers of a large power plant at a rate of about 30,000 gallons per second. How much water flows through the condenser in a day? Now compare the two calculations.

60. If a steam turbine is cooled with water from a cooling pond, about an acre of pond is required for each megawatt of power produced. Thus the pond area required for a 1000-megawatt plant is about 1000 acres. To get some feel for the size of 1000 acres, determine how many football fields are needed to make 1000 acres. A football field has an area of about 0.75 acre.

61. A cooling tower for a large electric power plant is about 400 feet in diameter. It is difficult to appreciate the size of these cooling units. Compare the area occupied by a tower 400 feet in diameter with the area of a football field. The area of a circle of diameter d is $A = 0.785\ d^2$. A football field is 100 feet wide and 300 feet long.

8

Atmospheric Problems

8.1 MOTIVATION

We have discussed the origin and the control methods for pollutants generated from energy conversion processes, and the environmental standards that have been established to ensure the protection of public health and welfare. Meeting the air quality standards relies heavily on natural air movement and precipitation. A serious situation can result if nature rebels and does not provide the atmospheric circulation for dispersing the pollutants. The possibility of both local and worldwide climatic changes stemming from the accumulation of particles and gases in the atmosphere is an added problem. All these effects involve visible, ultraviolet, or infrared radiation in some intimate way. These radiations have been mentioned in several cursory discussions. We must now

A dense midday haze shrouds this metropolitan area. Brought on by massive use of private automobiles, these hazy conditions are often compounded by weather conditions that prevent pollutant dispersal. (Photograph by Gene Daniels. Courtesy of the EPA.)

grasp some essential ideas in order to examine the atmospheric effects of interest. But our interest in radiation does not end here. When we study nuclear power in the next chapter, we shall discuss gamma radiation from radioactive wastes. Gamma radiation is very much akin to infrared, visible, and ultraviolet radiation. When we study the uses of solar energy in Chapter 10, we shall find that solar radiation contains components of infrared, visible, and ultraviolet radiation. A prospective nuclear fusion energy source that we shall study in Chapter 12 employs laser radiation.

8.2 WAVES AND PHOTONS

All of you have seen pianos and, if you were required to sketch one, you could provide a reasonable representation. You might even be able to depict the inside and explain how the sounds are produced. You might even eloquently describe mathematically the features of the sound waves. When you have finished your project, you might name it "A Model of a Piano." Had you never seen a piano but only heard its sounds, you surely would have arrived at a different model. But if the model explains your observations, that is what is important.

We seek a model for radiation. Bear in mind that you never see the radiation as you may see waves on the ocean. True, you see visible light radiation because of the nature of our eyes. But you cannot see infrared heat radiation nor can you see ultraviolet radiation from the sun. You see only the effects produced by the radiation when it interacts with matter or other radiation. Suntanning is an observable effect produced from ultraviolet radiation interacting with skin. From observations one tries to present some physical explanation or model for the radiation. Let us examine the wave aspects.

A wave is described as "a disturbance or oscillation propagated with a definite velocity from point to point in a medium." A sharp clap of the hands produces a sound disturbance (wave) that propagates through air. A rock dropped onto the surface of a still body of water such as a pond produces a disturbance that can be seen as it moves across the surface of the water (Fig. 8.1). If a float in the water is bobbed up and down continuously, a continuous train of disturbances having crests (peaks) and valleys moves across the surface. Figure 8.2 is representative of a "slice" in the direction the disturbance is moving. Counting the evenly spaced crests passing some position is like counting evenly spaced cars passing a position at the side of a busy highway. If a crest (or car) passes the position every $\frac{1}{10}$ of a second, we record the period, T, as $\frac{1}{10}$ of a second. The number of crests (or cars) passing each second would be 10 and we call this the frequency, given the symbol f. Frequency is often expressed in hertz, abbreviated Hz. A frequency of 60 cycles per second is recorded as 60 Hz. Frequency and period are related very simply.

$$f = \frac{1}{T}. \tag{8.1}$$

The distance from crest-to-crest (or valley-to-valley) is called the wavelength, denoted by the Greek lambda, λ. In a time equal to the period a crest travels a distance equal to one wavelength. Hence the speed of a crest is

$$\text{speed} = \frac{\text{distance}}{\text{time}},$$

$$v = \frac{\lambda}{T}. \tag{8.2}$$

In terms of the frequency, Eq. (8.2) becomes

$$v = f\lambda. \tag{8.3}$$

Equation (8.3) is a fundamental feature of wave propagation. Knowing the frequency and wavelength, it is a simple matter to compute the speed of the wave.

Observations of waves reveal several things. A water wave will bounce (reflect) off a shore. Sound echos are produced by sound waves bouncing off a building or a mountainside. A street lamp viewed through a window screen appears blurred because of light progressing around the wires forming the screen mesh. This effect, called diffraction, occurs also for sound waves. Waves may change their direction of propagation (refract) when entering a medium of different density. One has only to get in the path of an ocean wave rushing to the shore to know that a wave possesses energy. All these effects are well

FIGURE 8.1 A stone dropped into a still pond creates a circular water wave that propagates outward from the initial disturbance.

understood and can be described mathematically. Light radiation also reflects, refracts, and diffracts much like water waves that we can actually see. When we read a book, we see light reflected from the pages. We see a ray of light refract (or bend) when it passes from air into water. We see light diffract around the edges of a very narrow opening placed in its path. If we bathe in sun rays, we absorb energy and get a suntan. The speed of the radiation can be measured and a wavelength deduced from diffraction experiments. These measurements are consistent with the relation speed equals frequency times wavelength, which holds for observable waves. For these reasons we say that radiation is a wave phenomenon and use a wave model to describe it in these situations. However, some radiation releases electrons when impinging on the surface of many metals. This is called the photoelectric effect. (The photoelectric effect is utilized in some warning systems activated by crossing a beam of light.) A wave model fails to explain the photoelectric effect. We must resort to a branch of physics called quantum physics.

Quantum physics assumes that the energy associated with radiation is comprised of packets of energy or quanta, called photons. A photon has energy proportional to the frequency of the wave:

photon energy is proportional to frequency,

$$E \propto f.$$

$$E = hf, \tag{8.4}$$

where h is called Planck's constant. Planck's constant, like the speed of light, must be measured. It has a value of 6.63×10^{-34} joule · seconds. Electrons are able to escape from the surface of a metal in the photoelectric effect because all the energy of a photon is given to the electron initially bound to an atom. We will not need Planck's constant in our deliberations, but do remember that photon energy is proportional to frequency. As frequency increases, photon energy increases accordingly.

The wave aspects are apparent in the propagation of electromagnetic energy and when "waves" interact with "waves." The particle (quantum) nature emerges when electromagnetic waves interact with matter. Wave and particle aspects never occur together in the same experiment.

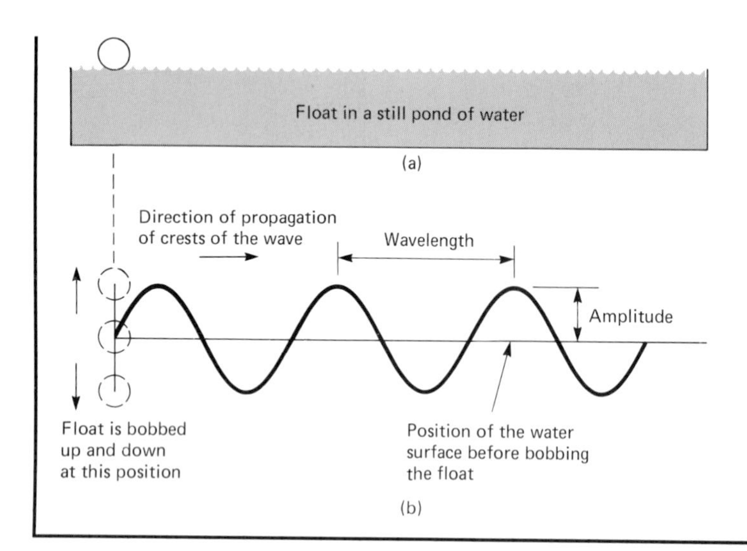

FIGURE 8.2 (a) A float such as a piece of wood rests on the surface of a still pan of water. When the float is oscillated up and down with your hand, circular waves spread out from the float. **(b)** "Snapshots" of a slice of a water wave along the direction of propagation. A complete cycle corresponds to taking the float up to its maximum height above the equilibrium position, back down to an equal distance below the equilibrium position, and then back to the equilibrium position. During the time *(T)* for one complete cycle of the bob a crest travels a distance of one wavelength (λ). Thus the speed of a crest is $v = \lambda/T$.

What is the disturbance associated with the waves? How are photons generated? These are fundamental questions. Atoms and molecules move in sound and water waves and therefore energy is associated with waves because particles are in motion. Through particle collisions, the waves transmit energy. Radiation waves possess electric energy and magnetic energy by virtue of oscillating electric and magnetic fields (Fig. 8.3). The sizes of the electric and magnetic fields at a particular position change with time much like the height of a water wave changes with time at a particular position. We call such a wave an electromagnetic wave. An electromagnetic wave transfers energy through the interaction of its associated electric and magnetic fields with electric charges that it encounters. For example, if a charge gets in the path of an electromagnetic wave, the charge interacts with the wave, and the charge experiences a force which may set it into motion.

A TV set receives energy from electromagnetic waves that interact with electrons in the antenna on the rooftop. The electromagnetic radiation from the antenna of a TV station is produced by accelerating electrons in the metallic conductors on the antenna. An electron not bound to an atom radiates photons when accelerated. The radiation from an ordinary incandescent light bulb is caused by ran-

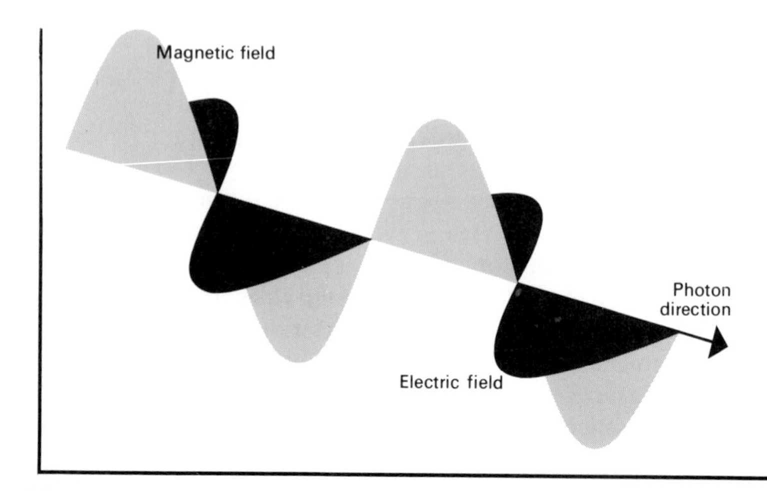

FIGURE 8.3 The oscillations of both the electric field (shown in black) and the magnetic field (shown in gray) are represented by a picture much like that for an ordinary water wave (Fig. 8.2b). The directions of the oscillations of the two fields are always at right angles to each other.

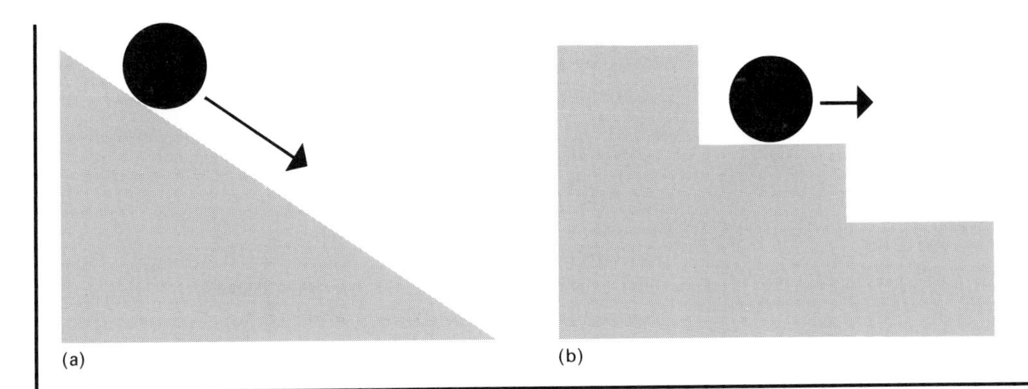

FIGURE 8.4 (a) A ball rolling down an inclined plane continuously loses potential energy and gains kinetic energy. **(b)** A ball rolling down a staircase loses potential energy in "jumps" each time it rolls over the edge of a step.

dom accelerations of electrons in the heated filament. This haphazard activity produces a broad, continuous spectrum of wavelengths. This continuous spectrum contrasts with the single wavelength provided by the controlled electron motion at the TV station antenna. Photons are also emitted when atoms, molecules, and nuclei make transitions from higher energy states to lower energy states. This is analogous to an energy transformation evolving from a ball making a movement from the top of an inclined plane (higher energy state) to the bottom (lower energy state). (See Fig. 8.4a.) But unlike the inclined plane example, the changes in energy occur discontinuously, more like those occurring with a ball rolling down a staircase (Fig. 8.4b). The radiation from a neon sign is not a continuous distribution of wavelengths (or colors) like that from an incandescent light bulb because the radiation is caused by transitions between atomic energy states in neon atoms. Such a discontinuous change in energy is accompanied by the emission of a photon with energy equal to the difference in energy of the initial and final states.* The energy given up by the system is transformed into electromagnetic energy carried by the photon. The frequency of the radiation is given by the difference in energy divided by Planck's constant:

$$f = \frac{E_2 - E_1}{h}. \tag{8.5}$$

Thus the greater the energy given up in the transition, the higher the frequency of the radiation. The range of energies covered by these radiations as well as the range of their associated frequencies (or wavelengths) is staggering (Fig. 8.5). Interestingly, the speed of all these waves in a space having no matter (vacuum) is 300 million meters per second *regardless of the frequency.* This speed is also called the speed of light. Infrared, ultraviolet, and visible radiation are all categories of electromagnetic radiation. They are distinguished by differences in wavelengths. X-rays and gamma rays are also categories of electromagnetic radiation.

Several experiments verify the photon concept of electromagnetic radiation. One is the photoelectric effect mentioned earlier. Another is the study of thermal radiation from a heated object.

* When the atom emits a photon it is analogous to a person throwing a ball while standing on a skateboard. A reaction to the force that makes the ball move one direction causes the person and the skateboard to recoil in the opposite direction. The person and the skateboard acquire kinetic energy as a result of the recoil. When a photon is emitted by an atom, the atom recoils and acquires kinetic energy. We have neglected this recoil energy because it is generally small compared to the energy of the photon.

8.3 THERMAL RADIATION

Any object at a temperature above zero kelvins emits electromagnetic radiation having a continuous spectrum of wavelengths. This radiation is called thermal radiation. The energy associated with each wavelength depends strongly on the temperature of

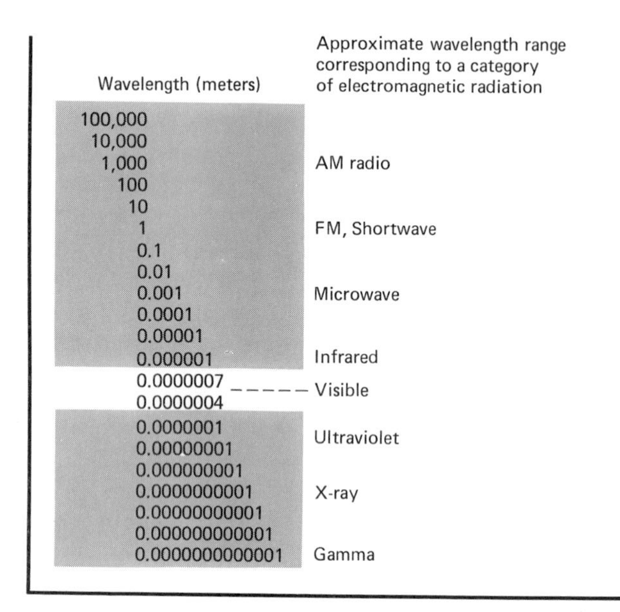

FIGURE 8.5 A portion of the spectrum of wavelengths of electromagnetic waves. An agreed upon range of wavelengths within the spectrum characterizes a certain class of radiation. For example, electromagnetic waves having wavelengths between 0.0000004 meters and 0.0000007 meters are classed as visible radiation. The boundaries between different types are not sharply defined.

the object. This is readily observed with the heating coils of an electric stove. In a dark room the unheated coils are invisible because our eyes are insensitive to the infrared radiation emitted by the coils. When an electric current is present in the coils, they warm and emit visible radiation. Your hand near the coils senses that the amount of radiation increases noticeably. Figure 8.6 shows measurements of the radiant power from a radiator maintained at different temperatures. Carefully note that all wavelengths (colors) are not emitted with the same intensity. The position of the peak along the horizontal axis denotes the dominant type of radiation emitted. For example, the peak position in the lowest curve is at about 1.6 micrometers, or 0.0000016 meters. This means that the dominant radiation for a temperature of 2000 K is infrared. If the measurement is repeated at different temperatures, similar results are obtained, but the peak in the curve shifts very systematically. A study of the results shows that if you multiply the wavelength corresponding to the peak position with the temperature expressed in kelvins, you always get the same answer. The result is

$$\lambda_m T = 2900 \text{ micrometer kelvins.} \tag{8.6}$$

As the temperature increases, Eq. (8.6) shows that the position of the peak in the thermal radiation curves shifts to smaller wavelengths. For example, if the temperature of two radiators were 300 K (about

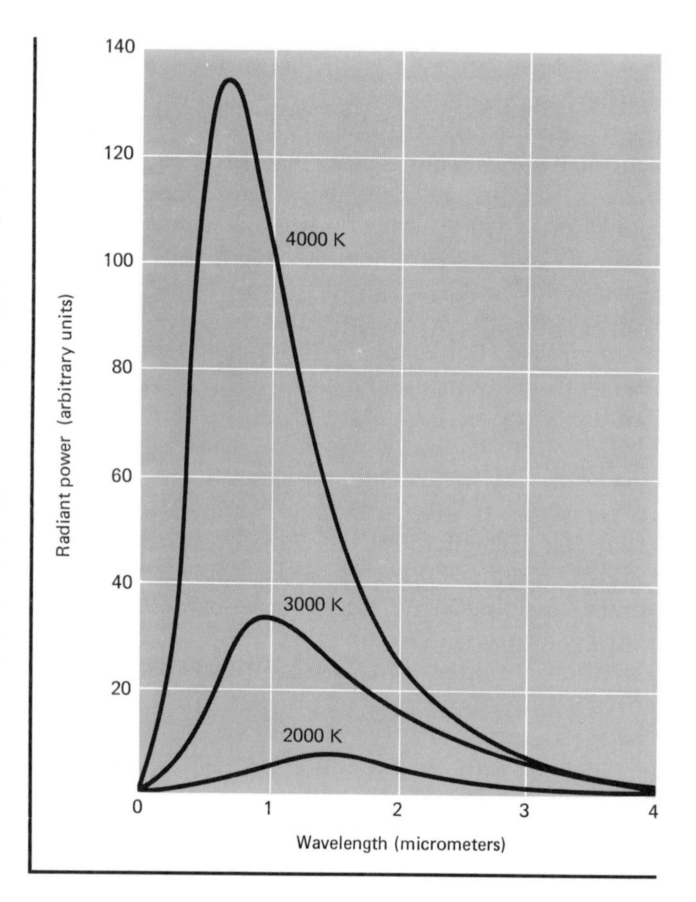

FIGURE 8.6 Illustration of how radiant power changes with temperature. The wavelength of the dominant type of radiation (given by the peak position) gets smaller as the temperature of the radiator increases. Note that if you multiply the wavelength of the dominant type and the temperature, you always get a number around 3000 micrometer • kelvins.

the temperature of the earth) and 6000 K (about the surface temperature of the sun), the peaks would occur at about 10 and 0.5 micrometers, respectively. A wavelength of 10 micrometers is in the infrared region of the electromagnetic spectrum and one of 0.5 micrometers is in the visible region. The earth and sun are not ideal radiators. Nevertheless, they both radiate at fairly constant temperatures and there is a predominant type of radiation given off that is determined by the radiating temperature. This difference in emission characteristics is a major factor in the "greenhouse" effect which we discuss in Section 8.7.

The amount of radiated energy depends very strongly on the temperature. Doubling the temperature produces a sixteen-fold increase in energy. This strong temperature dependence allows us to determine differences in temperature from measurements of the radiant energy. If the temperatures of interest are around a temperature you might encounter in a house, then the radiation would be mostly infrared, and so we would use a camera sensitive to infrared radiation. This type of camera functions much like an ordinary photographic camera. In black and white imagery, hot areas produce greater exposure and produce white areas on the recording. Figure 8.7 shows a thermal photograph of a building taken at night when the radiation is nearly all infrared. Areas of high heat loss are shown clearly. This infrared technique has become extremely useful for detecting heat leaks in buildings.

FIGURE 8.7 The building outlined in this photograph is located at the National Bureau of Standards. The picture appears to be a low resolution photograph taken with a conventional camera. However, the exposure is actually produced by infrared thermal radiation. Dark areas in the thermograph correspond to areas of high heat loss. (Photograph courtesy of the Department of Energy.)

8.4 TEMPERATURE INVERSIONS

One of the most serious recorded incidences of air pollution occurred in 1948 in the small industrial city of Donora, Pennsylvania. A weather condition prevailed that essentially put a lid over Donora and prevented the dispersal of pollutants spewed out from energy conversion processes in the city's industries. Many cities are harassed by these atmospheric lids that contribute to stagnant air conditions. Los Angeles experiences these "lids" for about 100 days of the year. Pollution problems from rapidly expanding industrialization in Denver, Colorado, are intensified by the existing atmospheric conditions. To a lesser but still significant extent the 25 easternmost states are victimized by intermittent stable weather conditions occurring for a variety of reasons. Many of the mechanisms causing them are very complicated, but some of the common ones are not difficult to understand. Because of their important role in the dispersal of by-products from energy production, we shall examine the common ones.

Anyone who has hiked up a mountain or ridden in a commercial jet airplane knows that the air temperature is lower at higher altitudes than it is on the ground. Outside a plane flying at 30,000 feet the temperature is about −50° F. The variation of temperature with altitude is measured by a quantity called the lapse rate, expressed as the change in degrees Celsius per kilometer (°C/km) or as the change in degrees Fahrenheit per mile (°F/mi). If the temperature decreases 4° C for each kilometer increase in altitude, we say the lapse rate is −4° C/

km. The negative sign reminds us the temperature decreases as altitude increases. Generally, the temperature decreases about 7° C for each kilometer increase in altitude up to a distance of about 13 kilometers (about 8 miles). This region from the ground up to about 13 kilometers is called the troposphere and contains about 80% of the entire mass of the atmosphere.

Warm air rises in the atmosphere or in a room if the air temperature decreases with altitude. The dispersal of pollutants from a smokestack relies on rising warm air to carry particles into the upper atmosphere. What if the temperature increases with altitude? This is called a temperature inversion because the lapse rate is inverted from normal. The warm air from the smokestack no longer rises (Fig. 8.8). Pollutants become more concentrated, fouling the air. Conditions are further aggravated if fog is present as it often is in London. The word smog, meaning smoke and fog, was coined for these conditions.

Two common types of temperature inversions are described as radiation and subsidence. Radiation inversions develop in the following way. On a sunny day the ground absorbs solar radiation and becomes warm. Air in contact with the ground also becomes warm. But remember, air is a very poor conductor, so heat is not readily conducted vertically to other parts of the atmosphere. Thus the temperature decreases from the ground up for several thousand feet (Fig. 8.9a). At night the solar energy source for the ground is removed, and the ground cools by emitting thermal radiation. Air in contact with the ground cools but again because air is a very poor conductor cooling is limited to air close to the ground. Air above the ground is warmer than air close to the ground. Consequently, the air temperature decreases from the ground up and we

FIGURE 8.8 An illustration of how smoke fails to rise because of a temperature inversion in the atmosphere. This type of inversion is often seen in the morning on a clear, cold day. After the sun is up for awhile, the inversion disappears and the smoke can be seen to rise.

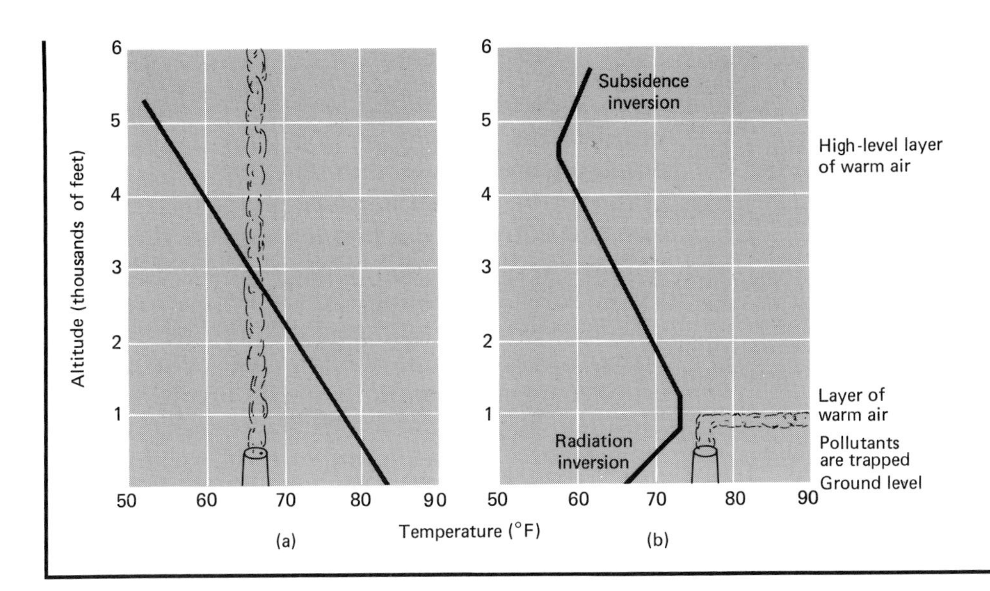

FIGURE 8.9 (a) Usually the temperature of the atmosphere decreases as the altitude increases, as shown in the left graph. These conditions allow warm air to rise and to be dispersed by air circulation. **(b)** The graph illustrates a radiation inversion at ground level and a subsidence inversion starting at about 4500 feet. Both types of temperature inversion can prevent warm air from rising.

say there is a radiation inversion (Fig. 8.9b). Usually, a radiation inversion is confined to altitudes of one to two thousand feet. Pollutants released during the night do not rise and tend to collect in this low "inverted" layer. Normally, the sun comes up, the ground warms, the situation reverts to normal, and the pollutants disperse. But if there is a cloud cover and no wind and the ground does not warm, the pollutants remain and the health concerns worsen. A radiation inversion prevailed at the time the photograph in Fig. 8.8 was taken. Radiation inversions are common on cold, clear, winter mornings, and they can be identified by observing smoke emanating from a chimney or smokestack.

The second common type of temperature inversion results from the phenomenon of *subsidence.** Air pressure, as well as temperature, decreases with increasing altitude. Under certain conditions, cool, high-level air sinks to a higher pressure level. This happens, for example, when a high-altitude wind loses speed. Once in the high-pressure area, the air compresses and warms. (Remember, this principle is used in a diesel engine, Section 6.7.) As a result, the air temperature is higher than normal in a local upper area and a temperature inversion results. Inversions of this type tend to form at altitudes of 1000–10,000 ft (Fig. 8.9b). Subsidence inversions are common on the West Coast and, because of the enclosing mountainous terrain, produce the stagnant air conditions around Los Angeles. Such inversions are not as common in other parts of the United States. The Donora incident resulted from an inversion of this type and conditions there were aggravated by the surrounding mountains. Weather forecasters anticipating the onset of conditions producing a subsidence inversion are often able to forewarn the public. Figure 8.10 shows how photochemical oxidant levels rose in Cincinnati, Ohio, during a temperature inversion and a stagnant air condition that lingered on for nearly two weeks. These conditions plague most major cities east of the Mississippi River during the summer months. Because most of the photochemical oxidants are a consequence of private automobiles, citizens are urged not to drive in the city until the weather changes and the pollutants are flushed.

* From the word *subside,* meaning to sink to a lower level.

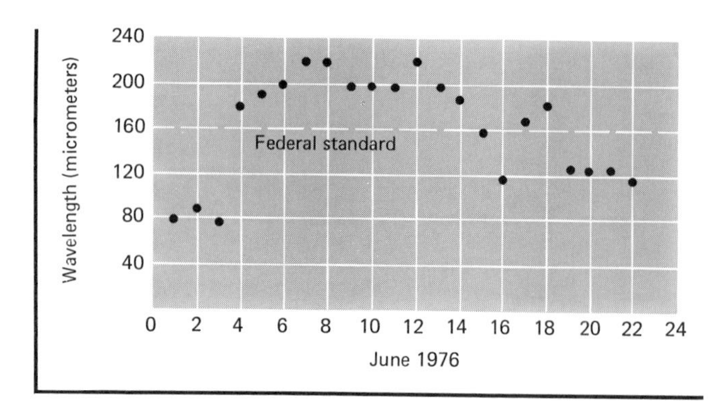

FIGURE 8.10 This plot is a record of the concentration of photochemical oxidants in Cincinnati, Ohio, during an extended stagnant air condition. The federal standard for photochemical oxidants was exceeded for eleven consecutive days. Rain and wind eventually flushed the oxidants from the area. Concentrations like these are common in the major cities east of the Mississippi River.

8.5 A STABLE ATMOSPHERIC CONDITION WITHOUT A TEMPERATURE INVERSION

A stable atmospheric condition preventing pollutant dispersal can occur without a temperature inversion. This happens when the prevailing lapse rate is less than a certain reference lapse rate. For example, if the reference lapse rate is $-4°$ C per kilometer and the actual lapse rate in the atmosphere is $-2°$ C per kilometer, then warm air will not rise even though there is no temperature inversion.

It is fun to play with a child's balloon filled with a gas "lighter than air." Sometimes it rises. Other times it sinks or perhaps remains at some position until a gust of air moves it. If it accelerates, there must be a net force in the direction of the acceleration. This is just Newton's second law. If the balloon is stationary, the net force on it is zero. Because gravity is always pulling downward on the balloon, then for the balloon to rise there must be a larger upward force. This upward buoyant force is a consequence of air pressure decreasing from the ground up. The pressure is greater at the bottom of the balloon than on the top because the bottom is at a lower altitude. Hence the force on the balloon is larger at the bottom than at the top thereby producing an upward force. If the buoyant force is larger than gravity, the balloon rises; if it is not, the balloon sinks. Archimedes first deduced that the buoyant force is equal to the weight of the air displaced by the object. For example, if the object has a volume of 100 cubic centimeters, then the upward buoyant force is equal to the weight of 100 cubic centimeters of air. So if the weight of a balloon is less than the weight of the air that it displaces, the balloon rises. The volume of the displaced air is, of course, the same as the volume of the balloon. Hence, the balloon's weight will be less than the air it displaces if its average density is less than the density of the air. Thus one can deduce whether a balloon rises or falls by comparing its density with that of the air it is placed in.

We can examine the behavior of a gas introduced into the air just as we examined the behavior of the balloon in air. We pick a certain volume of

Droplets of oil may be suspended in a mixture of isopropyl alcohol and water. The downward force of gravity is balanced by the upward buoyant force of the surrounding fluid, and the droplets remain motionless.

the gas released and compare its density with the density of the surrounding air. Three properties of a gas need to be recognized to understand its behavior in the atmosphere. First, a gas cools as it expands. You can observe this by letting gas escape rapidly from an inflated balloon. Second, the density of a gas increases as its temperature decreases. This is why cool, dense air at the ceiling of a room falls to the warmer floor thereby creating convection currents. Third, a gas is a very poor conductor of heat. When a parcel of gas is released into the air, it essentially does not exchange heat with the surrounding air. Because air is such a poor heat conductor the parcel behaves as if it is confined within the boundaries of a balloon. As it rises and expands into the lower pressure surroundings, it cools at its own internal rate. If when the gas (for example, smoke) is released its temperature is higher than the surrounding air, then its density is less so it starts to rise, cooling as it ascends. If it cools faster than the surrounding air, its temperature eventually falls to the temperature of the surroundings, the densities become equal, and the gas (smoke) stops rising (Fig. 8.11) producing a stagnant atmospheric condition.

8.6 ATMOSPHERIC EFFECTS DUE TO PARTICULATES

Few natural scenes rival the beauty of a red sunset or a clear, blue sky. And although the astronauts exploring the moon enjoyed many exciting views that we on earth cannot, they saw no red sunsets or blue skies. These effects are produced by encounters of light with atoms and molecules in the atmosphere, and there is no such atmosphere on the moon. When an atom glances off some particle such as a molecule, we say the atom is scattered. Similarly a ray of light is scattered when it glances off a tiny particle in the atmosphere. The degree of deflection of the ray depends on the color of the light; small particles scatter blue light much more than red light. Light from the sun is a mixture of the colors of the rainbow. When you look west toward the setting sun, the light you see penetrates the earth's atmosphere. The blue component is scattered away from your sight and you see what is left over, mainly the red component. The sky is blue because of blue light scattered toward your eyes.

Particulates in the atmosphere scatter and absorb radiation. The effects are often seen as dense

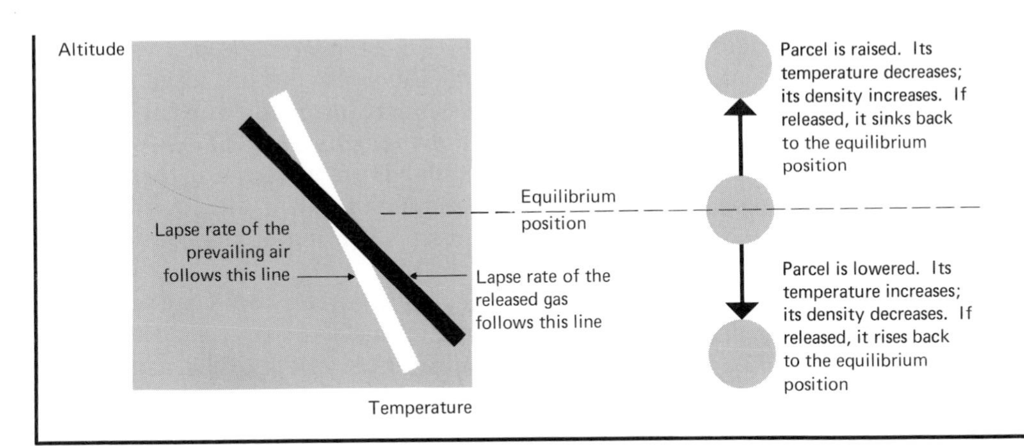

FIGURE 8.11 The white line represents a prevailing condition whereby the temperature decreases as one moves up into the atmosphere. The black line shows how the temperature of a gas might decrease with increasing altitude when the gas is released in the atmosphere. Initially, the gas is at a higher temperature than the air. However, it cools faster than the air as it rises. When the temperatures of the gas and air equalize, a stable condition results. Forces on the gas always tend to restore it to the stable position.

FIGURE 8.12 The same view of New York City recorded one day apart. The thick smog is the result of heavy smoke concentrated by a stagnant air condition. (Photograph courtesy of the *New York Daily News*.)

hazes in areas of high particulate concentration (Fig. 8.12). The hazes limit the maximum distance from which objects can be distinguished from their background. This maximum distance is called the visual range.

Optimum control of an aircraft at a municipal airport requires a minimum visual range of about 20 miles. In an atmosphere containing a particulate concentration of 100 micrograms per cubic meter of air, which is not unusually high, the visual range is reduced to about eight miles. Visual ranges of this order begin to curtail air-traffic control.

Reduced visibility from scattering and absorption of radiation means energy associated with the radiation is diminished. Turbidity* is a measure of this energy loss. It accounts for the type of radiation and the size and concentration of the scattering and absorbing particles; it is a number varying from zero to infinity. Zero means no radiation is removed; infinity means all radiation is removed. Solar radiation encounters a "thickness" of air (the atmosphere) in its path toward the earth. There will be a complex distribution of particulates mixed with the air in the atmosphere that varies for different positions on the earth. So the turbidity and energy loss also varies with position on the earth. There is, of course, some energy loss even if there are no particulates; particulates only enhance this loss.

Particulate concentrations are high enough in some areas that sunlight has been significantly reduced at the earth's surface. Analyses of the radiation received in these areas have led to the following conclusion: For concentrations varying from 100 $\mu g/m^3$ to 150 $\mu g/m^3$, where large smoke turbidity factors persist, in middle and high latitudes direct sunlight is reduced up to *one-third* in summer and *two-thirds* in winter. Furthermore, scientists have estimated that the total sunlight is reduced 5% for every doubling of the particulate concentration.

An astronaut orbiting the earth outside the earth's atmosphere is in a position to measure the amount of incoming solar energy. The astronaut would also observe some radiation returning back into space as a result of reflection and scattering in the atmosphere and on the earth. The energy returning back into space divided by the incoming energy is called the albedo. Measurements of the albedo average around 0.4, meaning that about 40% of the solar radiation returns to space. Particulates

* The word *turbidity* comes from *turbid* which implies cloudy, smoky, or hazy conditions.

contribute to the albedo by reflecting and absorbing radiation. If particulate concentrations are significant and if reflection predominates over absorption, then the earth would be deprived of solar energy and the temperature of the earth would decrease. If absorption predominates, the earth would warm. There is no real evidence that either effect is occurring. However, the decrease in worldwide air temperature since 1940 may be due to reflection of solar radiation by particles accumulating in the atmosphere.

8.7 THE GREENHOUSE EFFECT

Carbon dioxide (CO_2) is not toxic and does not harm plants or property. Superficially it would seem that it should be of no concern even though a 1000-megawatt coal-burning electric power plant at full capacity delivers about 29,000 tons of carbon dioxide into the atmosphere each day. Yet the long-term effects of carbon dioxide accumulation in the atmosphere may be disastrous. The concern is for a warming of the earth by what is popularly called the "greenhouse" effect. The warming of the earth by the greenhouse effect can now be explained using our understanding of radiation.

Solar radiation passes through carbon dioxide in the atmosphere with very little attenuation. After penetrating the earth's atmosphere, the radiation is absorbed and reflected by objects on the earth's surface. As for any object at a temperature above zero kelvins, the earth radiates energy, but because the temperature of the earth's surface is much lower than the temperature of the surface of the sun, the radiation is mostly infrared. Carbon dioxide strongly absorbs infrared radiation at selected infrared wavelengths. Thus radiation comes freely through carbon dioxide in the atmosphere, but is prevented from escaping upon reradiation by the earth. The energy is trapped and the temperature of the atmosphere and the earth would obviously rise if significant amounts of carbon dioxide were to accumulate.

It is not particularly difficult to forecast the amounts of carbon dioxide produced by the combustion of fossil fuels. However, all the carbon dioxide does not remain in the atmosphere. Significant amounts are taken up by the oceans and some carbon dioxide promotes plant growth via photosynthesis. These factors are difficult to estimate. So the only reliable measure is actual monitoring of carbon dioxide in the atmosphere. Sporadic measurements have been made for over a century. Systematic monitoring at the Mauna Loa Observatory in Hawaii shows global concentrations increasing about 7% between 1959 and 1979. The increase reflects the retention in the atmosphere of about half the carbon dioxide produced by combustion of fossil fuels.

Assessing the global impact of the accumulation of carbon dioxide in the atmosphere poses a unique problem to the human race. There is geological evidence that only a few degrees temperature variation differentiates an ice age and our present situation. Short-term temperature variations occur naturally. From actual measurements, it is known that in the last 100 years the mean global temperature rose about 0.6 K to 1940 and has fallen about 0.3 K in the interim. While we have measurements of the accumulation of carbon dioxide, we do not have experimental evidence of its impact on global temperature. The earth is a complex energy machine. Accordingly, physical models of the earth are intricate and are laden with uncertainties. Anticipated effects of the accumulation of carbon dioxide are generally framed in terms of the consequences of a doubling of the preindustrial (about 1800) concentration of about 290 parts per million. The concentration in 1980 was about 335 parts per million. A doubling could increase the average global surface temperature by about 3° C. But, importantly, the effect is not uniform over the earth's surface. The north polar region could experience 7°–10° C increases in winter. With such overall temperature increases it is likely that wind, ocean, and precipitation patterns would change and perhaps lead to shifts in agricultural patterns and major social, economic, and political impacts. At the extreme of the effects there could occur a melting of the West Antarctic ice sheet and a consequent rise of several meters in existing ocean levels. There is great worldwide pressure to accelerate the use of fossil fuels, especially coal. If this occurs, the carbon dioxide levels will accelerate. The doubling time depends sensitively on the growth of fossil fuel use. If we accept a 50% increase in carbon dioxide levels as tolerable, and if worldwide use of fossil fuels grows at a moderate rate of 2.5% per year in the 1980s, then the use of fossil fuels would have to taper off in the first quarter of the twenty-first century to avoid the possible consequences of a doubling of carbon dioxide concentrations. Whether necessary global energy planning can be implemented and enforced remains to be seen, but all indications are that a global greenhouse effect is a distinct possibility which may bring with it highly undesirable social consequences.

8.8 ACID RAIN

Changing a battery in a car, an unknowing man spills solution from the battery onto his trousers, and later finds them riddled with holes. This encounter introduces him to the notoriously corrosive property of acids. Acids have a greater than average concentration of positive hydrogen ions, that is, hydrogen atoms stripped of their electrons. Concentrated acids are highly reactive because the hydrogen ions in them are extremely chemically active. Consequently, battery acids eat away clothing. Chemists use a pH scale ranging from 0 to 14 to measure acidity. Neutral solutions having a very low concentration of hydrogen ions are assigned a pH of 7. Pure water has a pH of 7. As the hydrogen ion concentration increases, the solution becomes more acidic and the pH decreases. Conversely, as the hydrogen ion concentration decreases, the solution becomes more basic and the pH increases. A change of 1 on the pH scale denotes a factor of 10 change in acidity. Thus a solution having a pH of 4 is ten times more acidic than a solution having a pH of 5.

Carbon dioxide is a natural gaseous component of the atmosphere, and some carbon dioxide reacts with water vapor to produce liquid carbonic acid. Consequently, natural rain and snow are slightly acidic having a pH in the range of 5.6 to 5.7. Rain and snow also contain sulfuric acid (H_2SO_4) and nitric acid (HNO_3) originating primarily from human endeavors. The acidity of precipitation has increased rather dramatically in several parts of the world. The northeastern United States and southeastern Canada have experienced the greatest increases in North America. Figure 8.13 depicts the changes in the northeastern United States. The acid content of northeastern rain tends to be about 65% sulfuric, 30% nitric, and 5% other types. Acid rain on the far western coast tends to contain a greater proportion of nitric acid. Lakes in the Adirondacks have clearly deteriorated because of the increased acidity of their waters. Fish populations have vanished in some lakes because of acid-induced changes in their reproductive cycles. The effects of acid rain on forests, crops, and soils are much less understood, and there is no clear evidence that acid rain affects human beings. The origin of the acids is controversial, but there is strong circumstantial evidence that power plants and automobiles are involved. Sulfur oxides (SO_x), nitrogen oxides (NO_x), and water (H_2O) are ingredients for making sulfuric acid (H_2SO_4) and nitric acid (HNO_3). Power plants are prolific producers of sulfur oxides and nitrogen oxides. Automobiles spew out nitrogen oxides. Nature provides the water. But the precise atmospheric chemistry that converts the ingredients into products is not clear, and the pathways from sources to recipients are not completely understood. Wind in the United States and Canada moves naturally from west to east. Many power plants and industry in the Ohio River Valley and other midwestern areas have installed high smokestacks in an effort to reduce local sulfur oxide levels. Consequently the sulfur oxides enter the high-level wind system and migrate hundreds of miles. Conceivably the chemical formation of acids occurs in transit. Then the acids are deposited in rain or snow several hundred miles from the origin of the oxides. Evidence for implication of power plants is strong, but circumstantial. Mandates to limit emission or to control sulfur oxide for all power plants, not just new ones, would be expensive to obey. But if scientific research definitely identifies power plants as the cause of acid rain, those mandates will probably be made.

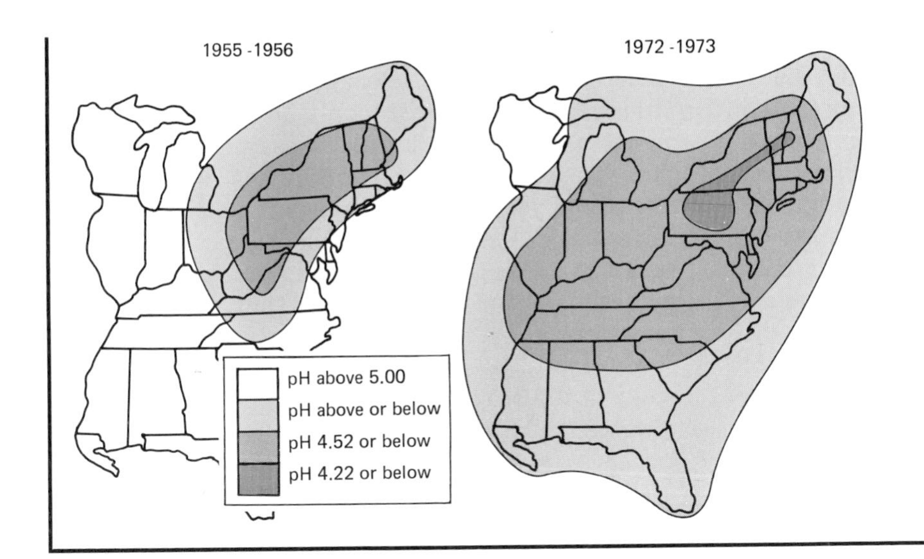

FIGURE 8.13 These two maps illustrate the changes in the annual average of the pH of precipitation in the eastern United States. Data for the maps were taken from the publication *Acid Rain*, United States Environmental Protection Agency, 1980.

8.9 DEPLETION OF THE OZONE LAYER

A suntan, admired and craved by many, is actually a mild form of skin damage produced by ultraviolet radiation from the sun. Overexposure to the sun's rays produces skin damage that in severe cases can lead to skin cancer. Nature provides a defense against gross effects from ultraviolet radiation by creating a band of ozone (the ozone layer) at an altitude of 20,000 to 30,000 meters. The ozone layer removes the great bulk of the potentially very damaging ultraviolet radiation emitted by the sun. Life as we know it would be drastically different without this ozone layer.

Scientists are concerned about the possibility of depleting the ozone layer through the interjection of manufactured products. The first of these anxieties was registered by opponents of the American commercial supersonic transport (SST) plane. The SST was intended to cruise in the midst of the ozone layer. There was fear that nitric oxide (NO) from the engine exhausts would react with and reduce the ozone concentration.

From the spectrum of wavelengths impinging on the ozone layer, the ozone molecules (O_3) selectively absorb short wavelength ultraviolet radiation. Absorption of energy may split an ozone molecule according to the chemical reaction

$$O_3 + \text{solar energy} \rightarrow O_2 + O.$$

The atomic oxygen (O) is very active chemically and we find the reverse process also occurring.

$$O_2 + O \rightarrow O_3 + \text{energy}.$$

Thus the ozone layer is normally a balanced dynamic system with equal rates of production and depletion of ozone. If nitric oxide were introduced into the ozone layer from the exhaust of an SST, we could have the loss of ozone according to the chemical reaction

$$NO + O_3 \rightarrow NO_2 + O_2$$

followed by

$$NO_2 + O \rightarrow NO + O_2.$$

Note that NO serves as a catalyst. It is regenerated in the pair of reactions and O_3 is depleted. A small amount of NO could conceivably deplete a large amount of ozone. If enough ozone were consumed, serious consequences could result.

A paint brush is a vehicle for carrying paint from a can to an area of interest. A good paint brush does not alter the color of the paint or mar the surface where the paint is applied. When the job is finished, the brush is cleaned to make ready for the next job. There are many needs for a vehicle to move paints, disinfectants, deodorants, insecticides, and so on, to hard to reach areas. Manufacturers often use compressed gases to propel liquids through an aerosol-producing nozzle. Ideally, no chemical reactions occur between the propellant and chemical compounds in the can. The aerosol is sprayed at the area of interest where the chemical remains to do its job and, hopefully, the propellant goes harmlessly into the atmosphere. There are no brushes to clean and all seems well. But the chemical inertness of the propellant, so attractive for having no effect on the chemical to be sprayed or the surface to be sprayed, has contributed to concerns for the ozone layer.

A popular type of aerosol spray propellant is termed chlorofluorocarbons. As the name suggests, chlorofluorocarbons are molecules having chlorine (Cl), fluorine (F), and carbon (C) as atomic constituents. Two of this type of aerosols of interest are called fluorocarbon 11, formula $CFCl_3$ and fluorocarbon 12, formula CF_2Cl_2. They are chemically inert and highly attractive as propellants in aerosol spray cans. Because these chlorofluorocarbons are chemically inert, their retention time in the atmosphere is very long and eventually they are thought to migrate to the ozone layer. There, solar energy is sufficiently intense and of the proper wavelength to separate these chlorofluorocarbon molecules. It is a classic case of adding energy to a bound system to free its constituents. Of particular concern is the liberation of chlorine (Cl) followed by formation of chlorine oxide (ClO). Thereafter the chemistry is complex but in simplistic terms there follows

$$ClO + O_3 \rightarrow ClO_2 + O_2,$$
$$ClO_2 + O \rightarrow ClO + O_2.$$

Here ClO functions as a catalyst that is regenerated to deplete more of the ozone. Scientists conceive of conditions that could produce a 40% decrease in the ozone layer by 1995. Predictions such as this have stimulated vigorous theoretical and experimental research programs to learn more about the potential problems, and have brought about a federal ban on the use of chlorofluorocarbons in aerosol spray cans. However, chlorofluorocarbons have other important uses. Most refrigerators and air conditioners use a chlorofluorocarbon called Freon™. In such use they are not released to the environment by design as are aerosol sprays, but there is still some concern over their accidental release to the ozone layer. Thus the monitoring and theoretical study of the ozone layer continues to be an important scientific endeavor.

REFERENCES

ATMOSPHERIC PROBLEMS

Introduction to Environmental Science, Joseph M. Moran, Michael D. Morgan, and James H. Wiersma, Freeman, San Francisco, 1980.

Environmental Pollution, 2nd ed., Laurent Hodges, Holt, Rinehart and Winston, New York, 1977.

"Coal: Climate and Health Hazards," Chapter 10 in *Energy: The Next Twenty Years,* Ballinger Publishing Company, Cambridge, Mass., 1979.

State of the Environment 1982, The Conservation Foundation, Washington, D.C. 20036.

THE GREENHOUSE EFFECT

"Atmospheric Carbon Dioxide: What to Do?" Gregg Marland and Ralph Rotty. In Lon C. Ruedisili and Morris W. Firebaugh (eds.), *Perspectives on Energy,* Oxford University Press, New York, 1982.

"Carbon Dioxide and the Climate: The Uncontrolled Experiment," C. F. Baes, Jr., H. E. Goeller, J. S. Olson, and R. M. Rotty, *American Scientist* **65,** 3 (May–June 1977).

Global Energy Futures and the Carbon Dioxide Problem, Council on Environmental Quality, January 1981. Superintendent of Documents, United States Government Printing Office, Washington, D.C. 20006.

Carbon Dioxide and Climate: A Second Assessment, National Academy Press, Washington, D.C., 1982.

"Carbon Dioxide and World Climate," Roger Revelle, *Scientific American* **247,** 35 (August 1982).

ACID RAIN

Acid Rain, Report Number EPA–600/9–79–036, Superintendent of Documents, United States Government Printing Office, Washington, D.C. 20402, National Acid Precipitation Assessment Plan, June 1982, Interagency Task Force on Acid Precipitation, Washington, D.C. 20006.

Proceedings of the Energy Outlook Meeting on Acid Rain, Illinois Energy Resources Commission, Springfield, Ill. 62706.

An Updated Perspective on Acid Rain, Edison Electric Institute, Washington, D.C. (November 1981).

WAVES AND PHOTONS

The wave-particle duality of electromagnetic radiation is discussed in nearly all elementary physics and physical science textbooks. Two such texts useful for the expansion of the material presented in Chapter 8 are *Physics in Perspective,* Eugene Hecht, Addison-Wesley, Reading, Mass., 1980, and *Conceptual Physics,* 4th ed., Paul G. Hewitt, Little, Brown, Boston, Mass., 1981.

REVIEW

1. Comment on the wave-particle duality of electromagnetic radiation.
2. What is a photon?
3. How does the dominant type of radiation from a heated source change when the temperature changes?
4. What is a temperature inversion?
5. Describe the conditions for the formation of radiation and subsidence inversions.
6. Why are temperature inversions of concern in the environment?
7. What is the meaning of lapse rate?
8. Explain how warm air rises in the atmosphere using the behavior of a balloon as an analogy.
9. How can a stagnant air condition result without a temperature inversion?
10. Why is there concern over accumulation of particulates in the upper atmosphere?
11. Describe the conditions leading to the greenhouse effect. Why is the greenhouse effect of concern?
12. What definitive statements can you make about the present-day accumulation of carbon dioxide and particulates in the atmosphere?
13. What is acid rain and what are the environmental concerns over it?
14. Describe the concerns for depletion of the ozone layer.
15. Waves are termed "disturbances propagating with a definite velocity." Disturbances as associated with electromagnetic waves involve
    a) movement of molecules.
    b) electric and magnetic fields.
    c) electrons and magnetons.
    d) magnets.
    e) magnetohydrodynamics.
16. A wave having a wavelength of two meters and frequency of 200 cycles per second travels with a speed of
    a) 0.01 meters/second.
    b) 800 meters/second.
    c) 80,000 meters/second.
    d) 400 meters/second.
    e) 100 meters/second.
17. If the wavelength of electromagnetic radiation increases, then
    a) the frequency of the wave will decrease.
    b) the speed of the wave will increase in direct relation.
    c) the frequency of the wave will increase.
    d) the speed of the wave will change.
    e) both (a) and (d) are correct.
18. Any object at a temperature above zero kelvins will emit electromagnetic radiation. If the wavelength of the dominant radiation is one unit when the temperature is 300 K, then the dominant radiation at a temperature of 6000 K is
    a) 6000 units. b) 20 units. c) $\frac{1}{20}$ units.
    d) 300 units. e) none of the above.
19. Object A emits mostly infrared radiation. Object B emits mostly visible radiation. The temperature of A is ______ the temperature of B.
    a) greater than
    b) twice
    c) less than
    d) four times
    e) equal to

20. Pick the graph below that describes a temperature inversion.

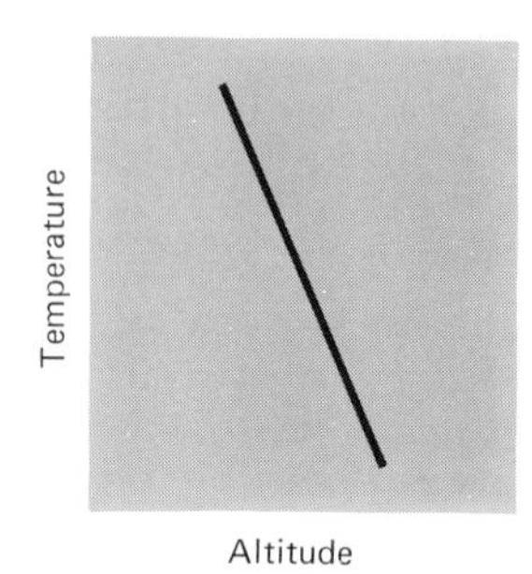

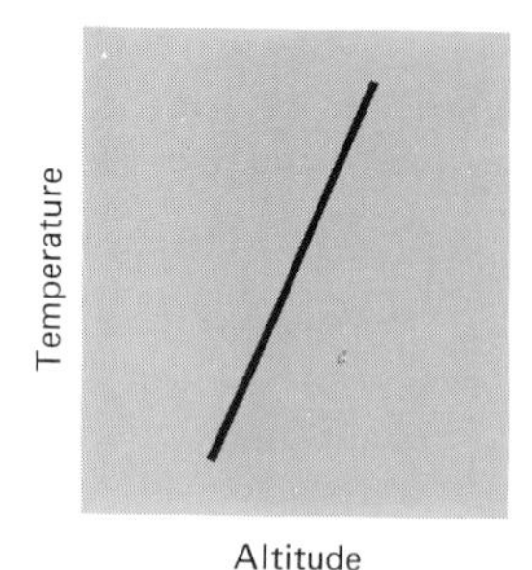

21. In order for smoke to rise in the atmosphere, it is necessary that
    a) the temperature increase with altitude.
    b) the temperature not change with altitude.
    c) the density of the air be greater than one g/cm^3.
    d) the temperature decrease with altitude.
    e) the smoke be at a temperature of at least 150° F.
22. Temperature inversions are of concern in the environment because
    a) they create a condition that prevents infrared radiation from escaping from the earth.
    b) they invert the distribution of carbon dioxide in the atmosphere.
    c) they accelerate the production of carbon dioxide in the atmosphere.
    d) they involve a condition whereby pollutants cannot rise up into the atmosphere and be dispersed.
    e) none of the above.
23. If we say the temperature of the atmosphere changes with altitude at a rate of $-7°$/km, we are discussing
    a) a synergism.
    b) the lapse rate of normal daytime conditions.
    c) a radiation inversion.
    d) a subsidence inversion.
    e) conditions that lead to increased pollution levels.
24. A subsidence inversion
    a) occurs primarily at night.
    b) takes place primarily near the ground.
    c) usually occurs at altitudes well above ground.
    d) occurs usually during daylight hours.
    e) occurs always in early morning hours in winter, and results in an accumulation of pollutants.
25. A setting sun is often particularly red when viewed through a polluted atmosphere. This is caused by
    a) absorption of solar radiation with subsequent reradiation of red light.
    b) preferential scattering of the blue colors out of the line of sight of the viewer.
    c) illumination of the red-colored pollutants.
    d) heat radiation from the molecules in the atmosphere.
    e) none of the above.
26. A certain inflated balloon has a volume of 1000 cm^3 and a mass of one gram. The balloon is released in air where the density is 0.0011 grams per cm^3. We can expect the balloon to
    a) rise. b) fall. c) remain stationary.
    d) explode. e) collapse.
27. Heat transfer by convection is often used to warm rooms in a building. The formation of convection currents is based on the principle that
    a) the density of air decreases when it is warmed.
    b) the density of air increases when it is warmed.
    c) cool air rises.
    d) air cools when compressed.
    e) the pressure of a gas is independent of its temperature.
28. We say that light sometimes behaves like a wave because
    a) we actually see wave motion of light just like we see wave motion of water.
    b) the first law of thermodynamics specifies that light is a wave.
    c) of the photoelectric effect.
    d) photons are waves and light is composed of photons.
    e) light has properties that are very much like water waves, for example.
29. About 40% of the radiation coming from the sun is returned by scattering from the earth's atmosphere. This is measured by a quantity called the
    a) coercivity.
    b) refractivity.
    c) scattering coefficient.
    d) albedo.
    e) photoeffect.
30. Heat leaks in homes are sometimes located by detecting differences in the emission of
    a) light.
    b) infrared radiation.

c) ultraviolet radiation.
d) gamma radiation.
e) air.

31. The greenhouse effect works because
    a) the invisible ultraviolet cannot penetrate the ozone layer.
    b) the invisible long wavelength infrared cannot penetrate the carbon dioxide in the atmosphere.
    c) the invisible long wavelength infrared cannot penetrate the ozone layer.
    d) the invisible short wavelength infrared cannot penetrate the ozone layer.
    e) visible radiation can get in through the atmosphere but can't get out again.

32. Fluorocarbons in the atmosphere present a major environmental problem because they
    a) are poisonous.
    b) may produce a blanket which may significantly reduce the radiation coming to the earth from the sun.
    c) may interact with and deplete the ozone layer.
    d) may enhance the ozone layer.
    e) are intimately involved in the greenhouse effect.

Answers

15. (b)	16. (d)	17. (a)	18. (c)
19. (c)	20. (b)	21. (d)	22. (d)
23. (b)	24. (c)	25. (b)	26. (a)
27. (a)	28. (e)	29. (d)	30. (b)
31. (b)	32. (c)		

QUESTIONS AND PROBLEMS

8.2 WAVES AND PHOTONS

1. What everyday experiences with sound suggest that it is a wave phenomenon?
2. If the wavelength of electromagnetic radiation increases, how does the frequency change?
3. If the energy of a photon increases, how does the frequency change?
4. Why don't our eyes detect individual photons from a light source?
5. Radio station WLW broadcasts electromagnetic waves with a frequency of 700,000 hertz. The speed of the waves is 300 million meters per second. What is the distance between crests of these waves?
6. Regardless of frequency, all electromagnetic waves travel with the same speed in a vacuum (v = 300,000,000 meters/second). Using Eq. (8.3) relating speed, frequency, and wavelength of a wave, compare the frequency of an AM radio wave and a gamma wave with a light wave. Values of the wavelengths are presented in Fig. 8.5.
7. Observing crests of a wave moving past some position is like watching evenly spaced cars move by some location on a highway. What wave property would be analogous to the distance between cars? If T is the time between the passing of two cars and V is the speed of the cars, how far will a car travel in time T? Point out the similarities between this result and Eq. (8.3) relating speed, frequency and wavelength of a wave.
8. If the energy of a photon is 0.00000000000000000002 joules, what type of radiation is it?
9. The energy of a photon is related to frequency by $E = hf$, where

   $h = 6.63 \times 10^{-34}$ joule · second

   Burning a pound of coal liberates about 10,000,000 joules of heat energy. Knowing that the frequency of visible light is 600,000,000,000,000 Hz, or 6×10^{14} Hz, how many visible photons are needed to have an energy equivalent to the heat energy liberated from burning a pound of coal?
10. A 100-watt light bulb produces about 10 joules of visible energy each second. Assuming the light emitted is green with a wave frequency of 600 trillion hertz (6×10^{14} Hz), how many photons are produced each second the bulb is lit?

8.3 THERMAL RADIATION

11. If all objects at a temperature greater than zero kelvins radiate electromagnetic waves, why don't we see objects in an unlighted room?
12. The human body radiates thermal energy. Why is it we do not continually cool down as we sit in a room of a house?
13. An ordinary 100-watt incandescent light bulb when lit is much too hot to handle with bare hands. Knowing that the dominant radiation is visible, explain why the temperature is necessarily high.
14. Why does the temperature of an auditorium tend to rise when filled with people?

15. If you were given the task of developing a camera that would delineate a warm area in a body of water, what special property is needed for the camera film?
16. A homeowner suspects that the builder of his home has neglected to insulate selected regions of the walls. Without disassembling the walls, how could he determine if the builder has cheated him?
17. As an energy conservation measure, it is important that the seals around the doors of a refrigerator fit properly. How would a thermal photograph of a refrigerator be a useful energy diagnostic tool? What areas of a refrigerator would you want to show up dark and light in the thermal photograph?
18. How could a thermal photograph of an automobile help a police officer determine if a car had been involved recently in a high-speed chase by police?
19. Light from an incandescent light bulb is due to thermal radiation from a metallic filament heated by electric charges flowing through it. What physical restriction limits severely the number of materials that can be used for the filament?
20. Why does an iron ingot in a foundry first turn red and then become a more "whitish" color when heated to the melting temperature?
21. The tip of a cigarette often becomes "red hot" when a smoker inhales. Using the relation between dominant wavelength and temperature for an ideal radiator, estimate the cigarette tip temperature knowing that red color has a wavelength of about 0.65 micrometers. Is it any wonder that an errant cigarette burns a hole in a carpet?
22. An object at room temperature radiates energy at a rate of about 0.3 watts per square inch of surface. The human body has about 2800 square inches of surface area. (You might like to verify this by assuming the body is a cylinder of your height (h) and diameter (d) of 12 inches. The formula needed is $A = 1.57d^2 + 3.14dh$.) What is the rate of emission of energy by the human body at room temperature? How does this rate of emission compare with a typical light bulb?

8.4 TEMPERATURE INVERSIONS

23. Why does smoke from a burning cigarette generally rise up into the air?
24. Two identical balloons are inflated to the same size. One is filled with air, the other with helium. Why will the balloon filled with helium rise more readily in the atmosphere?
25. If you inflate a balloon by blowing your breath into it, seal it by tying a knot in the open end, and then release it in the atmosphere, is it possible for the balloon to rise?
26. Why do you sometimes have to exert considerable force to hold an inflated beach ball under water?
27. A friend observes that the smoke emanating from the chimney of her fireplace floats toward the ground. What explanation of this phenomenon could you offer?
28. Sometimes smoke rising from a cigarette seemingly hits a barrier and stops rising. What is the reason for this?
29. As a result of a severe radiation temperature inversion occurring in the nighttime, pollutants are often trapped in a low-lying layer. Then when the sun comes out, the pollutants will migrate toward the ground. What temperature conditions from the ground up give rise to this effect?
30. A variation of temperature with altitude is shown in the accompanying graph. Pick out the portions of the plot indicating temperature inversions.

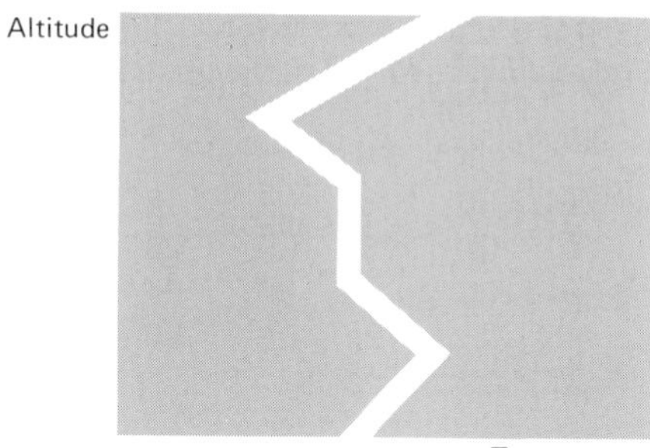

31. Why should there be concern over particulate emissions from power plants and industrial sources if only 10% of the total particulate emissions into the atmosphere come from manufactured sources?
32. An inflated balloon has a volume of 5000 cubic centimeters and a mass of three grams. The balloon is released in air having density of 0.0011 grams per cubic centimeter. Does the balloon rise or fall?
33. A small toy balloon has a mass of two grams. It is inflated to form a sphere of diameter 20 centimeters and is filled with $\frac{1}{4}$ gram of helium. When it is placed in air having a density of 0.001 grams per cubic centimeter, does the balloon rise or fall? The formula for the volume of a sphere of diameter d is $V = 0.52d^3$.
34. If the temperature is 27° C at ground level and the lapse rate in the atmosphere is −5° C/kilometer, what

is the temperature at an altitude of 1600 meters?

35. A commercial jet plane often cruises at an altitude of about 35,000 feet (about seven miles). If the lapse rate is −20° F/mile, what is the temperature outside the plane if it is 75° F at sea level?

36. At the top of a building 400 meters high, the temperaure is 19.7° C. If it is 21.5° C at ground level, what is the average lapse rate?

37. If the temperature is 80° F at the base of a mountain 10,000 feet high, what is the approximate temperature at the top under ordinary weather conditions?

38. The temperature of the air near the ground is 30° C and the temperature decreases 4° C for each kilometer of altitude. If the temperature of the gas released from a smokestack is 36° C and its temperature decreases 7° C for each kilometer as it rises, how high will it rise before stabilizing? One way of doing this is to make a plot of temperature versus altitude as is done in Fig. 8.11 and seeing where the two lines cross.

39. In a day's time a large coal-burning electric power plant may burn 3000 tons of coal. If the coal contains 3% sulfur, 180 tons of sulfur oxides are produced. Suppose that because of a temperature inversion all the sulfur oxides are uniformly distributed in a volume 80 km × 80 km × 0.5 km. What is the concentration of sulfur oxides in micrograms per cubic meter? How does this concentration compare with the primary standard given in Section 5.4? Useful information: 1 ton = 2000 pounds, 1 pound = 454 grams.

8.6 ATMOSPHERIC EFFECTS DUE TO PARTICULATES

40. Why does a sunrise or sunset appear particularly red when viewed through a polluted atmosphere?

41. Why does smoke rising from a cigarette in an ash tray often have a bluish color?

42. To a reasonable approximation the visual range measured in miles multiplied by the particulate concentration measured in micrograms per cubic meter always equals 750. Determine the visual range for particulate concentrations of 50, 100, 150, and 200 micrograms per cubic meter.

43. In 1883 the volcano Krakatoa in the East Indies erupted and spewed about 10 cubic kilometers of particulates into the atmosphere. The particulates were spread worldwide by atmospheric circulations and produced spectacular sunsets for more than two years. In 1982 all the coal-burning electric power plants in the United States released about one million cubic meters of particulates into the atmosphere. At the 1982 rate, how many years of operation by these power plants would be required to produce the particulate output of the Krakatoa explosion?

8.7 THE GREENHOUSE EFFECT

44. If the accumulation of carbon dioxide in the atmosphere proves to be a serious problem, what alternatives would there be to burning fossil fuels?

45. Nitrogen dioxide (NO_2) is a strong absorber of ultraviolet and visible blue radiation. If you view solar radiation after passing through an atmosphere containing nitrogen dioxide, what overall color would you expect to see?

8.9 DEPLETION OF THE OZONE LAYER

46. Why does it take energy to separate chlorine atoms from a chlorinated fluorocarbon molecule like $CFCl_3$? What is the source of energy that causes a chlorinated fluorocarbon to break up in the ozone layer?

47. The chlorinated fluorocarbons used in aerosol sprays are extremely inert, i.e., they do not react easily with other chemicals. Why is this property extremely useful in aerosol sprays? Why does it lead to the problems associated with depletion of the ozone layer?

48. Nature generates the same gaseous pollutants as human beings. Why, then, is there concern about depletion of the ozone layer by nitrogen oxides produced from human activities?

49. On a trans-Atlantic flight an SST is in the air about three hours less than a conventional jet transport. Using the rate concept (amount = rate × time), explain why this does not necessarily mean that the SST pollutes the atmosphere less than a jet transport.

50. Chlorine oxide (ClO) molecules, formed from chlorine atoms separated from chlorofluorocarbons in the ozone layer, enter into chemical reactions with ozone much as does nitric oxide (NO). Following the analysis presented in Section 8.9, fill in the steps in the equations below.

$$ClO + O_3 \rightarrow \underline{\qquad} + O_2,$$
$$\underline{\qquad} + O \rightarrow ClO + \underline{\qquad}.$$

What constitutes the catalyst in these reactions?

Nuclear Power

9.1 MOTIVATION

The rapid growth of technology in the Industrial Revolution that began sometime in the middle of the eighteenth century in England was triggered by the development of the steam engine. The steam engine powered industries, drove shovels for digging canals and mines, and performed a variety of other operations. When the first central electric generating station was built in New York City in 1881, the steam engine found new use—providing mechanical power for driving electric generators. Technology has changed dramatically since the Industrial Revolution. Jet planes whisk us between and across continents in a few hours. Communications through orbiting satellites puts us in nearly instant contact with all parts of the earth. Computers help instruct us,

A nuclear reactor vessel being installed in its protective containment structure. It is in the reactor vessel that energy is derived from nuclear reactions. The comparative size of the workmen and the vessel gives an indication of the magnitude of the engineering aspects. (Photograph courtesy of the General Electric Company.)

grade our exams, report our grades, send us bills, and so on. But our dependence on the steam engine, limited in performance by the second law of thermodynamics, has not changed. In spite of efforts to develop new technology for providing heat to vaporize water for steam turbines in large power plants, only two commercial schemes have emerged—burning a fossil fuel, usually coal, or fissioning nuclei in a nuclear reactor. Both systems involve trade-offs. If coal is used, we must cope with

1. disposal of particulates and scrubber sludge,
2. environmental effects of sulfur oxides, acid rain, and a certain amount of toxic heavy metals, and
3. the greenhouse effect.

If a nuclear reactor is used, we must deal with disposal of radioactive wastes and nuclear safety. We might argue against large, central generating stations employing large coal-burning units or uranium. We may argue that we should forestall using either coal or nuclear reactors, promote energy conservation, and wait for the development of more desirable technology. These are separate issues. But for new, large base load systems in this century, the choices are coal or uranium.

It is hoped that we have a grasp of the technology and problems of burning coal to produce electricity. It is not an easy engineering task to produce a system that burns 10,000 tons of coal a day. Yet we all have the ability to instigate a burning process similar to coal-burning in a power plant. The engineering of a nuclear power plant is far from trivial and unlike burning coal we do not all have the ability to instigate or see the type of energy conversion going on in a nuclear reactor. Yet we citizens are being asked to make decisions about nuclear energy. To be informed we must have some basic knowledge of nuclear reactors. This chapter on nuclear reactors is not intended to make nuclear engineers out of you. Rather it is intended to help you understand the wealth of material being written in newspapers and magazines and being aired on television and radio about nuclear power. We wish to answer questions such as

1. What is attractive and unattractive about the use of nuclear energy?
2. What is the origin of the energy?
3. What are the essential features of a nuclear reactor?
4. How is the nuclear reactor controlled?
5. What is radioactivity?
6. How are the radioactive wastes generated?
7. What are the options for disposing of radioactive wastes?

No single equation in physics has the universal prestige of the Einstein relation $E = mc^2$. In a strikingly simple way, it relates mass (m) to energy (E). The letter c denotes the speed of light or electromagnetic radiation in a vacuum (300,000,000 meters per second). To express energy in joules, m must be expressed in kilograms and c in meters per second. To illustrate, let us calculate the energy equivalent of a penny, which has a mass of about three grams (0.003 kg).

Energy = mass in kilograms multiplied by the square of the speed of light in meters per second.

$$= 0.003\,(300{,}000{,}000)^2,$$

$$= 270 \text{ trillion joules.}$$

Because heat from burning one ton of coal is about 27 billion joules, the potential energy in one gram of matter is nearly 3000 times the heat produced from burning a ton of coal (Fig. 9.1). Knowing that a potential energy source exists is one thing; exploiting it is another. However, our knowledge of the physics of the nucleus of the atom has led to a scheme involving reactions between atomic nuclei that converts mass to useful energy. Let us now look into these important principles.

9.2 PROPERTIES OF THE ATOMIC NUCLEUS

A model of the atom consisting only of a positively charged nucleus surrounded by negatively charged electrons is adequate for discussing molecules and

FIGURE 9.1 Atomic nuclei are a vast source of energy. A single truckload of nuclear fuel can supply the total electric energy needs of a city of 200,000 people for a year. (Photograph courtesy of the Department of Energy.)

chemical reactions because only interactions of the electrons are involved. Nuclear reactions depend on the details of atomic nuclei and this requires that the model be improved. Because both the atom and the nucleus are *bound* systems, many of the concepts used in the atomic model are also useful in the refinement.

The atom is bound together by attractive *electric forces* between the positively charged nucleus and electrons having negative charge. The nucleus is bound together by "strong" *nuclear forces*. In the hierarchy of fundamental forces, the strong nuclear force has the same status as the gravitational and the electromagnetic force. It is the strongest force of all and exists between protons and neutrons. A proton has positive charge equal in magnitude to that of an electron. The neutron is electrically neutral and its mass is slightly larger than the mass of the proton. The proton and neutron are about 2000 times more massive than an electron. Because the atom is electrically neutral and only protons in the nucleus have charge, an atom has equal numbers of electrons and protons.

The strong nuclear force has two features radically different from the electric force. First, it is enormously stronger. Consequently, nuclear energies tend to be much larger than electric energies. Using an energy unit called an electron volt,* the total energy, kinetic plus potential, of an electron about a nucleus is about 10 electron volts. The total energy of a proton in the nucleus is about 50,000,000 electron volts (or 50 MeV). Thus energy released in nuclear reactions tends to be millions of times larger than the energy released in a chemical reaction. Second, the nuclear force acts only over a distance about the "size" of a nucleus. This distance is roughly 0.0000000000001 cm and is about $\frac{1}{100,000}$ the diameter of an electron orbit about the nucleus.

* The electron volt (eV) is a favorite energy unit in atomic and nuclear physics. One eV is the energy acquired by an electron (or a proton) accelerated through a potential difference of one volt. Using the definition of potential difference ($W = QV$) it follows that $1\ \text{eV} = 1.602 \times 10^{-19}$ joules. It is not terribly important to be able to do calculations in terms of electron volts. It is important to remember the relative energy differences between atomic and nuclear systems.

The mathematical form of the electric force between two charges is quite simple. This simplicity allows fairly elementary calculations of the structure of atoms. The Bohr planetary model* of the atom evolves from such a calculation. The Bohr model incorporates electrons revolving about the nucleus in orbits much as planets revolve about the sun. The electrons exist in precise orbits, each characterized by a well-defined (quantized) energy. To move to a higher orbit, an electron must absorb just the right amount of energy. Conversely, an electron descending to a lower orbit releases a precise amount of energy. (See Fig. 9.2.) The concept of orbits is not particularly important. In fact, this notion does not enter in more sophisticated calculations. However, the idea of well-defined energies is a feature of all atomic models. And with modern computers, the energies of atoms and molecules can be calculated fairly well. However, the nuclear force is very strong and acts over a very short distance, and its precise mathematical form is still not completely understood. The search for a detailed understanding constitutes one of the challenges and major efforts of current research. There is no such thing as an elementary calculation for a model of the nucleus. However, like electrons around the nucleus, protons and neutrons (or nucleons as they are termed collectively) have kinetic energy due to their motion and potential energy associated with forces operating within the nucleus. Therefore, it is probably not surprising that the total energy of a nucleus, like the atom, assumes well-defined values. We say the energies are quantized (Fig. 9.3). Just as for an atom a nucleus must absorb just the right amount of energy to assume a higher energy state. And when a nucleus changes to a lower energy state, the change in energy is precise. For an atom the energy changes are expressed in electron volts. But for a nucleus the energy changes are thousands and millions of electron volts (Fig. 9.3).

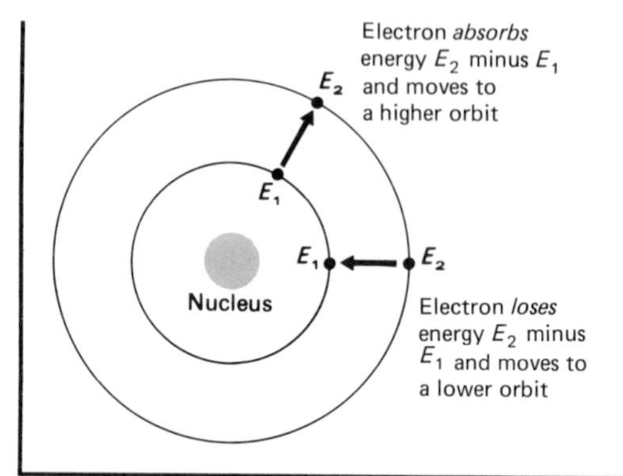

FIGURE 9.2 An atom in the energy state E_1 absorbs energy and the atom "moves" to a state with larger energy E_2. In the Bohr model of the atom, this corresponds to an electron moving to an orbit of larger diameter. An atom in an energy state E_2 loses energy and the atom "moves" to a state with less energy E_1. In the Bohr model of an atom, this corresponds to an electron moving to an orbit of smaller diameter.

An atom and its nucleus are labeled with the chemical symbol of the atom. When referring to the nucleus, it is customary to also affix the proton number (Z), the neutron number (N), and the nucleon number (A) to the chemical symbol (X) in the manner ${}^{A}_{Z}X_N$. The nucleon number is just the sum of the proton and neutron numbers, $A = N + Z$. The nucleus of the hydrogen atom is denoted ${}^{1}_{1}H_0$. This complete description is necessary in order to keep track of neutrons and protons in nuclear transformations. Those atoms with nuclei having the same number of protons but a different number of neutrons are called isotopes. The nuclei of the hydrogen, deuterium, and tritium atoms all have one proton but 0, 1, and 2 neutrons, respectively (Fig. 9.4). The most abundant isotope of carbon has 6 protons and 6 neutrons. This isotope is assigned a mass of exactly 12 atomic mass units (amu) on the atomic mass scale. All other nuclear and atomic masses are measured relative to this isotope of carbon. For example, the most abundant isotope of oxygen, ${}^{16}_{8}O_8$, has a mass of 15.994915 amu on the atomic mass scale.

The ideas that mass can be converted to energy and, conversely, that energy can be converted to mass seem to belong in a science fiction discussion. Yet there is a wide body of evidence supporting both

* Named for Niels Bohr, Danish physicist, 1885–1962.

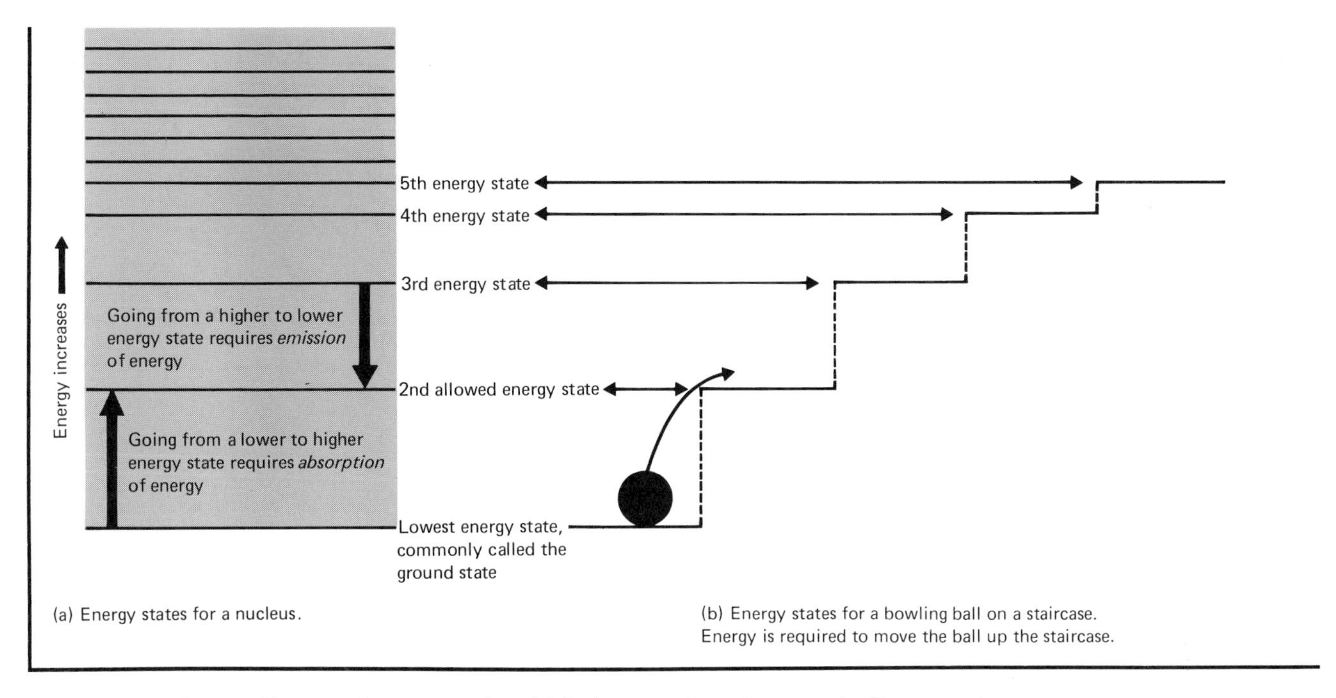

FIGURE 9.3 Energy diagrams for systems in which the energies take on only discrete values.

transformations. Mass energy is seldom included in chemical reactions because the changes in mass are imperceptible. But in nuclear reactions the forces involved are so strong that mass–energy changes are observable.

When we say neutrons and protons are bound together into some stable configuration, we mean that energy must be added to the nucleus to separate it into its constituents. It is somewhat like saying a marble at the bottom of a can is bound and you must add energy to the marble to free it. Energy added to a nucleus to free a constituent is converted to mass. Thus the sum total of the masses of the freed nuclear constituents is always greater than the mass of the bound nucleus. The deuterium nucleus consists of a single proton bound to a single neutron. To separate the two, energy must be absorbed by the deuterium nucleus. Thus the mass of a neutron added to the mass of a proton exceeds the mass of the deuterium nucleus. Conversely, when a deuterium nucleus is assembled from a proton and a neutron, energy is liberated. Always remember, the mass of a stable nucleus is always less than the combined masses of its freed constituents. Total mass is not conserved when a nucleus is formed from neutrons and protons. But total energy is conserved when we include mass energy.

9.3 NUCLEAR STABILITY

Water atop a dam has gravitational potential energy that is converted to kinetic energy when the water flows over the dam. The fall of the water leads to a conversion of energy. A carbon atom reacting with an oxygen molecule to form carbon dioxide liberates energy because the final rearrangement of atoms is less energetic than the original. There is no change in the numbers or types of atoms, only rearrangement. Similarly, nuclei of atoms may participate in nuclear reactions liberating energy. Not only is there a rearrangement of nuclear constituents but in some cases an actual transformation of chemical ele-

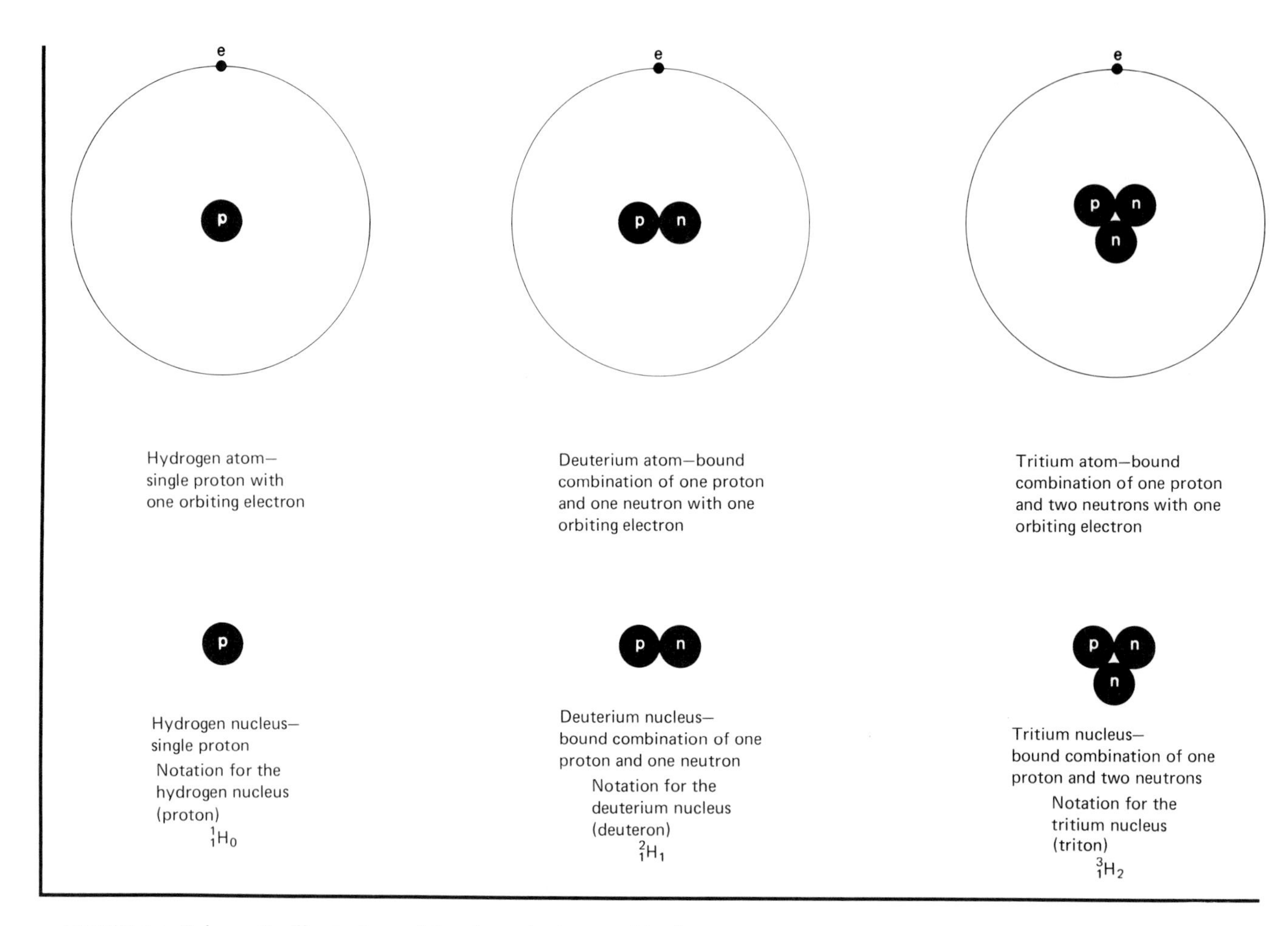

FIGURE 9.4 Schematic illustration of the three isotopes of hydrogen.

ments. In rearranging water in a dam system, atoms in chemical reactions, or nuclei in nuclear reactions, energy is conserved. The forms of energy may change but there is never a reaction having more total energy at the end than at the beginning. In any reaction, mass energy through $E = mc^2$ must be considered as a form of energy. Thus a spontaneous energy change in a nucleus imparting kinetic energy to the reaction products must derive energy from a loss of mass in the reaction.

A marble resting inside a bowl is said to be stable because, if it is moved and released, the forces acting on it tend to restore it to its stable position at the bottom of the bowl. A ball sitting at the edge of a stair step is in an unstable position. Nuclei are subject to forces that promote breakup. Some nuclei are stable against these "provocations." Others are induced to nuclear transformations. Stable nuclei with few nucleons have nearly equal numbers of neutrons and protons. For example, $^2_1H_1$, $^3_2He_1$, $^4_2He_2$, $^6_3Li_3$, $^9_4Be_5$, $^{10}_5B_5$, $^{12}_6C_6$, $^{16}_8O_8$, and $^{17}_8O_9$ are all stable. As the number of nucleons in a nucleus increases, the neutron number gets progressively larger than the proton number. For example, the one stable isotope of cesium has 55 protons and 78 neutrons.

9.4 NUCLEAR REACTIONS

Alpha Particle Emission

A nucleus having two protons is labeled helium. Two protons and two neutrons identify the nucleus as ^4_2He. This nucleus is also called an alpha particle. Thus the ejection of an alpha particle by an unstable nucleus is just the emission of a bound configuration of two protons and two neutrons. Clearly, if two protons are removed, the leftover is chemically different. $^{238}_{92}U$ that has 92 protons and 146 neutrons is a natural alpha particle emitter. Removing two protons and two neutrons from $^{238}_{92}U$ leaves a nucleus with 90 protons and 144 neutrons. This nucleus is an isotope of thorium, $^{234}_{90}Th$. Symbolically,

$$^{238}_{92}U \rightarrow {}^{234}_{90}Th + {}^4_2He.$$

At the time of emission, the $^{238}_{92}U$ nucleus is at rest and has no kinetic energy. Both the $^{234}_{90}Th$ nucleus and the alpha particle have kinetic energy after the emission. The kinetic energies are a consequence of converting part of the mass of $^{238}_{92}U$ into kinetic energy.

It appears that any nucleus having at least two protons and two neutrons could spontaneously emit an alpha particle. But remember, energy considerations dictate the outcome. If the emission of an alpha particle necessitates that the totality of the end products have more mass than the emitting nucleus, then the emission will not happen. This requirement generally limits alpha particle emitters to very massive nuclei such as $^{238}_{92}U$ and $^{239}_{94}Pu$.

Beta Decay

Beta decay is an example of the weak nuclear force, the fourth and final natural force we shall encounter. Beta decay is very intriguing because its explanation seems to contradict the neutron–proton model of the nucleus. There are two types of beta particles which differ from each other only in the signs of their electric charge. The negative variety is indistinguishable from an electron, so we give it the symbol $_{-1}e$. The positive counterpart is called a positron with the symbol $_{+1}e$. Beta particles originate in the nucleus of an atom but in reality they do not exist as entities in the nucleus that wait for weak nuclear forces to provide the right set of conditions for ejection of the particle. Rather, under the influence of weak nuclear forces, the beta particle is created at the moment of emission and we find after electron emission one more proton and one fewer neutron in the nucleus. In positron emission we find one fewer proton and one more neutron in the nucleus. A neutrino accompanies positron emission and an antineutrino accompanies electron emission. Neutrinos and antineutrinos are not important in commercial nuclear power but they are very important for the physics of beta decay. Like photons, they possess no electric charge and no mass. They interact extremely weakly with matter. But, importantly, they have energy and if for no other reason are needed to conserve energy.

Tritium (^3_1H) is an electron emitter and $^{11}_6C$ is a positron emitter. At some moment for ^3_1H, having one proton and two neutrons, an electron and antineutrino are ejected. The nuclear aftermath now has two protons and one neutron, labeling the residual as helium. Symbolically and pictorially,

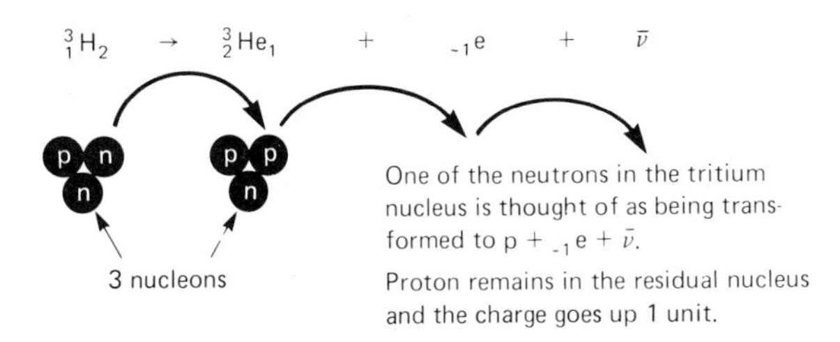

At some moment for $^{11}_6C$, having six protons and five neutrons, a positron and neutrino are ejected. Symbolically and pictorially,

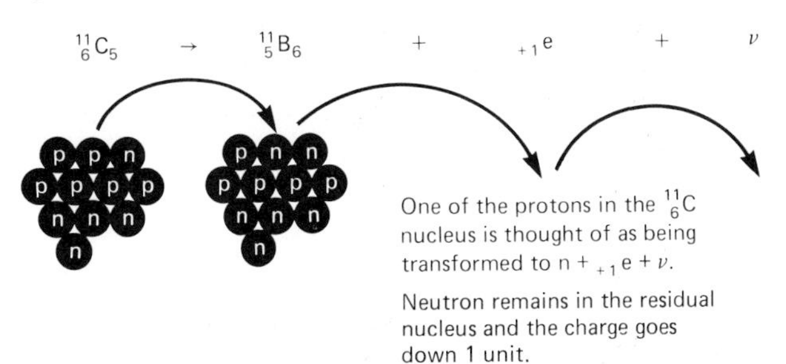

Note that in both cases a new chemical is formed—^3_2He in the beta decay of ^3_1H—and $^{11}_5B$ in the beta decay of $^{11}_6C$. Gaseous ^3_2He is actually produced on a commercial basis by collecting ^3_2He atoms from the beta decay of tritium. In Chapter 12 we shall find that some types of nuclear reactor fuel are actually produced from the end products of beta decay nuclear reactions.

Gamma Radiation

The potential energy of a bowling ball increases when it is lifted from the floor to a table top. If it rolls off, the potential energy decreases; the larger the table height, the larger the decrease. When an electron in an atom is "lifted" to a higher orbit, its potential energy increases. The energy it loses when returning to the lower orbit is converted to electromagnetic energy carried off by a photon having frequency

$$f = \frac{E_{\text{higher}} - E_{\text{lower}}}{h}$$

where h is Planck's constant (Section 8.2). Depending on the energy changes, the radiation may be infrared, visible, ultraviolet, or X-rays. Very similar energy changes occur in nuclei. However, the energy changes and hence the frequency of the photons are some one million times larger than for an atomic process. Photons having energies greater than about 100,000 electron volts are called gamma rays. Infrared radiation, visible radiation, ultraviolet radiation, X-rays, and gamma rays are all members of the family of electromagnetic radiation. In a vacuum, they all travel with the same speed. The types are distinguished by frequency or, equivalently, wavelength. Going from infrared radiation to gamma radiation, the frequency and photon energies increase. Because the wavelength equals the speed of light (3×10^8 meters/second) divided by frequency, the wavelength decreases in going from infrared radiation to gamma radiation.

Nuclear Fission

A nucleus is a reservoir of neutrons and protons. Some nuclei are small reservoirs; others, like the uranium nucleus, are fairly large. Alpha particle emission is a nuclear reaction in which two protons and two neutrons escape as a unit from a nuclear reservoir. All sorts of nuclear reactions involve shuffling the protons and neutrons of the reservoirs of the interacting nuclei. But no perceived rearrangement is possible unless the principles of energy conservation are obeyed.

Nuclear fission reactions play the key energy role in a nuclear reactor. To fission means to split. Nuclear fission involves the splitting of a nucleus. In a nuclear reactor, splitting is instigated by neutrons. A neutron interacting with certain nuclei induces fission. The fissioning nucleus in most nuclear reactors is $^{235}_{92}U$. Symbolically,

$$^{235}_{92}U + n \rightarrow X + Y + N + \text{energy}$$

in which n represents the impinging neutron and N represents neutrons liberated in the reaction. X and Y symbolize the aftermath of the splitting. They represent nuclei formed from the reservoir of 144 neutrons and 92 protons provided by $^{235}_{92}U$ and the impinging neutron. These 144 neutrons and 92 protons are distributed in the reaction products X, Y, and N. One possibility is for X to be ^{143}Ba and for Y to be ^{90}Kr (Ba and Kr are the symbols for barium and krypton). There are many other possibilities but in all cases energy is conserved as are the total numbers of protons and neutrons. The energy liberated comes from the conversion of mass. In accordance with energy conservation, we find the combined masses of the impinging neutron and $^{235}_{92}U$ to be greater than the combined masses of the reaction products. The energy freed from the loss of mass is perhaps ten million times the energy freed in a single chemical reaction. Viewed another way, it takes about ten million chemical reactions to free as much energy as a single nuclear fission reaction can release. This is the basic energy appeal of nuclear fission reactions.

We will say more about nuclear fission reactions when we discuss nuclear reactors in Section 9.6. For the present, understand that

1. neutrons and protons may be rearranged several ways in a nuclear fission reaction but there

is no change in the total number of each after the rearrangement,

2. the total energy remains constant throughout the reaction but we must include mass energy ($E = mc^2$), and
3. there is a decrease in mass in the reaction and a conversion of mass energy to kinetic energy of the reaction products.

When a neutron impinges on a nucleus there is no guarantee that nuclear fission occurs. There is a wide variety of possible nuclear rearrangements of which nuclear fission is but one. The general improbability of inducing fission severely limits the number of possibilities for practical sources of nuclei for fissioning. Nature provides only one—$^{235}_{92}U$.

9.5 HALF-LIVES OF RADIOACTIVE NUCLEI

A pile of tree leaves decays and disintegrates from natural processes, and the end products bear little resemblance to the leaves on a tree. An unstable nucleus emitting a charged particle changes its chemical character and we say the parent nucleus disintegrates. Emission of photons does not produce a change in the chemical character of the parent, but there are generally large energy changes in the process. Knowing the disintegration rate of a collection of unstable nuclei is very important in assessing its potential dangers. If a sample of unstable nuclei disintegrates at a rate of 37 billion disintegrations per second, we say the disintegration rate is one curie, abbreviated Ci. Normal living exposes us all to radioactivity from building materials and cosmic rays (Section 9.12). These sources are likely to have strengths in trillionths of curies. We call a trillionth of a curie a picocurie, abbreviated pCi. A picocurie source corresponds to about two disintegrations per minute.

Tritium ($^{3}_{1}H$) and $^{14}_{6}C$ are both unstable and disintegrate by emitting negative beta particles. If you had collections of equal numbers of tritium atoms and $^{14}_{6}C$ atoms, you might expect the disintegration rates of the two samples to be identical. Interestingly, they are not. The tritium disintegration rate is initially nearly 500 times larger than the $^{14}_{6}C$ disintegration rate. For an analogy, visualize one pan containing popcorn and another pan having the same number of kernels of popcorn but of inferior grade. When popped with the same heat source, the popping rates are very different even though both samples are popcorn and eventually all kernels pop. Tritium and $^{14}_{6}C$ are both radioactive nuclei but they are structured very differently. Hence, their radioactive features are very different and we should not expect samples of them to disintegrate at identical rates. With a pan of popcorn we cannot say when a particular kernel is going to pop. Similarly we cannot say when a particular radioactive nucleus is going to disintegrate. However, we can determine how many radioactive nuclei will disintegrate in a specified time period and how the disintegration rate changes as time passes. When kernels of popcorn explode in a pan, it is easy to perceive from the sound that the rate of popping decreases as time progresses, and eventually, all the kernels pop. So it is with the disintegration rate of a collection of unstable nuclei. The time for a 50% decrease in the disintegration rate is called the half-life. If a radioactive material started disintegrating at a rate of 500 per second and the rate fell to 250 per second in five minutes, we record its half-life as five minutes. After five more minutes the rate drops to 125 per second. Every five-minute interval witnesses a 50% drop in the disintegration rate. This halving at regular intervals is called an exponential decrease.

It is a rather simple laboratory exercise to measure the half-life of some radioactive materials. The experiment (Fig. 9.5) requires a clock, an appropriate detector, and a way of recording the number of disintegrations. The disintegration rate is the number of particles detected divided by the time interval for the measurement.

$$R = \frac{N}{t}.$$

Atoms of barium-137 emitting gamma rays are a convenient radioactive source. Table 9.1 contains data recorded in a student laboratory exercise. Figure 9.6 graphically displays disintegration rate and time of the measurement. In a time of 2.6 minutes,

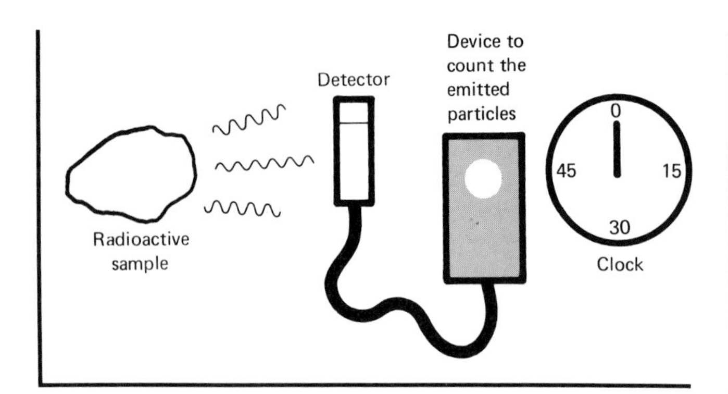

FIGURE 9.5 An elementary setup for measuring the decay rate of a radioactive sample. One records the number of particles or photons interacting with the detector in a measured time interval. The measurement is repeated for several points in time.

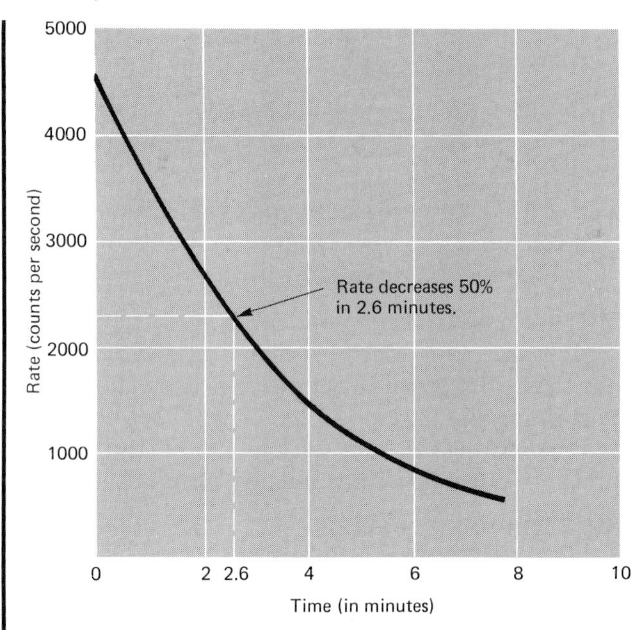

FIGURE 9.6 Plot of the ^{137}Ba decay rate data shown in Table 9.1. Initially, the decay rate was 4617 decays/sec. After a lapse of 2.6 minutes the decay rate decreased to 2300 decays/sec—a decrease of 50%. After a lapse of 5.2 minutes the decay rate decreased to 1190 decays/sec—a decrease of 50% more. Regardless of what time you pinpoint, the decay rate will be 50% less 2.6 minutes after that time.

the disintegration rate drops by 50%. Accordingly, the half-life is 2.6 minutes. Note that we could pick any rate, for example, 3000 per second, and the rate drops by 50% in a time equal to 2.6 minutes.

Experimentally, half-lives are determined from measurements of disintegration rates. Theoretically, the half-life may be discussed in terms of the disintegration of numbers of nuclei. If there were 500,000 unstable nuclei in a sample and 250,000 of them disintegrated in 13 days, we record the half-life as 13 days. Another 13 days hence half of the 250,000 disintegrate and there remain 125,000 unstable nuclei waiting to disintegrate. A generalized graph illustrating this idea is shown in Fig. 9.7.

No two isotopes are structured the same. Consequently, no two unstable isotopes have identical half-lives. Not only do the half-lives vary but they vary enormously. Some are so short that extremely sophisticated techniques are required for measurement. The unstable nucleus $^{8}_{4}Be$ has a half-life of about a tenth of a millionth of a billionth (10^{-16}) of a second. $^{10}_{4}Be$ has just two more neutrons than $^{8}_{4}Be$, but it has a half-life of 2.7 million years. The half-lives of some unstable nuclei of interest in nuclear power plants are shown in Table 9.2.

TABLE 9.1 Data for the decay rate of ^{137}Ba.

Time (minutes)	Particles detected/second
0	4617
0.6	3963
1.2	3332
1.8	2904
2.4	2542
3.0	2114
3.6	1820
4.2	1541
4.8	1279
5.4	1151
6.0	960
6.6	789
7.2	728
7.8	603

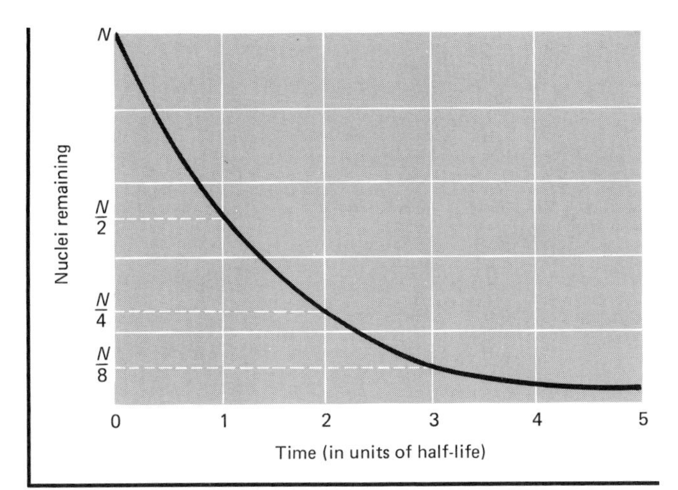

FIGURE 9.7 A generalized plot for radioactive decay. *N* denotes the number of radioactive nuclei at the beginning. *N* was one million in the example given in the text If one starts out with *N* radioactive nuclei, then after one half-life (designated 1) has elapsed, there remains *N*/2 nuclei that have not decayed. After two half-lives (designated 2) there will remain *N*/4 nuclei, and so on.

9.6 NUCLEAR REACTORS

General Features

Although we never see the energy-producing fire in the cylinders of an automobile engine, it is not difficult for us to envision it because we see fires in all manners of circmstances. Unlike visualizing the fire in an automobile engine, it is very difficult to envision the energy-producing reactions in a nuclear reactor because we have no basis for comparison. For all its mystery, a nuclear reactor is a source of heat, and in an electric power plant the heat from a nuclear reactor vaporizes water to make steam for a turbine that powers an electric generator (Fig. 9.8). Let us draw on our experience with coal-burning electric power plants to try to remove some of the mystery about a nuclear reactor in a nuclear power plant.

A campfire requires a fuel, usually wood, a containment for the fuel, an appropriate arrangement of the fuel, and a supply of oxygen. The fire is controlled either by regulating the amount of fuel in the fire or by regulating the oxygen supply, or both. Adding fuel increases the energy output and depriving the fire of oxygen by covering it with dirt decreases the output. The same ideas pertain to a nuclear reactor. The fuel is generally $^{235}_{92}U$ atoms arranged appropriately in a reactor vessel. Neutrons interacting with and fissioning the nuclei of $^{235}_{92}U$ atoms liberate energy. The energy output may be controlled either by regulating the fuel, or by regulating the neutron supply, or both. Generally, the neutron supply is regulated.

Nuclear reactors providing sources of neutrons for research are found on many campuses. The heat generated by these reactors is usually more than 100,000 times less than that produced in a nuclear power plant. These research reactors are often submerged in an open pool of pure water that looks much like a swimming pool. There is little to see when viewing the pool. All you can see is a neat, somewhat cubical arrangement of fuel, and cylindrical control rods that can be moved up and down in strategically arranged spaces in the fuel assembly. To begin the production of energy and the liberation of neutrons from nuclear fission reactions, the reactor operator arranges to pull certain rods out of the fuel package. The same principles and materials are used in much larger reactors in nuclear power plants. But in nuclear power reactors, the goal is to produce and contain energy. Therefore the entire operation is housed in a containment vessel that is concealed from view. Any reactor, large or

TABLE 9.2 Half-lives of some nuclei of concern in nuclear power plants.

Nucleus		Half-life
^{3}H	(tritium)	12.26 years
^{90}Sr	(strontium)	28.8 years
^{137}Cs	(cesium)	30.2 years
^{131}I	(iodine)	8.05 days
^{85}Kr	(krypton)	10.76 years
^{133}Xe	(xenon)	5.27 days

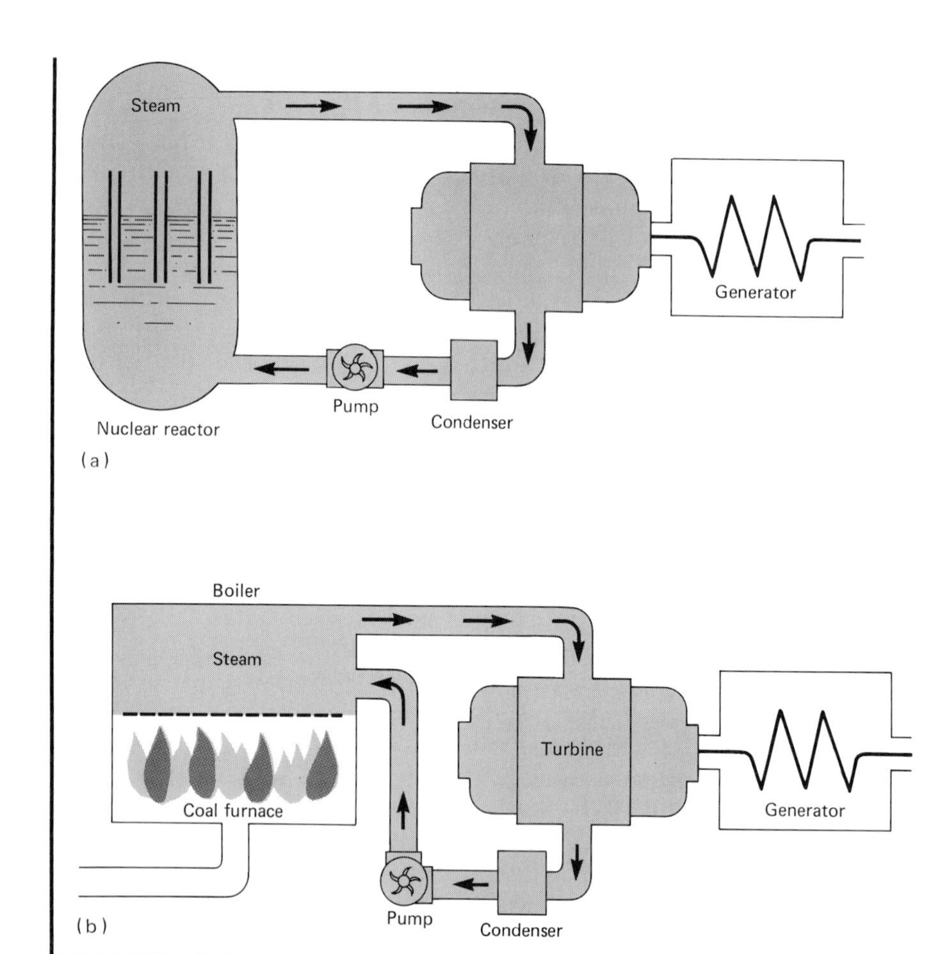

FIGURE 9.8 Schematic illustrations of **(a)** the nuclear-fuel and **(b)** fossil-fuel electric generating systems. Both use a steam turbine to provide power for an electric generator. Heat for vaporizing the water is provided by a nuclear reactor in the nuclear system and by a flame boiler in the fossil-fuel system.

small, has four aspects that we must understand.

1. the role of the fuel,
2. the role of neutrons in producing energy,
3. the role of the water surrounding the fuel, and
4. the function of the control rods moving amongst the fuel.

First let us examine the principles and then the practicalities of nuclear reactors.

Principles

A chain reaction is a situation in which an event triggers a sequence of identical events. If a student at the rear of a line of students in a cafeteria pushes the person in front of her and that person in turn pushes the student in front of him, and so on, a chain reaction develops. If a student in a crowd were to push two students and thereafter each pushed student in turn pushed two more, an expanding chain reaction results. This chain reaction idea is central in the operation of a nuclear reactor. It all starts with a neutron fissioning a nucleus. Neutrons liberated in the fission in turn fission other nuclei, and so on (Fig. 9.9). Each fission releases energy and unless there is some purposeful control, the energy release proceeds uncontrolled. Whether controlled or uncontrolled the chain reaction does not propa-

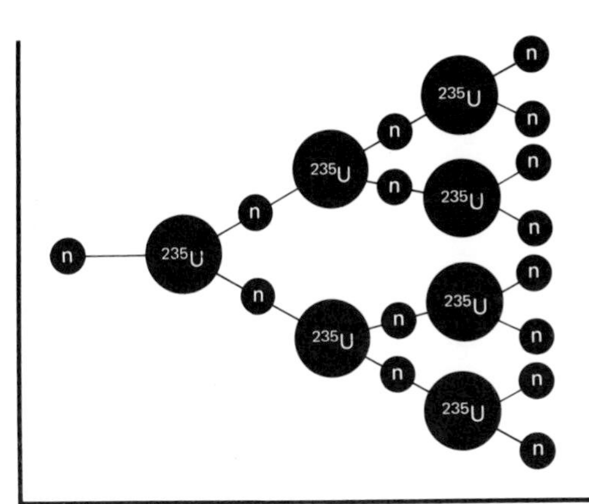

FIGURE 9.9 Schematic illustration of a nuclear chain reaction. The neutrons released when $^{235}_{92}U$ fissions as a result of interacting with a neutron are used to instigate other fission reactions and keep the process going.

gate unless

1. the fuel is properly arranged and
2. there is sufficient likelihood that a neutron induces fission.

This is just like saying that a chain reaction in a cafeteria line does not propagate unless

1. the students are arranged properly and
2. there is sufficient likelihood that students will push others in front of them.

Only three neutrons differentiate $^{235}_{92}U$ from $^{238}_{92}U$ and yet the likelihood of neutron-induced fission of $^{238}_{92}U$ is so low that $^{238}_{92}U$ cannot be used as a reactor fuel. With sufficient ingenuity, $^{235}_{92}U$ can be used as a reactor fuel. $^{235}_{92}U$ fissions most readily when the speed of the impinging neutrons is low. By low speed, we mean something like the speed of a molecule in the air of the room you are in. Neutrons produced in nuclear fission reactions are traveling very, very fast. Thus to optimize the chain reaction, the speed of reaction neutrons must be reduced. Water surrounding the fuel helps slow the neutrons. Any moving object loses speed in a collision with another object initially at rest. The loss in speed is greatest if the masses are the same. This is easily demonstrated by sliding a penny into another penny on a smooth surface. Protons are good targets for neutrons, and water is a good source of protons because each water molecule has two hydrogen atoms, and each atom has a single proton in its nucleus. Thus a neutron produced in a nuclear fission reaction bounces around in a pool of water losing energy with each collision with a proton. After its energy is reduced sufficiently, or moderated, it may react with a $^{235}_{92}U$ nucleus to induce fission.

Control rods, as their name suggests, control the chain reactions. They are made of materials that are more likely than the fuel to absorb neutrons. Generally, the materials are boron and/or cadmium. Neutrons in the vicinity of the control rods and the fuel are more likely to be captured by the rods thereby depriving the fuel of neutrons needed to induce fission. To decrease the energy output, the control rods are inserted closer to the fuel. To increase the energy output, the control rods are pulled away from the fuel. Boric acid, H_3BO_3, a boron-containing liquid that strongly absorbs neutrons, is often included in the water moderator to provide further control of the energy output of the reactor.

To summarize the four reactor principles:

1. the fuel is both an energy and a neutron source,
2. neutrons induce nuclear fission reactions in the fuel,
3. water surrounding the fuel moderates neutron speeds, and
4. control rods regulate the energy output of the reactor (Fig. 9.10).

Practicalities

The path to electricity from a nuclear power plant begins with uranium atoms embedded in the earth's crust. Uranium, like iron, is a metal. It is found naturally as a constituent of chemical compounds in minerals such as pitchblende, which is a very important uranium ore. Uranium ore is mined much like coal; open pits are used to mine shallow deposits, shaft mining is used to extract deeper deposits. Commercial ores yield 3–5 pounds of uranium compounds per ton of ore. To avoid transportation of

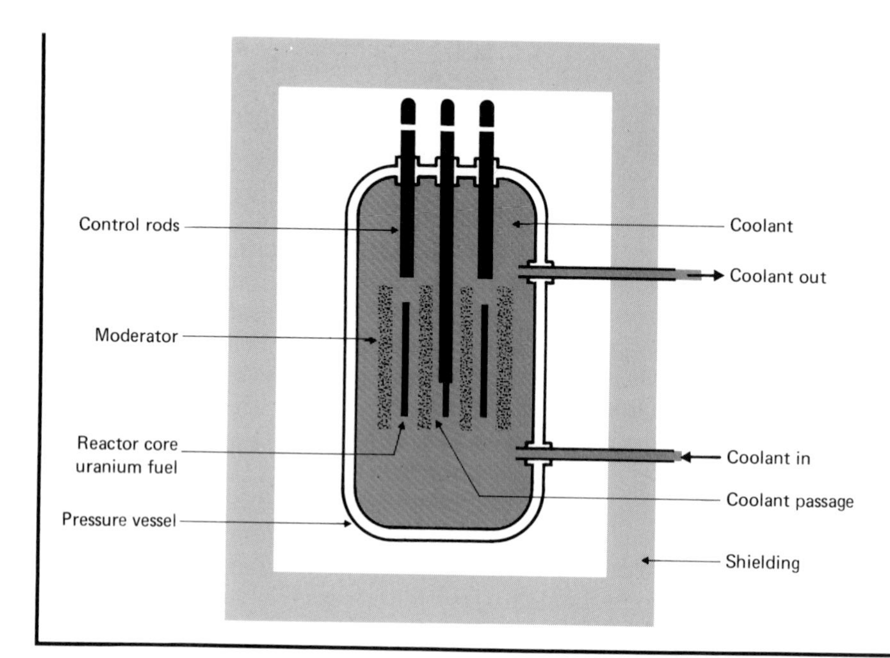

FIGURE 9.10 Schematic cutout view of a nuclear reactor.

large amounts of material having a small amount of uranium, the ore is refined near the mines to produce a material called yellowcake that is rich in the uranium compound U_3O_8. About seven of every 1000 uranium atoms in the yellowcake is of the reactor variety, $^{235}_{92}U$.

Some types of reactors (Section 9.9) use this natural uranium that has 0.7% $^{235}_{92}U$ and 99.3% $^{238}_{92}U$, but United States reactors that use water moderators require the proportions be changed to about 3% $^{235}_{92}U$ and about 97% $^{238}_{92}U$. To effect this enrichment, the solid uranium compounds in the yellow-cake are converted to gaseous uranium hexafluoride UF_6 (Section 9.7). Following enrichment, gaseous UF_6 is converted to solid uranium oxide (UO_2) for fabrication of fuel elements for a nuclear reactor. The physics of enrichment is covered in Section 9.7.

Uranium oxide fuel is formed into solid pellets about $\frac{3}{8}$ inches in diameter and $\frac{1}{2}$ inch long (Fig. 9.11). The fuel pellets are stacked in tubes about 12 feet long. Neutrons must be able to penetrate the tube walls, called the cladding, to interact with the uranium, and the walls must withstand the temperature required for heating water for a steam turbine. Stainless steel or a material called zircaloy is used for the cladding. Some 200 tubes are packaged into a fuel bundle. About 175 appropriately arranged fuel bundles form the reactor core where the energy is produced. The core is housed in a cylindrical steel vessel about 20 feet in diameter and about 40 feet high. Energy and neutrons are liberated from nuclear fission reactions within the fuel pellet. About 83% of the energy is in kinetic energy of the fission fragments. The fuel pellets warm when the fission fragments lose energy through collisions with atoms in the fuel element. Heat is conducted from the pellets by water circulating around the fuel cladding. Thus water has two very important roles—conducting heat from the reactor core and moderating neutron speeds.

Imagine an assembled reactor core with water surrounding the core and the control rods in their lowest positions. There are about 100 tons of uranium in the fuel bundles but there is no chain reaction, no energy produced, and no neutrons liberated from nuclear fission reactions. Energy production starts by raising the control rods and inserting a

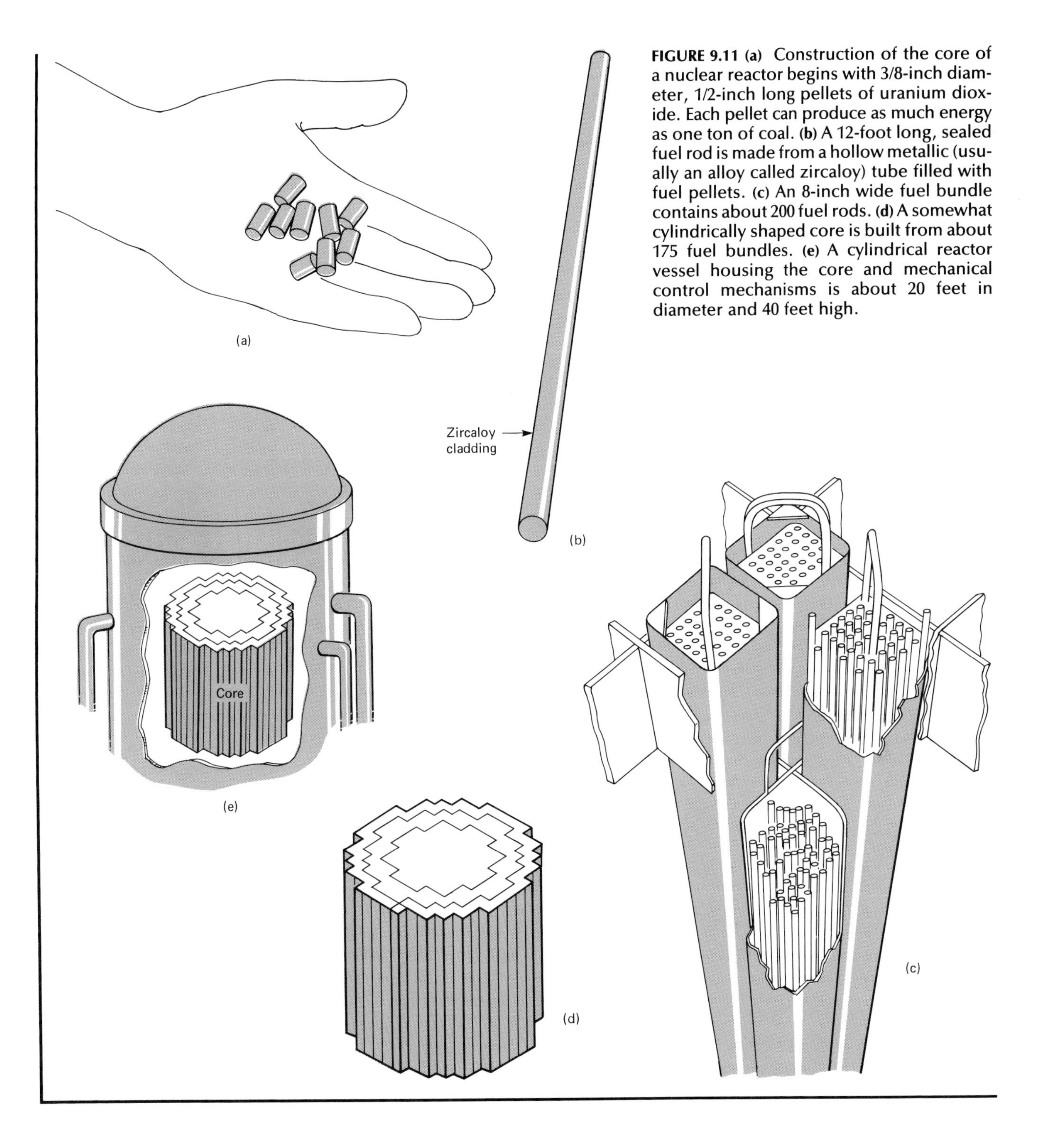

FIGURE 9.11 (a) Construction of the core of a nuclear reactor begins with 3/8-inch diameter, 1/2-inch long pellets of uranium dioxide. Each pellet can produce as much energy as one ton of coal. **(b)** A 12-foot long, sealed fuel rod is made from a hollow metallic (usually an alloy called zircaloy) tube filled with fuel pellets. **(c)** An 8-inch wide fuel bundle contains about 200 fuel rods. **(d)** A somewhat cylindrically shaped core is built from about 175 fuel bundles. **(e)** A cylindrical reactor vessel housing the core and mechanical control mechanisms is about 20 feet in diameter and 40 feet high.

source of neutrons. Nuclear fission reactions begin liberating energy and neutrons. Neutrons escape from the fuel elements into the water where they lose speed and change direction with each collision with a proton. After losing sufficient speed, a neutron may reenter a fuel element and induce a $^{235}_{92}U$ nucleus to fission to help propagate the chain reaction. If on the average one neutron from a nuclear fission reaction goes on to produce another nuclear fission reaction, the chain reaction is self-sustaining, and there is a steady rate of energy production. Raising the control rods further allows more nuclear fission reactions and more energy production. Higher, but steady, energy production occurs at the new control rod position.

Realize that there is an atmosphere of neutrons in the confines of the reactor vessel. Like molecules of air in your room, neutrons are moving and colliding with atoms comprising the interior of the reactor vessel. Clearly, the desired controlled chain reaction depends on

1. the arrangement of the fuel elements,
2. how well the moderator functions,
3. the amount of fissionable $^{235}_{92}U$,
4. the neutron energy required for a high probability for the fission reactions, and
5. the shielding material surrounding the reactor core used to minimize the escape of neutrons.

Although protons are very efficient neutron moderators, they also efficiently capture neutrons to form bound proton–neutron pairs called deuterons. To offset this, reactors using ordinary water for the moderator use fuel in which the proportion of $^{235}_{92}U$ is increased from its natural abundance of 0.7% to about 3.0% in order to compensate for the neutrons lost in the deuteron production. This process of increasing the $^{235}_{92}U$ concentration is called enrichment. Enrichment is a very sophisticated bit of technology.

9.7 FUEL ENRICHMENT

Commercial schemes for enriching the $^{235}_{92}U$ component in uranium take advantage of the slight mass difference of $^{235}_{92}U$ and $^{238}_{92}U$. The oldest and most widely used method is called gaseous diffusion. The principle is straightforward. The average kinetic energy of the molecules in the room you are in depends only on the temperature of the room. To have the same kinetic energy ($\frac{1}{2}mv^2$), a light molecule must have more speed than a heavy molecule. Accordingly, a light molecule makes more collisions with the walls of the room than a heavy molecule because, on the average, it is traveling faster. If both light and heavy molecules have the same chance of penetrating the walls, then the light ones are more likely to succeed because they have more opportunities. Gaseous uranium hexafluoride (UF_6) having both $^{235}_{92}U$ and $^{238}_{92}U$ atoms is confined in a container with one semiporous wall. The relative concentration of the lighter $^{235}_{92}U$ atoms versus $^{238}_{92}U$ atoms in the gas making it through the semiporous wall is higher than the concentration of $^{238}_{92}U$ atoms. Thousands of passes through semiporous walls are required to achieve the desired 3% $^{235}_{92}U$ concentration. United States gaseous diffusion plants are located at Oak Ridge, Tennessee; Portsmouth, Ohio; and Paducah, Kentucky.

Another separation scheme exploiting the slight mass difference of $^{235}_{92}U$ and $^{238}_{92}U$ is called the centrifuge process. The centrifuge operates in principle like the cyclone particulate separator discussed in Chapter 4. A closed cylindrical container of a gas having molecules of varying mass is set into rapid rotational motion. Molecules move away from the center of the cylinder and the gas compresses near the walls of the container. Thermal motion of the molecules tends to keep the molecules uniformly distributed throughout the volume occupied by the gas. The gas pressure is not sufficient to change the direction of the velocity of the heavier molecules so they tend to accumulate at the walls. A "gas-scooping" device is used to collect the more massive molecules. As with gaseous diffusion the separation of $^{235}_{92}U$ from $^{238}_{92}U$ begins by converting uranium to a gas composed of uranium hexafluoride (UF_6) molecules. Those molecules having a $^{238}_{92}U$ atom are slightly more massive than those containing a $^{235}_{92}U$ atom. Three pilot centrifuge separation plants are operated by a consortium consisting of the United Kingdom, West Germany, and the Netherlands. Expansion of these centrifuge separation facilities is underway. A demonstration plant in the United States is operating at the Oak Ridge National Laboratory in Tennessee.

Fuel enrichment by gaseous diffusion and centrifuge technologies requires substantial amounts of energy and equipment. A technology in the development stages requiring smaller plants and much less energy is termed laser isotope separation. Radiation from lasers induces chemical reactions involving uranium atoms. Because of slight differences in the atomic properties of $^{235}_{92}U$ and $^{238}_{92}U$, laser-induced chemical reactions can differentiate between the two species of uranium. Reaction products containing $^{235}_{92}U$ atoms can then be collected and used as raw material for reactor fuel elements.

9.8 BOILING WATER AND PRESSURIZED WATER REACTORS

Nearly all United States reactors use water to moderate neutrons and to remove heat. But there are two basic designs labeled BWR (for boiling water reactor) and PWR (for pressurized water reactor). Schematic diagrams of these two reactor types are shown in Figs. 9.12 and 9.13. Water comes to a boil producing steam inside the containment vessel of a boiling water reactor. Water is kept under pressure and does not boil inside the containment vessel of a pressurized water reactor. Keeping the water from boiling allows higher temperatures in the reactor vessel. Steam for the turbine in a nuclear power plant using a pressurized water reactor comes from water brought to a boil in a steam generator garnering heat from water circulating around the reactor core. Steam for a turbine in a nuclear power plant using a boiling water reactor comes from the reactor vessel. But in a system using a pressurized water reactor, the steam comes from a steam generator physically separated from the reactor vessel. The pressure and temperature in a boiling water reactor are about 1000 pounds per square inch and 545° F. These numbers increase to about 2250 pounds per square inch and 600° F in a pressurized water reactor.

9.9 HTGR AND CANDU REACTORS

Two other reactor designs are labeled HTGR for High Temperature Gas-cooled Reactor and CANDU for CANadian Deuterium Uranium. The HTGR uses solid graphite (carbon) for a neutron moderator and

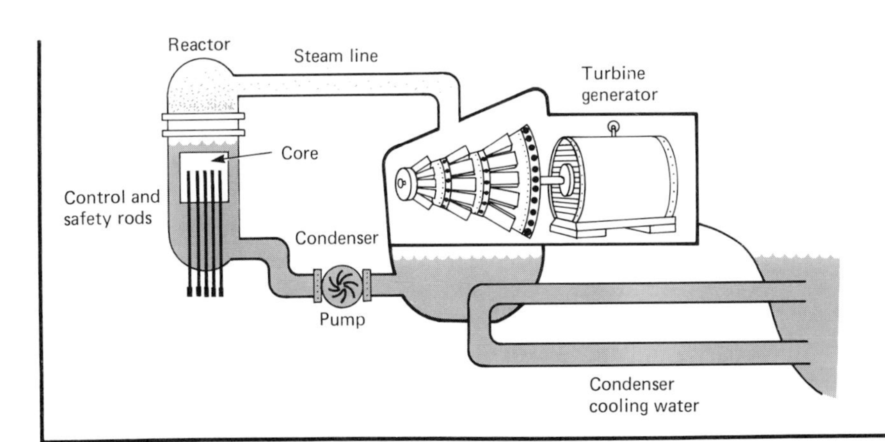

FIGURE 9.12 The water-steam circuit in a boiling water reactor. Water is vaporized in the reactor and the steam is channeled directly to the turbine. Water condensed after passing through the turbine is pumped back into the reactor. The cycle is closed and neither the steam nor the water come into contact with anything exterior to the system. The condenser cooling water never comes into contact with the water circulating in the reactor.

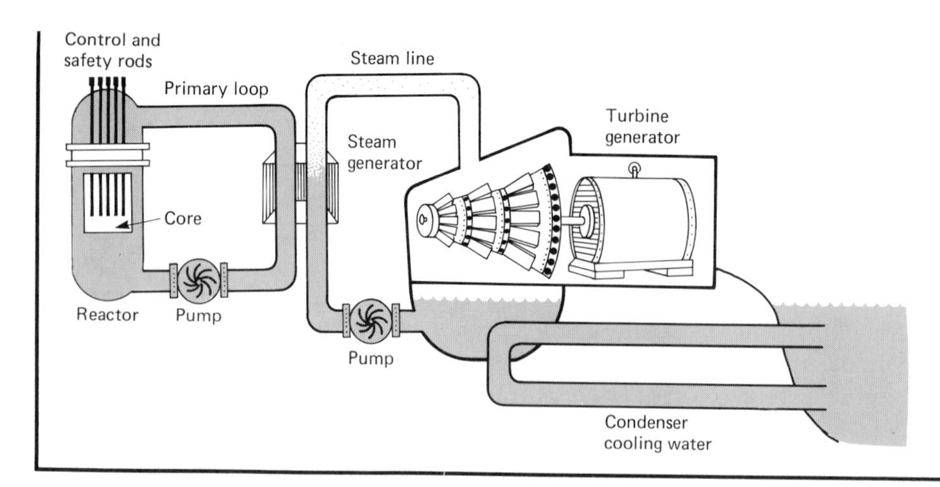

FIGURE 9.13 The water-steam circuit in a pressurized water reactor. Water circulating through the reactor in the primary loop is kept under pressure and never comes to a boil. Heat is transferred from this loop to a steam generator that provides steam for the turbine. Water in the primary loop never comes into direct contact with water in the steam turbine circuit. In principle, the steam turbine circuit is identical to that in a boiling water reactor.

helium gas to remove heat from the reactor. Uranium fuel enriched to 93% ${}^{235}_{92}U$ is dispersed in the solid graphite moderator. Control rods moving among the fuel regulate the power output of the reactor. Because temperatures of about 1430° F can be achieved with a HTGR, the thermodynamic efficiency of the steam turbine is higher than in boiling water and pressurized water reactors. The Colorado Public Service operates a HTGR at the Fort Saint Vrain Nuclear Generating Station about 35 miles north of Denver.

Deuterium is a hydrogen isotope having a proton and a neutron in its nucleus. Unlike ordinary hydrogen which has a single proton as its nucleus, deuterium does not readily capture neutrons. Water made with deuterium atoms is called heavy water. Heavy water is a poorer neutron moderator than ordinary water but if heavy water is used, then natural uranium can be used for fuel thus avoiding the enrichment process. This is the ploy in the CANDU reactor.

9.10 NUCLEAR WASTES

When a new nuclear reactor is fueled, the only radioactivity in the nuclear power plant comes from the uranium fuel. This radioactivity is relatively small because the half-lives of both ${}^{235}_{92}U$ and ${}^{238}_{92}U$ are very long. Radioactive wastes evolve once the reactor is set into operation.

These wastes are broadly categorized as low level and high level. The first comes from routine operation of the reactor and fuel processing plants. Contributions come from such things as leaks in the cladding of the fuel elements and irradiation of the coolant and air by neutrons. This radioactive material is collected by various filtering systems and released to the environment under controlled conditions. Gaseous radioactive wastes are normally vented to the atmosphere. Solid and liquid low-level wastes are buried. Disposal is done within a framework of regulations administered by the Nuclear Regulatory Commission (NRC). Disposal of the low-level wastes is considered a low-hazard process.

High-level wastes come from the spent fuel. There are two primary sources. One is from radioactive fission fragments. The other, called actinides, is due to radioactive nuclei produced from the bombardment of ${}^{235}_{92}U$ and ${}^{238}_{92}U$. For example, plutonium-239 (${}^{239}_{94}Pu$) with a half-life of 24,390 years can be made by bombarding ${}^{238}_{92}U$ with neutrons (see Section 12.2). There are about 35 fission fragments and 18 actinides in the spent fuel element. Some

TABLE 9.3 Typical radioactive fission products and actinides in high-level radioactive wastes. Half-lives were taken from *Nuclear Data Sheets*, published by Academic Press, Inc., Orlando, Fla.

Fission fragments			Actinides		
Isotope	*Half-life (years)*	*Radiation*	*Isotope*	*Half-life (years)*	*Radiation*
Strontium-90	28.9	beta	Radium-226	1600	alpha, gamma
Zirconium-93	950,000	beta	Thorium-229	7340	alpha, gamma
Technetium-99	210,000	beta, gamma	Plutonium-238	87.8	alpha, gamma
Cesium-135	2,300,000	beta	Plutonium-239	24,390	alpha, gamma
Cesium-137	30.1	beta, gamma	Americium-241	433	alpha, gamma
			Curium-244	17.9	alpha, gamma
			Curium-245	8700	alpha, gamma

important ones from the hazard standpoint are presented in Table 9.3. These wastes are analogous to the hot ashes left over in a coal-burning stove or furnace. Hot ashes can be cooled by dousing with water. Radioactive "hot ashes" cannot be cooled by any external process. They rid themselves of the excess energy on a time scale determined by the half-lives of the nuclear constituents.

Isotopes with half-lives less than a few months long disappear rapidly on a human time scale so they do not figure into long-term radioactive waste disposal concerns. Isotopes with very long half-lives, say greater than a million years, emit their radiation at a negligible rate and so they are not of great concern. It is the isotopes with half-lives somewhere between these broad limits that must be reckoned with in radioactive waste disposal schemes. The levels of radioactivity involved are so large that hundreds of years must elapse before the leftover radioactive by-products can be considered harmless. The long-term stability required of an appropriate disposal system poses a unique engineering and social problem. Several alternatives have been proposed. We will examine these options in Section 9.13.

9.11 REPROCESSING NUCLEAR WASTES

Each fission of a $^{235}_{92}U$ nucleus in a fuel pellet produces a pair of nuclei chemically different from uranium. As time progresses, a variety of chemical species accumulates in the pellets. Some of the nuclei compete with $^{235}_{92}U$ for neutrons. So severe is the competition that about one-fourth of the fuel elements must be replaced annually leaving substantial amounts of $^{235}_{92}U$ in the spent fuel elements. Accompanying the valuable $^{235}_{92}U$ in the spent fuel is fissionable $^{239}_{94}Pu$ formed from nuclear reactions of neutrons with $^{238}_{92}U$. Recovery of fissionable $^{235}_{92}U$ and $^{239}_{94}Pu$ from the spent fuel for use in new fuel elements is termed reprocessing.

Reprocessing of spent fuel is controversial. Commercial attempts at reprocessing spent fuel from nuclear power plants have been fraught with technological problems. Additionally, $^{235}_{92}U$ and $^{239}_{94}Pu$ are used for nuclear bombs and there is concern for diversion of these recovered materials, especially $^{239}_{94}Pu$, for criminal purposes. Consequently, commercial reprocessing in the United States has waned for reasons both technological and political. When, or if, commercial reprocessing will be reimplemented is an open question. But if nuclear breeder reactors (Chapter 12) are developed, then reprocessing is essential. In places such as France, in which the nuclear breeder concept has been adopted, reprocessing is done as a matter of necessity.

9.12 BIOLOGICAL DAMAGE DUE TO RADIATION

Fundamental Mechanisms for Biological Damage

Many of our discussions used the notion of a bound system built from some basic constituent(s). Human tissue can be viewed within this same framework

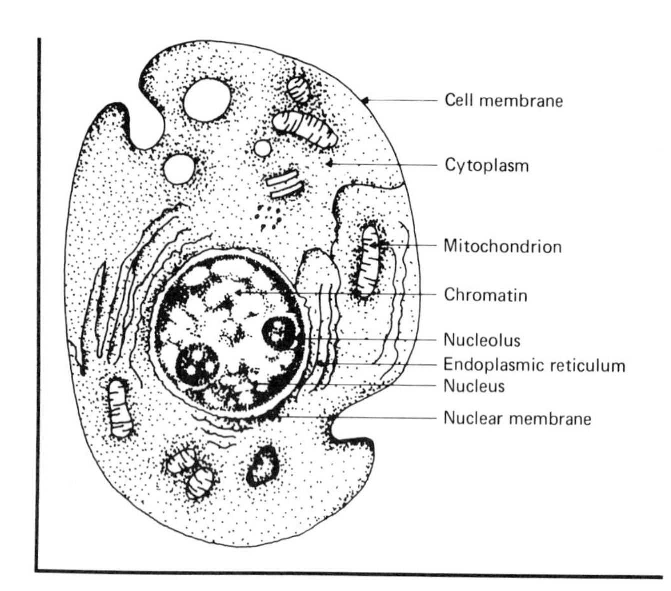

FIGURE 9.14 Schematic depiction of a cell showing its main components.

with cells as the basic units (Fig. 9.14). Cells are responsible for both metabolic processes (movement, respiration, growth, and reaction to environmental changes) and reproduction. Cells controlling metabolism are termed somatic; those controlling reproduction are termed genetic. Control of these cell functions is exercised by the nucleus of the cell. Cells reproduce in tissue through a process of division called mitosis. The hereditary aspects of life are passed on from the parent cell to the divided ones (daughters) in chromosomes that are rich in a substance called deoxyribonucleic acid (DNA). The growth of a cell and its reproduction are governed by biochemical reactions involving many complex molecules. Damage to the molecules involved in these processes *may* alter the function of a cell to the extent that

1. it dies outright and is removed from life processes,
2. it is damaged so that it cannot reproduce and therefore it eventually dies, or
3. it is still able to divide but the functioning of the new cells is altered (mutated).

Some mutations may be inconsequential. However, some can produce somatic effects such as uncontrolled cell reproduction resulting in cancerous growths. Mutations in genetic cells can be passed on to succeeding generations.

Any mechanism that can supply the energy necessary to break a chemical bond, ionize an atom, or alter the chemistry of cells is capable of producing biological damage. Roughly it takes 10–30 electron volts to damage a molecule. The alpha, beta, gamma, and neutron radiations from radioactive nuclei typically have energies in the range of millions of electron volts. Thus they are capable of damaging hundreds of thousands of cells. The actual damage produced depends on the physics of the interaction that, in turn, depends on the type and energy of the particle involved.

A particle loses energy by forces acting on it through some distance (work). The energy the particle loses is transformed (or converted) to other forms. For example, the brakes of an automobile exert a force on the wheels that decelerates the automobile. The kinetic energy of the car is converted into heat. A charged particle, like a beta or an alpha particle, loses its energy primarily through the electric interaction with the charged particles associated with atoms. Atoms may absorb energy and be excited to higher energy states or, if enough energy is transferred, electrons may be completely removed from the atoms. The rate of energy loss by a particle depends on the energy, charge, and mass of the radioactive particle. An alpha particle, being about 7000 times as massive and having twice the charge of an electron, loses energy at a much greater rate. For example, a one million electron volt alpha particle cannot penetrate a sheet of paper, but it takes many sheets of paper to stop a one million electron volt beta particle. Because an alpha particle deposits most of its energy in a small region of space, it produces much more local damage than a beta particle of the same energy.

A neutron, by virtue of having no charge, is very difficult to stop. It loses energy by collisions with nuclei the same way billiard balls lose energy in collisions. If a neutron makes a collision and loses its energy, it can do considerable damage. Because a neutron does not lose energy continuously as does a charged particle, it is not meaningful to talk about a thickness required to stop a neutron of given energy. Rather one talks about a thickness required to reduce the intensity of a neutron source by some given amount. Concrete is often used as a material to attenuate neutrons. It takes ten inches of concrete to reduce the intensity of a 10 million electron volt neutron source by 90%.

A gamma ray has no mass and no charge, but it does have electromagnetic energy and experiences electromagnetic forces when traversing matter. A gamma ray loses energy by three mechanisms:

1. It may scatter from an electron and impart some energy to it. This is called the Compton effect.
2. It may be absorbed by an atom whereupon energy is relinquished to one of the electrons surrounding the nucleus. This is called the photoelectric effect.
3. It may interact with an atom and create an electron ($_{-1}e$) and a positron ($_{+1}e$). This is called pair production. Pair production requires a minimum of 1.022 million electron volts of energy to compensate for the masses of the electron and positron.

Like neutrons, gamma rays do not lose energy continuously and again we talk about a thickness required to reduce the intensity of a beam of gamma rays. Lead is often used to attenuate a source of gamma rays. It takes about four centimeters of lead to reduce the intensity of a beam of 10 million electron volt gamma rays by 90%.

Radiation Units

A nine-ounce baseball and a nine-ounce chunk of glass moving 60 miles per hour have the same kinetic energies. To a physician, who has to repair the damage inflicted on a person having the misfortune to intercept these particular objects, they produce quite different effects. The situation for nuclear radiation is quite similar. Different effects are produced by different types of radiation.

For some situations, we need to know only how much energy is deposited in some type of matter. This is referred to as a *dose*. The earliest unit for quantifying this concept was the roentgen (abbreviated R). It is a useful measure for X-rays and gamma rays but is not a unit appropriate for radiations in general. A more appropriate unit is the rad, meaning *radiation absorbed dose*. One rad is defined as 0.00001 joules of energy absorbed per gram of substance. When the substance is human tissue, the rad and roentgen are essentially the same and the terms are often used interchangeably. The rem, meaning *roentgen equivalent man,* is a biological unit accounting for the biological damage produced. It is much less precise than a physical unit like a rad because of the many factors entering into biological damage. The energy of the particle is a major consideration. For example, both visible light and gamma rays are electromagnetic radiation. However, a photon of visible light lacks a sufficient amount of energy to be biologically dangerous. Intense local tissue damage can be more serious than diffuse damage. For this reason, an energetic fission fragment that travels a very short distance in tissue is more dangerous than 250-keV X-ray photons, that distribute their energy over a much longer distance. Also, some body organs are much more susceptible to radiation damage than others. Because the amount of radiation damage is directly proportional to the dose received, the rem and rad are related to each other. But, because some radiation is more effective than others in producing biological damage, the proportionality factor, called the relative biological effectiveness (RBE), depends on the type and energy of the particle. Stated formally,

$$\text{dose equivalent in rems} = \text{RBE times dose in rads},$$
$$DE = \text{RBE} \cdot D.$$

It is customary to use the damage done by a whole body irradiation of 250-keV X-rays as the norm and assign an RBE of unity to this radiation. The RBE for any other particle is then measured relative to this standard. Table 9.4 lists some representative RBEs for various radiations of interest.

TABLE 9.4 Some representative RBEs.

Radiation	Biological effect	Approximate RBE
X-rays, gamma rays, and beta rays (photons and electrons) of all energies above 50 keV	Whole-body irradiation, hematopoietic system critical	1
Photons and electrons, 10–50 keV	Whole-body irradiation, hematopoietic system critical	2
Photons and electrons below 10 keV, low-energy neutrons and protons	Whole body irradiation, outer surface critical	5
Fast neutrons and protons, 0.5–10 MeV	Whole-body irradiation, cataracts critical	10
Natural alpha particles	Cancer induction	10
Heavy nuclei, fission particles	Cataract formation	20

Background Radiation

Knowing that radiation can produce very undesirable biological damage, it is only natural to seek protection by avoiding radiation exposure. It is, however, impossible to avoid all radiation because everyone is routinely exposed to a variety of natural and manufactured radiation sources. As inhabitants of the earth we are continually bombarded with nuclear radiations coming from unstable nuclei in the earth (Table 9.5) and from radiations produced by the interaction of cosmic rays with elements of the atmosphere. Cosmic rays are primarily very high-energy protons and gamma rays of extraterrestrial origin. There is considerable variation in the intensities of these radiations depending on geographic location and altitude. The cosmic background radiation at Boulder, Colorado, may be a factor of two larger than it is in an eastern city that is at a substantially lower elevation. The water from wells in Maine has some 3000 times more radium content than water from the Potomac River. On the average, each person in the United States receives an annual exposure of 50 mrem* from cosmic radiation and 50 mrem from the earth and building materials. An additional 25 mrem is received internally fron inhalation of air (5 mrem) and isotopes formed naturally in human tissue (20 mrem). From diagnostic X-rays and radiotherapy, a person receives about 60 mrem and from the nuclear industry, television, radioactive fallout, etc., 5 mrem. Thus the grand total comes to about 190 mrem/year.*

Radiation Doses for Various Somatic Effects

Human beings will always be exposed to some uncontrolled radiation. Some risk will always be taken when there is exposure to additional amounts. Naturally, we want to know how much radiation is required to produce a given effect. This is a very difficult question to answer because

1. controlled experiments cannot be done on human subjects, and
2. damage to genetic cells shows up only in succeeding generations.

Information must be obtained from controlled experiments on animals, accidental (and unfortu-

* The symbol for millirem, which is one-thousandth of a rem, is mrem.

* See, for example, "Ionizing-Radiation Standards for Population Exposure," Joseph A. Lieberman, *Physics Today* **24,** 11 (November 1971): 32.

TABLE 9.5 Primary radionuclides from the earth that contribute to background radiation.

Isotope	Half-life (years)	Type of radiation
Radium-226 (^{226}Ra)	1622	alpha, gamma
Uranium-238 (^{238}U)	4,500,000,000	alpha
Thorium-232 (^{232}Th)	14,000,000,000	alpha, gamma
Potassium-40 (^{40}K)	1,300,000,000	beta, gamma
Tritium (^{3}H)	12.3	beta
Carbon-14 (^{14}C)	5730	beta

nate) exposures to human subjects, and scientific judgment. The most intense radiation exposures to large numbers of people occurred in the World War II nuclear bombing of Japan and in the Marshall Islands following the hydrogen bomb testing at Bikini Atoll in 1954. Studies of these and other victims have yielded considerable information on somatic effects. Some of these results are shown in Table 9.6. Most of the effects given in Table 9.6 are for one-time, whole-body exposures. Larger exposures are required to produce the same effects if the net dose is accumulated over a period of time. This

TABLE 9.6 X-ray and gamma ray doses required to produce various somatic effects.

Dose (rads)	Effect
0.3 weekly	Probably no observable effect
60 (whole body)	Reduction of lymphocytes (white blood cells formed in lymphoid tissues as in the lymph nodes, spleen, thymus, and tonsils)
100 (whole body)	Nausea, vomiting, fatigue
200 (whole body)	Reduction of all blood elements
400 (whole body)	50% of an exposed group will probably die
500 (gonads)	Sterilization
600 (whole body)	Death (likely)
1000 (skin)	Erythema (reddening of the skin)

is because somatic cell damage is repairable to some extent. A sunburn, for example, produces considerable cell damage that is normally repaired by metabolic processes.

Nuclear radiation has greatest biological effect in areas of the body that have rapidly dividing cells. That is why, for example, radiation attacks hair follicles and skin. Because cancer also involves rapid cell growth, nuclear radiation is used to control and, in some cases, stop the growth. It is ironic, but understandable, that radiation has the ability both to induce and to cure cancer.

Radiation Standards

From our knowledge that risks are involved in exposure to nuclear radiations and that a certain amount of radiation will be released to the environment in the routine operation of a nuclear reactor, the question arises, "Who will decide the maximum allowable exposure for the general public?" In the United States, this job is given to the Environmental Protection Agency (EPA). It is the function of the Nuclear Regulatory Commission (NRC) to set emission standards consistent with the guidelines of the EPA. These standards have changed throughout the years but since the mid-1950s the maximum allowable *average* exposure for the general population has been set at 170 mrem/year above background. This figure is based on the recommendation that a person should receive no more than 5 rems of radiation above background in a 30-year period. The recommendation is based on studies of the genetic effects of radiation by a committee established by the United States National Academy of Science–National Research Council and the United Kingdom Medical Research Council in the mid-1950s. If we add this value of 170 mrem/year to that of the approximate 190 mrem/year that a person normally receives from natural and manufactured sources, we find that his or her total could amount to about 360 mrem/year.

The 170 mrem/year exposure for the general population is controversial. That radiation induces human cell damage and forms of cancer such as leukemia is unquestioned. In fact, for exposures above 100 rads, experimental evidence indicates a direct linear relationship between incidences of cancer and radiation exposure (Fig. 9.15). Reliable data for exposures below 100 rads is lacking. Data for these low exposure levels are of extreme importance because they apply to levels of radiation in the realm of background exposures and those which could be achieved under present EPA guidelines. Any estimate of cancer incidences at low dosages must be based on an extrapolation of this curve. The feeling is that if the incidence rate increased as the exposure decreased, then it would be observed; it has not been. There are two other alternatives commonly offered. The first is that no effect occurs until some level of exposure, say around 100 mrad, is obtained. Then the cancer incidence rate increases with exposure as observed. This is referred to as the threshold hypothesis. If this were true, it would be the best alternative. The second proposal is that the observed linear relationship continues down to zero exposure. This is the linear hypothesis. In the absence of conclusive contrary evidence, the linear

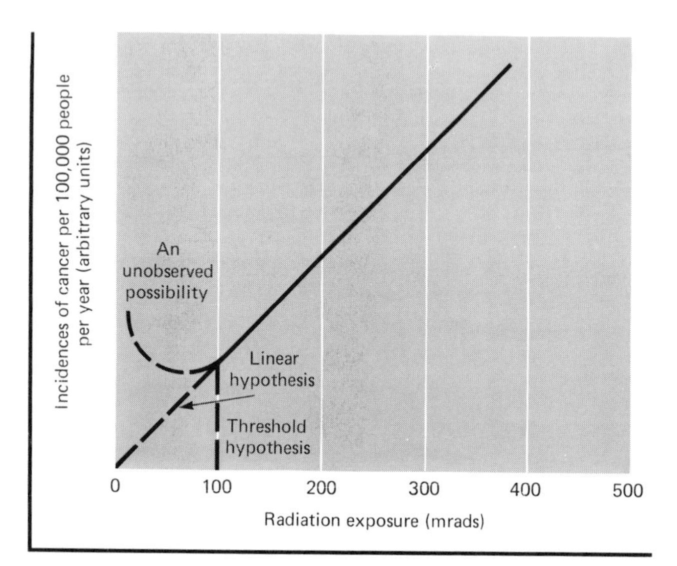

FIGURE 9.15 The induction of cancer by nuclear radiation is proportional to the dose received for doses greater than about 100 millirads. Experimental information is lacking for smaller doses. The graph illustrates three possible ways of extrapolating the cancer incidence–radiation behavior for low-dose exposures.

hypothesis is the most conservative assumption. Based on this assumption, it has been estimated* that an added exposure of 170 mrem per year to the entire population for 30 years would produce 16,000 cancer cases annually. These estimates are probably realistic *if* the entire population did, in fact, get 170 mrem/year. However, the law† requires that at the site of any nuclear power plant the average annual exposure to an individual cannot exceed 5 mrem. In fact, annual exposures during routine operation are much less than 5 mrem. Thus public concerns for radiation exposure focus more on accidental releases of large amounts of radioactivity rather than routine releases.

9.13 DISPOSAL OF THE HIGH-LEVEL RADIOACTIVE WASTES

Spent fuel is highly radioactive and must be properly protected (Fig. 9.16). When first removed from a reactor, it is stored in pools of water at the plant site. In about 120 days enough short-lived unstable nuclei disintegrate to reduce the radioactivity by over 90%. If not reprocessed, the spent fuel will probably be stored in such a way that the valuable fissionable material could be recovered at a later time. If reprocessed, uranium and plutonium will be chemically separated and the solid radioactive leftovers taken to a permanent repository. The reprocessed uranium and plutonium can be used separately for producing new fuel elements or a mixture of uranium and plutonium can be fabricated into a mixed-oxide fuel.

Disposal of wastes from a coal-burning electric power plant is a nontrivial proposition (Section 5.4). So it is with high-level radioactive wastes from nuclear power plants. Some radioactive wastes of concern are shown in Table 9.3. While nuclear waste disposal technology is fairly well developed, there are some public misconceptions about it. Reprocessing technology that is fairly well developed leaves

---

* See, for example, "Radiation Risk: A Scientific Problem?" *Science* **167** (6 February 1970): 853.

† *Federal Register* **36,** 111 (June 9, 1971).

the high-level radioactive wastes in liquid form. After about five years of isolation in underground tanks, the radioactivity declines to a sufficiently low level for solidification of the wastes. The wastes may become an integral part of either a glasslike material or are embedded into a ceramic. Thus the radioactive atoms are held rigidly in a solid binder. The solidified wastes enclosed by a stainless steel canister can then be taken to a permanent disposal site. A site covering ten acres (a football field is about $\frac{3}{4}$ of an acre) could handle all solid wastes through the year 2010. Options for disposal sites include deep ocean trenches, polar ice caps, and outer space. But the most likely site is in underground formations of salt or rock (Fig. 9.17).

Salt formations have received the most attention of the geologic formations considered. The United States is underlaid with about 400,000 square miles of salt beds. This is roughly 10% of the entire area of the country. The beds are located in areas not subject to active geological processes. The areas are devoid of water; otherwise the salt would not be there since it is water soluble. Salt is easily deformed, has good thermal properties, so that heat is readily conducted away from the wastes, and is easily mined.

Although salt formations are attractive, there is not uniform agreement that they are the best locations for waste disposal. Rock formations, made of granite or limestone, in arid areas are also promising sites. Sweden, at least, has decided on granite formations as the permanent disposal site.

FIGURE 9.16 Spent nuclear fuel elements are transported in massive, thick-walled casks to ensure that the radioactive contents do not escape to the environment. (Photograph courtesy of Atomic Industrial Forum, Inc.)

TABLE 9.7 Projected radioactivity levels for high-level radioactive wastes for the year 2000.

Fission fragments		Actinides	
Isotope	*Radioactivity (millions of curies)*	*Isotope*	*Radioactivity (millions of curies)*
Strontium-90	13,200	Plutonium-238	214
Zirconium-93	0.392	Plutonium-239	0.739
Technetium-99	2.92	Americium-241	64.9
Cesium-135	0.0786	Curium-244	891
Cesium-137	19,000	Curium-245	0.318

Source: "Managing Radioactive Wastes," *Physics Today* **26,** 6 (August 1973): 36.

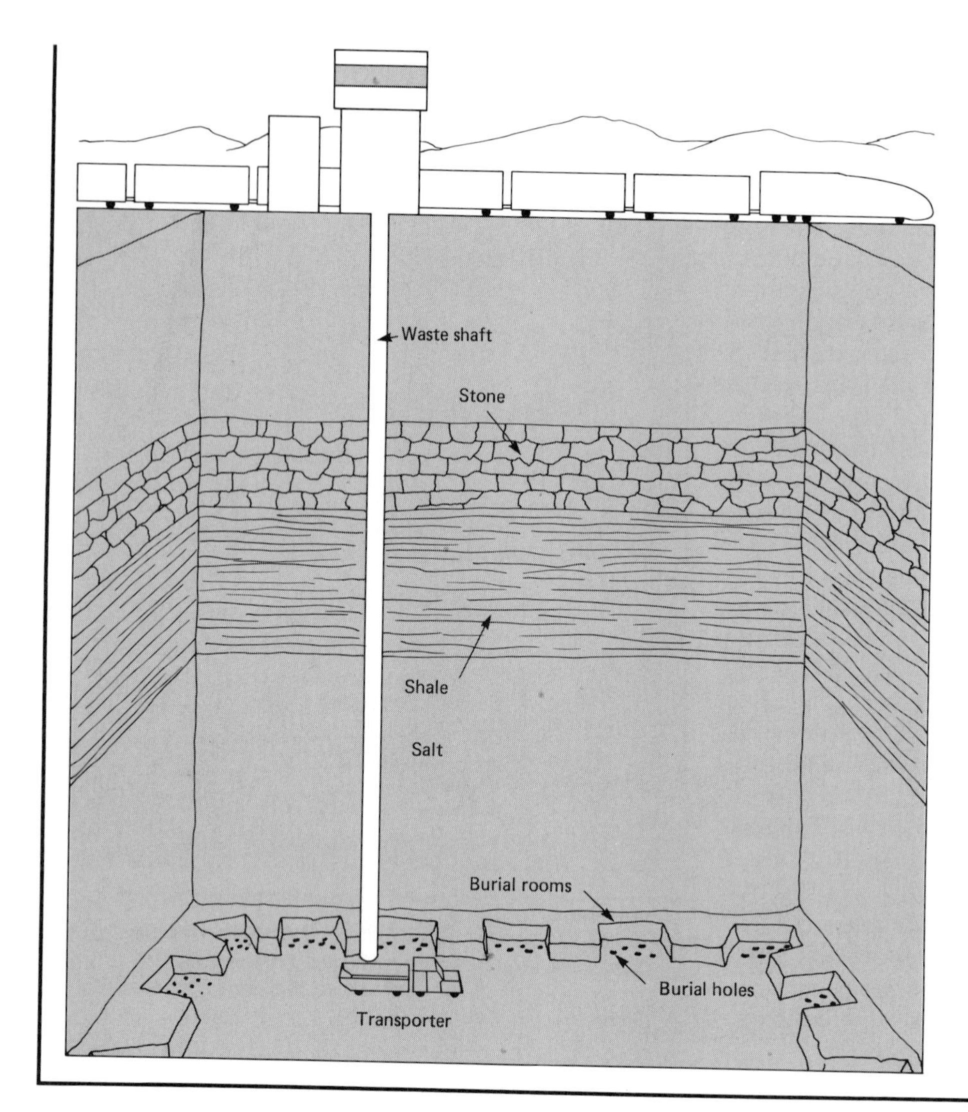

FIGURE 9.17 The salt formation concept for disposal of radioactive wastes. Solidified wastes would be permanently sealed in burial holes located several hundred feet below ground.

While radioactive waste disposal is an oft-debated issue, there is general agreement that it poses a much smaller potential hazard than nuclear accidents, diversion of weapons grade material, and nuclear proliferation. A 1978 study by the American Physical Society* concluded that technology exists for permanent disposal of solidified wastes in geological formations.

9.14 REACTOR SAFETY

When a man driving a car spots a stalled vehicle in his lane, information is fed to his brain that causes him to respond by applying brakes to the car. This is an example of the notion of feedback. Feedback is a widely used concept in many disciplines, but

* "Report to the American Physical Society by the Study Group on Nuclear Fuel Cycles and Waste Management," *Reviews of Modern Physics* **50,** 1, Pt. II (1978).

especially in physics and engineering. Basically, it means that something from the output of a system is fed back to its input (Fig. 9.18). The "something" can be information, as in the example cited, or such things as energy and electrical signals (voltage or current). For instance, some of the electric energy from a coal-burning power plant is fed back to the plant to run the electrostatic precipitators. The feedback may tend to either increase or decrease the output and thus is referred to as being either positive or negative feedback. The feedback in the car example is negative because it results in a decrease in output, namely, the speed of the car. In a nuclear bomb, the idea is to have a positive feedback mechanism so that the fission reaction proceeds uncontrolled. This is achieved by

1. using no control mechanisms such as cadmium rods,
2. using nearly 100% pure $^{235}_{92}U$ or other fissionable materials, and
3. forcing and holding together two pieces of the fissionable material to form a critical mass.

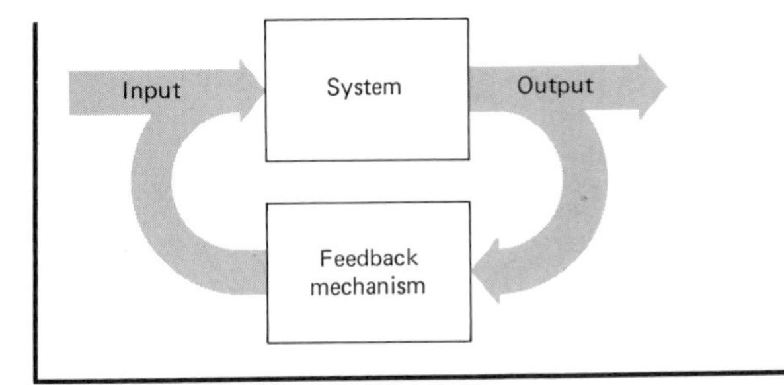

FIGURE 9.18 Schematic representation of a system with a feedback mechanism. If the system is the earth, the input might be carbon dioxide from burning fossil fuels and the output might be heat from warming by the "greenhouse" effect. Warming might release carbon dioxide trapped in polar ice—accelerating the "greenhouse" effect. This is an example of positive feedback.

Because a nuclear reactor employs the fission process, the question "Is it possible that these conditions could be achieved accidentally?" naturally arises. The answer is no for the slow neutron reactors considered here. The reasons are as follows. Neutrons emitted at the time fission occurs are termed prompt. Using nearly pure $^{235}_{92}U$ (or $^{239}_{94}Pu$) a nuclear bomb can achieve criticality using only prompt neutrons. Although the $^{235}_{92}U$ is enriched above its natural 0.7% abundance, it is still only about 3% of the uranium used. Thus it is some 30 times less concentrated than the fuel needed for bombs. A nuclear reactor cannot achieve criticality using only prompt neutrons. Rather, the reactor employs delayed neutrons emitted from nonfission reactions some time after the emission of the prompt neutrons. If an uncontrolled reaction does start, the intense heat developed will melt the uranium containers and the uranium will separate, tending to decrease the fission reactions. Finally, there is a multitude of fast-acting, negative feedback mechanisms that control the fission process. The control rods are the only moving parts in the reactor. If an "excursion" from normal is sensed, these rods are slammed into a position in which the reactor cannot possibly operate. Again, the time involved in delayed neutron emission is crucial for success of the control operation. Irregular excursions are indicated by such things as abnormal variations in the pressure and temperature of the reactor and in the flow rate of the coolant water. If the water supply fails or is purposely shut off, the neutron-moderating mechanism is removed. This drastically reduces the reaction probability for fission.

A nuclear reactor and its auxiliaries involve a complex array of pumps, pipes, and valves for circulating water into and steam out of the reactor core. Operation is conducted from an elaborate central control room. Minor malfunctions of the components are inevitable and routine maintenance of this system is inescapable. Apart from this there is concern over the possibility of the loss of water in the main cooling of the reactor core. This is referred to as a loss of cooling accident, abbreviated LOCA. Such an accident could occur from faulty construction, improper maintenance, natural disaster, sabotage, etc. The chain reaction fission process shuts down once the coolant stops flowing because of the need for a moderator. However, the reactor core still produces heat at a level of 5–7% of its rated power because of the high-level radioactive fission fragments and actinides in the core. Mounting steam and water pressure is first relieved by valves venting the reactor vessel. This material would be somewhat radioactive because it would have been in contact with fuel rods that are not sealed perfectly. This radiation would be contained by structures surrounding the reactor vessel. Two containment procedures are currently used. One employs a large spherical or cylindrical steel vessel that essentially surrounds the entire reactor. This is the function of the spherical structure seen at many older nuclear power plants (Fig. 9.19). In the second type, the reactor vessel is located in a steel containment tank that is surrounded by thick layers of high-density concrete (Fig. 9.20). The entire structure is built partially underground. Because a contemporary reactor generates about 3000 megawatts of thermal power when operating normally, the radioactive products produce about 150–200 megawatts of power at the time the reactor is shut down. Although the power level drops following shutdown because many of the radioactive products decay rapidly, there is sufficient thermal power to begin melting the core in about 30 seconds. Unless checked, this molten radioactive mass could melt its way through the bottom of the containment facilities. A set of conditions could be imagined that would release enormous quantities of radiation into the environment. Emergency core-cooling systems (ECCS) must operate within this 30-second time period to cool the core and prevent its melting. Several independent cooling mechanisms are built into the reactor system to cool the deactivated reactor. Knowing that the chance of an accident releasing radioactivity

FIGURE 9.19 The Yankee Atomic Electric Power Plant in Rowe, Massachusetts, built in 1960. This reactor uses a spherical radiation containment structure. (Photograph courtesy of Yankee Atomic Electric Company.)

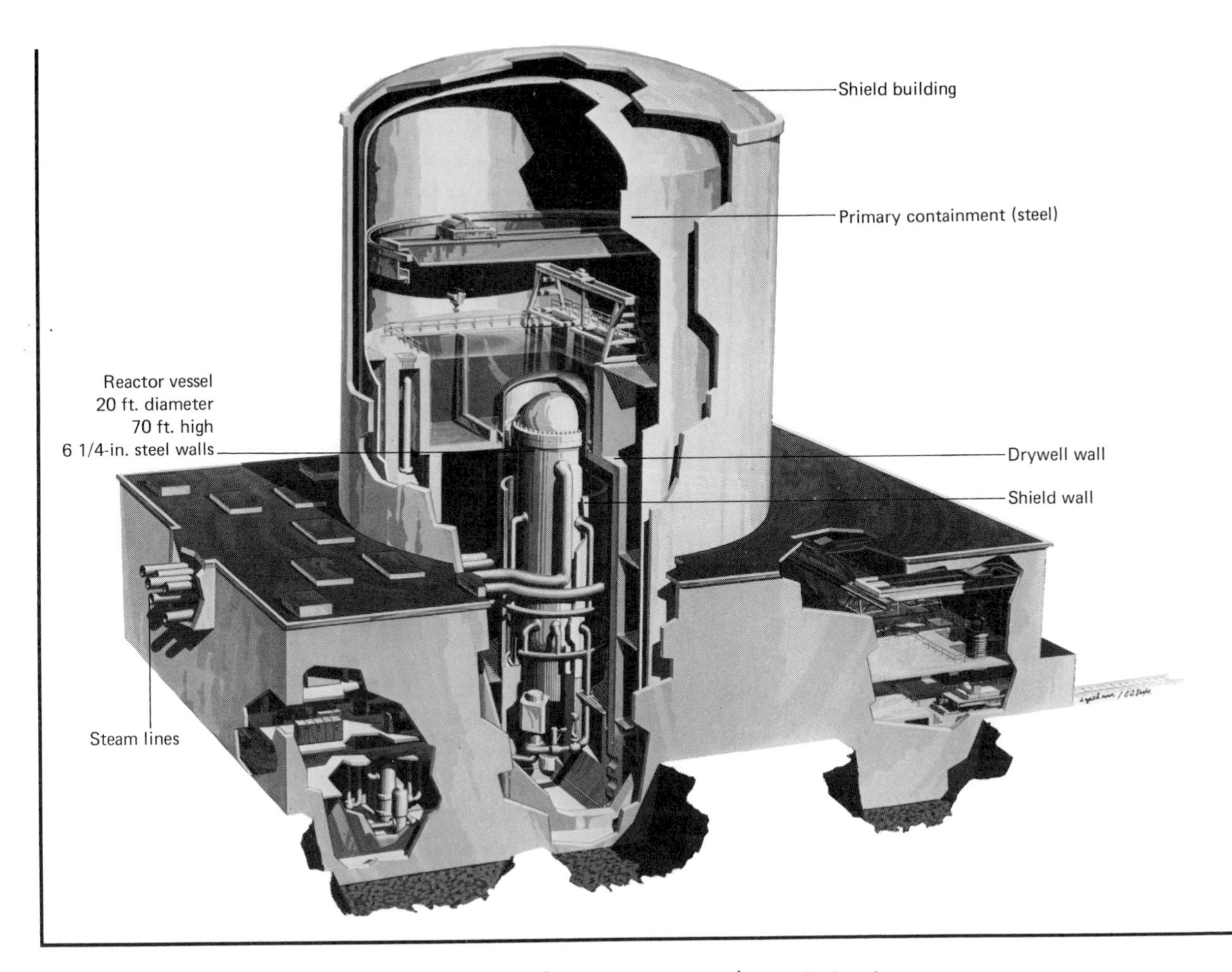

FIGURE 9.20 Drawing of a boiling water nuclear reactor containment structure using steel, reinforced concrete, and earth barriers to contain radiation in the unlikely event of an accident. (Courtesy of the General Electric Company.)

cannot be made zero, and knowing that exposure to nuclear radiation whether from medical X-rays or a nuclear power plant poses a biological threat, citizens naturally want to know the risks of living near a nuclear power plant.

Risk Assessment

Two factors figure into risk assessment.

1. the likelihood of the accident and
2. the consequences of the accident.

It seems the first factor could be assessed experimentally by depriving a full-scale reactor system of its cooling water and then testing the emergency cooling systems. Some testing of this sort is done but for the most part the likelihood of an accident is estimated by calculations because of the engineering magnitude of the system and the lack of sufficient operating experience.

A group of 60 specialists headed by Dr. Norman C. Rasmussen, Professor of Nuclear Engineering at the Massachusetts Institute of Technology, estimated the risks associated with contemporary nuclear power plants. Their report entitled "Reactor Safety Study" and other risk assessments are discussed in the literature.* The Rasmussen group envisioned sequences of failures and from probabilities of failure of components, including reactor operators, they pieced together an overall probability for a meltdown of the reactor core. From a judgment of the consequences of a core meltdown, they computed the risk. For example, the likelihood of a loss of cooling for a single reactor was assessed to be one chance in 2000 for each year of operation. Failure of the emergency cooling systems with subsequent core meltdown was assessed at one chance in ten per year of operation. Thus the annual likelihood that a single reactor would experience a meltdown of its core is one chance in 20,000. The odds of a core meltdown releasing sufficient radioactivity to cause 1000 deaths was assessed at one chance in 100. Thus the overall risk involving the chance of a meltdown of a single reactor core (one in 20,000) and the consequences (one in 100) is one in 2,000,000. Were there 100 reactors this risk is proportionally larger, that is, one in 20,000 per year of operation. Less severe accidents have correspondingly higher likelihoods.

Many citizens are comforted by the low risks from nuclear accidents as presented in the Reactor Safety Study. But many other citizens question the analysis. Studies by the American Physical Society, the prime organization of professional physicists, the Environmental Protection Agency, and the Union of Concerned Scientists identified omissions, errors, and sources of uncertainty in the Reactor Safety Study. Subsequently a Risk and Assessment Review Group concluded that the odds reported by the Reactor Safety Study should be revised upward. The group also maintained that the risks are still small. Regardless of how detailed the analyses, nuclear reactors or automobiles, for example, will not be proven either safe or unsafe. We can talk only in terms of risk. Some 50,000 Americans are killed annually in automobile accidents. Thus each person in a United States population of about 230,000,000 accepts an annual risk of one in 4000 of being killed in an automobile accident. Given the margin of errors in the risk assessment of nuclear power plants, the risks are compatible with many other accepted risks in our society.

* *Risks Associated with Nuclear Power: A Critical Review of the Literature,* National Academy of Sciences, Washington, D.C., 1979.

Three Mile Island

The Three Mile Island nuclear facility near Harrisburg, Pennsylvania, employs two pressurized water reactors. At 4 A.M. on March 28, 1979, the pump failed in the line returning water to the heat exchanger transferring heat from the reactor to the steam turbine in Unit 2. Emergency pumps functioned but before reaching full capacity, pressure in the reactor vessel increased to the point that a relief valve opened. The behavior of the emergency pumps and the pressure relief valve are design features in the system. However, two unanticipated events followed. A valve between the emergency pumps and the heat exchanger was found to be closed. The closed valve was discovered and opened about eight minutes following the pump failure. Having rectified the cooling water problem, pressure in the reactor vessel dropped and a signal alerted the operators that the pressure relief valve had closed. While the actuating mechanism for the valve did function, the valve did not close completely and it remained open for 2.5 hours. The operators thought there was too much water in the reactor when in fact there was too little. Water in the reactor boiled away exposing the reactor core leading to conditions oxidizing the zirconium fuel cladding and liberating hydrogen that formed a bubble in the reactor vessel. Explosion of the hydrogen was not possible because of a lack of oxygen. But the formation of the hydrogen was unanticipated and its presence caused grave concern at the time of the accident. Eventually, the accident was brought under control but not before major damage was done to the reactor core and not before substantial amounts of radioactive water escaped from the reactor vessel to the containment structures. While some radioactivity did escape to the environment, the average exposure to the surrounding population amounted to only two millirems per person. The maximum exposure to any individual was 100 millirems.

The trauma of the Three Mile Island incident cannot be denied. The accident shows that unanticipated events like the liberation of hydrogen can occur, that human errors are a factor, and that partial core meltdown can occur. But the accident also shows that in the face of a serious problem, radiation containment schemes held public exposures to a minimum. Whatever technological and social adjustments are made as a result of the Three Mile Island incident, we still cannot say whether nuclear power plants are safe or unsafe. But we would be remiss not to reduce the risks from nuclear power plant accidents from the lessons learned.

9.15 THE PRICE-ANDERSON ACT

In the event of a major nuclear accident, a nuclear power industry could sustain significant financial losses through liability claims. In order to protect the nuclear industry against liability claims and to encourage the development of nuclear energy, the Price–Anderson Act was enacted in 1957. The act requires a nuclear power industry to purchase the maximum liability insurance available from private insurance companies. In 1957, this maximum amount was $60 million. In 1983 it was $125 million. In addition to the private insurance, the government guaranteed an indemnity equal to the difference between $560 million and the amount obtained privately. A utility pays a premium for the government's liability coverage. In any accident, no more than $560 million can be paid out in liabilities. Thus, if a given accident produced $1120 million in liabilities, claims would be settled on the basis of 50 cents for each dollar claim. The 1957 act covered a ten-year period. The act was extended for a third ten-year period beginning in August 1977. Amendments to the 1977 renewal provide for a gradual phaseout of government indemnity through a system of private insurance. Opponents of nuclear power claim that the act proves the "unsafety" of nuclear power plants. Proponents argue that the act provides nothing more than what is done for other industries where lack of insurance experience prevails.

9.16 NUCLEAR PARKS AND OFFSHORE SITING

A bright future has been forecast for nuclear electric power since its feasibility was recognized in the early 1940s. Yet the actual development has consistently lagged behind the forecasts. One reason for this is that every nuclear power station has been designed to accommodate a particular site. Environmental considerations have been justifiably strict and this has increased the delays. The time from "decision to build" to "production of electricity" is often more than ten years. The industry is striving for standardization so that the construction time can be reduced. One way to standardize is to cluster several plants at the same site. Presently, it is fairly common to build two or three 1000-megawatt plants at the same site (Fig. 9.21). Nuclear parks envisioned for the future would have a combined output of 40,000–50,000 megawatts. With a capacity of this

FIGURE 9.21 An economic advantage accrues from conglomerating nuclear electric power plants. The Calvert Cliffs Nuclear Power Plant located in Lusby, Maryland, utilizes two pressurized water reactor systems, each producing 850 megawatts of electric power. (Photograph courtesy of the Baltimore Gas and Electric Company.)

magnitude, nuclear parks could have on-site chemical reprocessing and fuel fabrication and possibly even waste disposal. Nuclear parks would avoid most of the problems associated with transportation of radioactive materials. The isolation of the parks from urban centers would give the public added protection in the event of a nuclear accident. However, nuclear parks would be very vulnerable targets in the event of war.

A still more advanced concept calls for siting nuclear plants in the coastal waters offshore from the mainland. There are several attractive features to this idea. About half of the United States electrical demand is within 300 miles of the shores of the country. Nuclear stations located about three miles offshore could transmit electric power to the bulk of the users within reasonable transmission distances. Advantage could be taken of the ability of the oceans to assimilate heat and wastes. Offshore siting would also isolate the plants from populous areas. The much-sought-after standardization would be relatively easy to obtain because of the uniformity of the siting areas. This would tend to reduce the initial costs of the power plants. The safety risks in an offshore plant would be much the same as those in a land-sited plant, but there would be added concern for the possibility of releasing radioactive products into the ocean streams in the event of a large-scale accident.

There are several possible designs for offshore plants. One advanced design requires that plants be built on land in the form of barges that are then towed to and anchored at the site. After anchoring, the facility would be surrounded with a protective breakwater.

9.17 LIFETIME OF URANIUM FUEL SUPPLIES

When discussing energy sources of the magnitude of the reserves of a country, it is appropriate to define an energy unit equal to a billion-billion (10^{18}) Btu. This is symbolized by Q, which denotes both quintillion and a common symbol for heat. The total quantity of the energy derivable from fossil fuels in the United States amounts to about six to eight Q. Depending on projected estimates of consumption rates (Chapter 3), this supply may last for 100–150 years. The estimated fission energy available from thorium and uranium in the United States is about $900Q$. It would appear that if nuclear energy becomes an accepted energy source, our energy resources would be extended by hundreds of years. And it can, but not with light-water reactors. The reason is that although the $900Q$ is available, it includes ^{238}U and ^{232}Th that will not work in a light-water reactor because of the low fission probability for slow neutrons. Because the useful material, ^{235}U, constitutes only about 0.7% of natural uranium, the amount of energy available using nuclear reactor technology like that discussed in this chapter would be comparable to that available from fossil fuels. The ^{235}U resources could last, perhaps, 25 years. These facts create somewhat of a misunderstanding between the public and the promoters of nuclear energy who sometimes lead us to believe that the energy available from nuclear sources is nearly limitless. What may not have been explicitly stated is that although the source *is* nearly limitless, the present technology is not available to exploit it. There are, indeed, genuine prospects, but they are possibly ten to fifteen years from fruition. This and other advanced nuclear technologies are the order of business in Chapter 12.

REFERENCES

NUCLEAR POWER

The basic physics and the technology of nuclear reactors are presented in the paperback book, *Nuclear Energy: Its Physics and Its Social Challenge,* David R. Inglis, Addison-Wesley, Reading, Mass., 1973.

A historical account of nuclear fission, nuclear reactors, and nuclear bombs is presented in Chapter 16, "Fission and Fusion," in *Physics in Perspective,* Eugene Hecht, Addison-Wesley, Reading, Mass., 1980.

Energy, W. H. Freeman, San Francisco, 1979. A collection of *Scientific American* articles. Chapter III, "Nuclear Energy," includes:

The Necessity of Fission Power
The Disposal of Radioactive Wastes from Fission Reactors
The Reprocessing of Nuclear Fuels

Nuclear Power, Nuclear Weapons, and International Stability

Energy in Transition: 1985–2010, Final Report of the Committee on Nuclear and Alternative Energy Systems, National Research Council, W. H. Freeman, San Francisco, 1979. Chapter 5 is entitled "Nuclear Power."

Energy: The Next Twenty Years, Report by a Study Group Sponsored by the Ford Foundation and administered by Resources for the Future, Ballinger, Cambridge, Mass., 1979. Chapter 12 is entitled "Nuclear Power."

"The Future of Nuclear Energy," Alvin Weinberg, *Physics Today,* March 1981, p. 48.

"World Uranium Resources," Kenneth S. Deffeyes and Ian D. MacGregor, *Scientific American* **242,** 66, January 1980.

"Gas-cooled Nuclear Power Reactors," Harold M. Agnew, *Scientific American* **244,** 55, June 1981.

DISPOSAL OF RADIOACTIVE WASTES

"High-Level and Long-Lived Radioactive Waste Disposal," Ernest E. Angino, *Science* **198,** 885 (2 December 1977).

"The Disposal of Radioactive Wastes from Fission Reactors," Bernard L. Cohen, *Scientific American* **236,** 21 (June 1977).

"An Analysis of Nuclear Wastes," League of Women Voters Education Fund. In Lon C. Ruedisili and Morris W. Firebaugh, (eds.) *Perspectives on Energy,* Oxford University Press, New York, 1982.

"Report to the American Physical Society by the Study Group on Nuclear Fuel Cycles and Waste Management," *Supplement to Reviews of Modern Physics* **50** (1), 58 (1978).

"Geologic Disposal of Nuclear Wastes: Salt's Lead Is Challenged," Richard A. Kerr, *Science* **204,** 603 (11 May 1979).

SAFETY OF NUCLEAR REACTORS

"The Safety of Nuclear Reactors," Harold W. Lewis, *Scientific American* **242** (3), 53, 1980.

"Environmental Liabilities of Nuclear Power," John Holdren. In Lon C. Ruedisili and Morris W. Firebaugh (eds.), *Perspectives on Energy,* Oxford University Press, New York, 1982. From an original article in *Social and Ethical Implications of Recent Developments in Science and Technology,* Charles P. Wolff (ed.), Plenum, New York, 1981.

"Three Mile Island: The Accident That Should Not Have Happened," Ellis Rubinstein. In Lon C. Ruedisili and Morris W. Firebaugh (eds.), *Perspectives on Energy,* Oxford University Press, New York, 1982. From an original article in *IEEE Spectrum* **16** (11), 33 (1979).

"The Need for Change: The Legacy of TMI," Kemeny Commission. In Lon C. Ruedisili and Morris W. Firebaugh (eds.), *Perspectives on Energy,* Oxford University Press, New York, 1982. From an original article in *Bulletin of the Atomic Scientists* **36** (1), 24 (1980).

Reactor Safety Study, WASH–1400, United States Atomic Energy Commission, Washington, D.C., August 1974.

"Risk with Energy from Conventional and Nonconventional Sources," H. Inhaber, *Science* **203,** 718 (1979).

"Ultrasafe Reactors Anyone?" *Science* **219,** 265 (21 January 1983).

ENRICHMENT OF URANIUM FUEL

"Supplying Enriched Uranium," Vincent V. Abajian and Alan M. Fishman, *Physics Today* **26** (8), 23 (August 1973).

"The Reprocessing of Nuclear Fuels," William P. Bebbington, *Scientific American* **235,** 30 (December 1976).

RADIATION EFFECTS

Living with Nuclear Radiation, Patrick M. Hurley, The University of Michigan Press, Ann Arbor, Mich., 1982.

"Impacts of the Nuclear Energy Industry on Human Health and Safety," Bernard L. Cohen, *American Scientist* **64** (5), 550 (1976).

Report of the Committee on Biological Effects of Ionizing Radiation (BEIR), National Research Council, National Academy of Sciences, Washington, D.C., 1979.

"The Biological Effects of Low-level Ionizing Radiation," Arthur C. Upton, *Scientific American* **246,** 41 (February 1982).

"Catastrophic Releases of Radiation," Stephen A. Fetter and Kosta Tsipes, *Scientific American* **244,** 41 (April 1981).

PRICE–ANDERSON ACT

"Price–Anderson: Exploring the Alternatives," Hubert H. Nexon, *Nuclear News* **17** (4), 56 (March 1974).

OFFSHORE SITING OF NUCLEAR POWER PLANTS

"Offshore Nuclear Power Stations," Michael W. Golay, *Oceanus* **17,** 46 (Summer 1974).

REVIEW

1. What role does the Einstein equation $E = mc^2$ play in nuclear energy considerations?
2. Name some similarities and differences in the structure of atoms and nuclei.
3. What is the meaning of the symbol ${}^{4}_{2}He_2$?
4. What is an isotope? Name two isotopes of hydrogen.
5. Describe the role of mass-energy considerations in determining the stability of nuclei.
6. What are the emissions of interest from unstable nuclei and how are they produced?
7. How is energy liberated in a nuclear reaction?
8. Describe the fission of a heavy nucleus by a neutron.
9. What is the nuclear meaning of disintegration?
10. What is a curie?
11. How is the half-life of an unstable nucleus measured?
12. List the energy conversion steps in the production of electricity in a nuclear power plant. How does the process differ from a coal-burning power plant?
13. What is a chain reaction?
14. Summarize the four principles involved in making a nuclear reactor work.
15. Trace the path of uranium from the earth's crust to a nuclear reactor.
16. How is steady energy production achieved in a nuclear reactor?
17. What is meant by fuel enrichment and how is it achieved?
18. Describe the salient features of a boiling water reactor and a pressurized water reactor. Name some comparative advantages of each type.
19. Distinguish between low-level and high-level nuclear wastes.
20. Why are unstable nuclei with very short or very long half-lives relatively "safe"?
21. What is the motivation for reprocessing nuclear wastes and why is reprocessing controversial?
22. What are fission fragments and actinides and what role do they play in plans for disposal of radioactive wastes?
23. List three possible damaging effects on living cells induced by nuclear radiations.
24. How can it be that nuclear radiation both causes and cures cancer?
25. How do somatic radiation effects differ from genetic radiation effects?
26. What are the terms rad, rem, and RBE used for?
27. What are the sources of background radiation?
28. What characteristics of salt and rock formations make them attractive as sites for disposal of radioactive wastes?
29. What prevents a nuclear reactor from exploding like a nuclear bomb?
30. What could cause a "loss-of-cooling" accident and how might an "emergency core-cooling system" come to the rescue?
31. What factors enter into the computation of a risk, and the assessment of the risk?
32. Name some attractive features and disadvantages of siting a cluster of reactors into a "nuclear park."
33. It is appropriate to think of a nucleus as a bound state of _____ held together by _____ forces.
    a) electrons, electric.
    b) atoms, atomic.
    c) molecules, molecular.
    d) nucleons, nuclear.
    e) protons and neutrons, electric.
34. If the nucleon number A is unchanged, but Z changes, then
    a) the weight is essentially unchanged.
    b) the number of neutrons has changed.
    c) a new or different element is produced.
    d) the number of protons has changed.
    e) all of the above are correct statements.
35. Two nuclei are designated ${}^{90}Sr$ and ${}^{91}Sr$. These nuclei would be termed
    a) isobars. b) nucleons. c) isotopes.
    d) beta particles. e) curies.
36. The conversion of potential energy to some other form of energy always involves the rearrangement of something. That which is "rearranged" in a nuclear reaction involving the conversion of nuclear potential energy is
    a) protons and positrons. b) neutrons only.
    c) protons only. d) electric charge only.
    e) nucleons.
37. The spontaneous emission of an alpha particle by

^{239}Pu can be represented by $^{239}Pu \rightarrow \alpha + {}^{235}U$. A check of the masses of the products would reveal that

a) the combined masses of α and ^{235}U are greater than the mass of ^{239}Pu.
b) there is no change in the mass during the transformation.
c) the combined masses of α and ^{235}U are less than the mass of ^{239}Pu.
d) the mass of ^{239}Pu equals the mass of ^{235}U.
e) the alpha particle, like a photon, has no mass.

38. Of the several distinct types of emission from radioactive materials, the one which has much the same nature as ultraviolet radiation is

a) alpha particles b) beta particles (negative).
c) gamma rays. d) positrons.
e) fission products.

39. At 9 A.M. on Monday a radioactive sample contained four million nuclei. At 9 A.M. on Friday three million of these nuclei had decayed. The half-life of this sample is

a) 5 days. b) 4 days. c) 3 days.
d) 2 days. e) 1 day.

40. The disintegration of a sample of radioactive material is shown in the graph. The half-life of this material is ____ and after three hours ____ nuclei will not have disintegrated.

a) 2 hours; 7.5 million
b) 1 hour, 15 million
c) one-third of an hour, 15 million
d) 3 hours, 15 million
e) 1 hour, 105 million

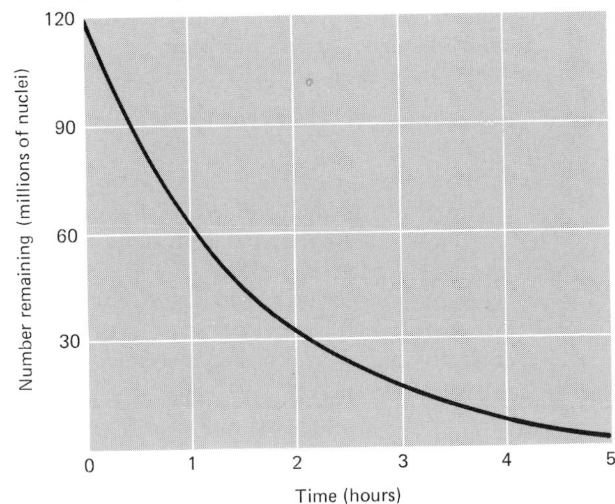

41. When you say a nucleus is radioactive, you mean the nucleus

a) is energetically unstable.
b) is an isotope.
c) actively emits radio waves.
d) actively absorbs radio waves.
e) necessarily emits gamma radiation.

42. Two radioactive samples have the same numbers of atoms but have different half-lives. The sample having the longer half-life will

a) have the greater rate of disintegration.
b) have the smaller rate of disintegration.
c) have the same rate of disintegration as the one with the shorter half-life.
d) probably be an alpha emitter.
e) probably undergo spontaneous nuclear fission.

43. A friend of yours reading an article concerning nuclear energy asks you to explain the meaning of curie. You would tell her (correctly) that a curie is

a) a measure of the half-life of a radioactive material.
b) the name of a type of radioactive material.
c) a measure of the penetrating ability of radiation.
d) a measure of the disintegration rate of a radioactive sample.
e) the plural of the word curious.

44. In physics, conservation means something of interest does not change. For example, when we say "energy is conserved" in energy conversion it means there is no change in energy. When ^{239}Pu is formed according to the reaction

$$^{239}Np \rightarrow {}^{239}Pu + {}_{-1}e,$$

something that is conserved is the number of

a) nucleons. b) protons. c) neutrons.
d) beta particles. e) protons, neutrons, and nucleons.

45. A student studying a radioactive material measures the rate at which the radioactive atoms disintegrate. It would be appropriate for her to express the results in terms of

a) half-life b) curies c) rems
d) remspergram e) half-livespersecond.

46. Textbooks often depict the fissioning of ^{235}U as

$$n + {}^{235}U \rightarrow Kr + Ba + 3n + \text{energy}$$

where Kr and Ba stand for krypton and barium,

respectively. In reality,

a) this is indeed the only possible reaction that could occur.
b) there are many other possible reactions producing products different from Kr and Ba.
c) this reaction is used because it is the only one producing radioactive products.
d) this reaction is used because it is the only one liberating energy.
e) this reaction is used because Kr and Ba are the only radioactive products of environmental concern.

47. The three figures below depict a head-on collision between a moving steel sphere and a stationary steel sphere. Which moving object loses the most speed?

48. A nuclear chain reaction
    a) means that uranium atoms must be positioned like links in a chain.
    b) means that a nuclear fission reaction produces radioactive nuclei with a chainlike structure.
    c) means that the products from a reaction instigate more reactions of the same type.
    d) is involved in a nuclear bomb but not in a nuclear reactor.
    e) is instigated by proton bombardment of uranium in a nuclear reactor.

49. Fuel pellets in a contemporary nuclear reactor are
    a) highly radioactive and must be handled by remote control.
    b) about 98% ^{235}U.
    c) a good source of bomb material because of their ^{238}U.
    d) about 3–5% ^{235}U.
    e) about 0.5 inch in diameter and 12 inches long.

50. Criticality in a nuclear reactor means the
    a) reactor has achieved a condition whereby it will explode.
    b) reactor has achieved a self-sustaining operating condition.
    c) radioactivity produced has reached a critical limit.
    d) steam pressure produced has reached a critical value.
    e) the news media have publicized design errors.

51. The function of a moderator in a nuclear reactor is to
    a) make the level of radioactivity more moderate.
    b) reduce the speed of neutrons.
    c) moderate the uranium fuel.
    d) moderate the effects of a loss of cooling.
    e) control the rate of nuclear fission reactions.

52. By using water made with deuterium for a neutron moderator, Canadian reactors are able to use natural uranium rather than the enriched uranium used in United States reactors. This is because
    a) the Canadian reactors operate at a much lower power.
    b) deuterium is less likely to capture neutrons than ordinary hydrogen.
    c) the Canadian reactors do not use control rods.
    d) deuterium is especially plentiful in Canada.
    e) all of the above are incorrect; Canadian reactors use plutonium for fuel.

53. The fraction of ^{235}U in natural uranium is ____; enrichment processes increase ^{235}U to ____ for light water reactors and ____ for nuclear weapons.
    a) less than 1%; about 3%; more than 90%.
    b) about 10%; 15%–30%; more than 90%.
    c) less than 1%; about 15%; about 15%.
    d) about 15%; about 50%; 99%.
    e) about 15%; more than 90%; more than 90%.

54. Water has two major roles in the operation of a nuclear reactor. These are as
    a) neutron absorber and heat transfer liquid.
    b) neutron moderator and heat transfer liquid.
    c) neutron moderator and neutron absorber.
    d) neutron absorber and lubricant.
    e) neutron moderator.

55. A nuclear-fueled electric power plant
    a) works on the principle of nuclear fusion.
    b) differs from a coal-fueled electric power plant mainly in the way the electricity is generated.
    c) produces exactly the same pollutants as a coal-fueled electric power plant.
    d) converts nuclear energy to heat in one stage of its operation.
    e) both (b) and (d) are correct answers.

56. The power level of a present day commercial nuclear power plant is controlled by the insertion or withdrawal of cadmium or boron rods (control rods). Cadmium or boron is used because

a) it is a good neutron absorber.
b) anything will do but cadmium and boron are cheap materials.
c) it doesn't react chemically with the uranium fuel.
d) it doesn't become radioactive when exposed to neutrons.
e) it doesn't rust in water.

57. A nuclear electric power plant differs from a coal-burning electric power plant only in the way that heat is provided to produce steam for the turbines. Also, we have learned that there are two types of nuclear reactors—boiling water (BWR) and pressurized water (PWR). We know that
    a) water is converted to steam inside the BWR and PWR reactors.
    b) there is only one company in the United States that makes both BWR and PWR reactor vessels.
    c) water is not allowed to boil in the PWR reactor.
    d) only the BWR system like that being built near Cincinnati requires cooling towers.
    e) all of the options above are reasonable.

58. Somatic radiation damage produces
    a) damage only to the surface layers of the skin.
    b) a physiological effect on the body.
    c) damage only to the reproductive organs.
    d) effects that are almost completely uninterpretable by experts in the field.
    e) none of the above.

59. For identical doses of radiation, X-rays will produce 20 times less biological damage than fission fragments. Hence the relative biological effectiveness of X-rays is ____ that of fission fragments.
    a) twenty times less than b) twenty times greater than c) equal to d) unrelated to
    e) none of the above.

60. Everyone receives a certain amount of nuclear radiation from cosmic rays, building materials, X-rays, etc. This so-called background radiation amounts to about 200 mrem/year. If you lived near a nuclear power plant, then you could expect to receive, from the routine operation of the plant, an additional amount of about
    a) 100 mrem/year. b) less than 5 mrem/year.
    c) about 200 mrem/year. d) much more than 200 mrem/year. e) none.

61. If a conventional nuclear reactor suddenly loses its cooling water, then the
    a) reactor will explode spontaneously.
    b) nuclear fission chain reactions will terminate.
    c) core of the reactor will melt in a matter of seconds.
    d) control rods will replace the water as the moderator.
    e) all of the above.

62. The main concern with a nuclear reactor core melt down is a
    a) possible nuclear bomblike explosion.
    b) release of chemically explosive substances such as liquid sodium.
    c) release of large quantities of highly radioactive deuterium in the cooling water.
    d) release to the environment of highly radioactive fission fragments produced by the core (during routine operation).
    e) release of extremely radioactive uranium to the environment.

63. At the accident at Three Mile Island in 1979, there was much concern about removing heat from the reactor that still existed several days after the accident first started. This heat energy had its origin in
    a) radioactive atoms in the core of the reactor.
    b) nuclear fission reactions that continued until several days after the accident started.
    c) hydrogen-producing reactions involving the zircaloy cladding.
    d) nuclear fusion reactions that began once the control rods were inserted.
    e) the melting of the reactor control rods.

64. A convincing argument for saying that a conventional nuclear reactor cannot explode like a nuclear bomb is
    a) the fuel mixture is too dilute in a nuclear reactor.
    b) nuclear bombs work only with ^{239}Pu while nuclear reactors use ^{235}U.
    c) nuclear bombs require a moderator as part of their structure.
    d) the basic energy conversion mechanism is different in a nuclear reactor.
    e) nuclear reactors do not use ^{235}U.

65. When one refers to radioactive waste disposal for the nuclear power industry, the radioactivity is due mainly to
    a) the control rods.
    b) the moderator.
    c) the casings of the fuel elements.
    d) the actinides.
    e) the actinides and fission fragments.

66. Of the radioactive waste disposal schemes listed below, the one most likely to be used in the near future is
    a) nuclear transmutations.
    b) dumping in the major rivers in the United States.
    c) disposal in the polar ice caps.
    d) rocketing into space.
    e) disposal in salt formations.
67. Following the reactor accident at Three Mile Island, the caption of a newspaper article read "Citizens Learn How to Measure Rems." Citizens are interested in a rem because it is a unit of radiation
    a) reflecting the half-life of a radioactivity substance.
    b) that assesses the amount of radiation absorbed as well as the biological effect.
    c) that is used interchangeably with the curie radiation unit.
    d) that stands for radiation en masse.
    e) named for Ralph E. Miller.

Answers

33. (d)	34. (e)	35. (c)	36. (e)
37. (c)	38. (c)	39. (d)	40. (b)
41. (a)	42. (b)	43. (d)	44. (a)
45. (b)	46. (b)	47. (c)	48. (c)
49. (d)	50. (b)	51. (b)	52. (b)
53. (a)	54. (b)	55. (d)	56. (a)
57. (c)	58. (b)	59. (a)	60. (b)
61. (b)	62. (d)	63. (a)	64. (a)
65. (e)	66. (e)	67. (b)	

QUESTIONS AND PROBLEMS

9.1 MOTIVATION

1. Using the Einstein equation, $E = mc^2$, show that the energy of one gram of matter is 90 trillion (i.e., 9×10^{13}) joules.
2. About 100,000,000 Btu of heat is needed annually for heating an average house in the central United States.
    a) If this energy were obtained from coal having a heat content of 10,000 Btu per pound, how many pounds of coal are required?
    b) The complete fissioning of one pound of ^{235}U produces 50,000,000,000 Btu of energy. If all the fission energy could be converted to heat, how many pounds of ^{235}U would be required to provide the 100,000,000 Btu for space heating?

9.2 PROPERTIES OF THE ATOMIC NUCLEUS

3. A newspaper article concerning nuclear energy makes reference to Pu-238. What does this notation mean?
4. An elevator stops only at floors. Could you associate an energy with each elevator stop? How is this situation like the energy states of an atom or nucleus?
5. If a nucleus of an atom were the size of an orange (about three inches in diameter), about how many lengths of a football field away would an electron be? A football field is 300 feet long.
6. A proton has a mass 1836 times greater than the mass of an electron. What percentage of the mass of the hydrogen atom is contained in its nucleus?

9.4 NUCLEAR REACTIONS

7. Why do mass considerations prevent an isolated proton from spontaneously disintegrating into a neutron, a beta ($_{-1}e$) particle, and a neutrino?
8. Water behind a dam has potential energy by virtue of an advantageous position. The potential energy is converted to kinetic energy when the water is rearranged as it plummets to the bottom of the dam. What is being rearranged when the potential energy of a nucleus is converted to kinetic energy?
9. The accompanying sketch shows a set of weighing scales with before and after products of the fission of ^{235}U by a neutron. Why are the scales tipped as shown?

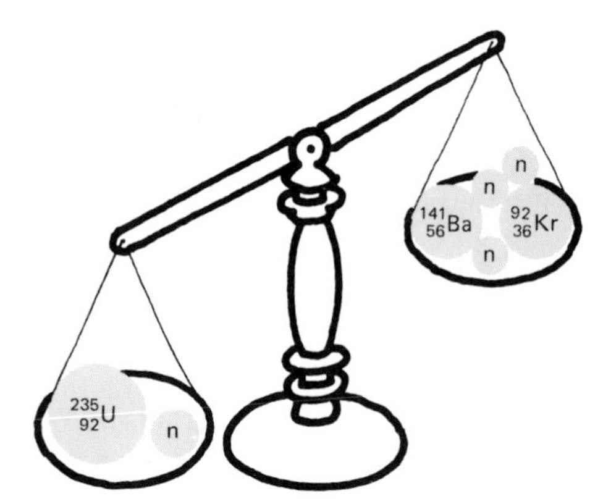

10. How is a nuclear fission chain reaction similar to a collision chain reaction of closely spaced cars on a highway?
11. What would you conclude about the reaction products in two nuclear fission reactions if one reaction produced three neutrons and the other reaction produced four neutrons?

12. If the spontaneous fission of ${}^{235}_{92}U_{143}$ produces only two nuclei, one of which is ${}^{91}_{36}Kr_{55}$, what is the other nucleus?

13. Two nuclei interact to form a single nucleus having a mass 0.01 amu less than the combined masses of the two interacting particles. What happens to this 0.01 amu of mass?

14. Each of the nuclear reactions listed below cannot happen because a fundamental physics principle is violated. Name the principle for each reaction.

$${}^{235}_{92}U_{141} \rightarrow {}^{90}_{36}Kr_{54} + {}^{142}_{56}Ba_{86}.$$
$${}^{3}_{1}H_{2} \rightarrow {}^{3}_{2}He_{1} + {}_{+1}e + \nu.$$
$${}^{11}_{5}B_{6} \rightarrow {}^{11}_{6}C_{5} + {}_{+1}e + \bar{\nu}.$$

15. One possible way that ${}^{235}U$ can be fissioned by a neutron is

$${}^{235}U + n \rightarrow {}^{141}Ba + {}^{90}Kr + 5n.$$

Using the masses given below for the nuclei involved in this reaction, show that 163 MeV are released in the fission. Remember, one amu is equivalent to 931 MeV of energy.

Nucleus	Mass (amu)
${}^{235}U$	235.0439
n	1.0087
${}^{141}Ba$	140.9141
${}^{90}Kr$	89.9198

9.5 HALF-LIFE OF RADIOACTIVE NUCLEI

16. The time required for one-half of a sample of radioactive atoms to disintegrate is called the half-life. In principle, how long does it take for a radioactive material to disintegrate completely?

17. Carbon-14 (${}^{14}_{6}C_{8}$) is a beta particle (${}_{-1}e$) emitter with a half-life of 5730 years. It is commonly used to radioactively date carbon-based relics in our environment. Write the nuclear reaction for the decay of ${}^{14}_{6}C_{8}$.

18. A certain radioactive nucleus has a half-life of ten days. What fraction of the sample decays in 30 days? What fraction of the sample remains after 30 days?

19. A radioactive material has a half-life of ten years and an initial strength of 640 curies. If it can be safely disposed of when its strength has diminished to 10 curies, how many years elapse?

20. After 15.2 minutes the disintegration rate of a radioactive sample is only 1/1024 of its initial value. What is the half-life of this radioactive sample?

21. A measurement of the disintegration rate of a radioactive sample was taken at 2 P.M. on Monday. At 2 P.M. on Friday of the same week, the disintegration rate dropped to one-fourth the Monday value. What is the half-life of this radioactive sample?

22. Picocurie radioactive sources are commonly used for demonstration purposes in classrooms and student laboratories. How many disintegrations occur in one minute with a one picocurie source?

23. Student measurements of the disintegration rate of ${}^{131}I$ yielded the data shown in the table. Show that these measurements are consistent with the known half-life of 8.05 days for ${}^{131}I$.

Disintegration rate (disintegrations per second)	Time (days)
3400	0
2600	3
2000	6
1600	9
1200	12
950	15
720	18

9.6–9.11 NUCLEAR REACTORS

24. The chemical reaction

$$C + O_2 \rightarrow CO_2$$

releases 4.1 electron volts of energy. A nuclear fission reaction liberates about 200 million electron volts of energy. How many of these chemical reactions does it take to deliver the same amount of energy as a nuclear fission reaction?

25. A city of one million people requires about 2000 million watts of electric power.
    a) How many joules of electric energy are required each day?
    b) The complete fissioning of one kilogram of ${}^{235}U$ produces about 82 trillion (82×10^{12}) joules of energy. If the fission energy is converted to electric energy with an efficiency of 30 percent, show that about seven kilograms of ${}^{235}U$ would provide the daily electric energy needs of a city of one million people.

c) A cube of pure ^{235}U one centimeter on a side has a mass of 0.0187 kilograms. What is the length of a side of a cube of ^{235}U if its mass is seven kilograms?

26. If several dominoes are standing on end in a line and the first is tipped over, then a chain reaction of falling dominoes may develop. Why is a critical arrangement required to sustain the chain reaction?

27. The overall efficiency for generating electric energy is nearly the same for a coal-burning and a nuclear-fueled electric power plant. Yet a nuclear power plant releases 12–15% more heat to the cooling water for the condenser than does a coal burning power plant. Trace the energy conversions in both systems and explain why the nuclear power plant releases more heat to the cooling water.

28. The capture of a neutron by a proton is depicted as

$$n + p \rightarrow {}^{2}_{1}H_{1} + \gamma$$

where γ represents a gamma photon. From mass–energy considerations, why is the inclusion of the gamma photon reasonable?

29. There are some neutrons in the "rain" of radiation from cosmic sources. What prevents a cosmic neutron from striking a ^{235}U nucleus in the ground and triggering a fission chain reaction?

30. Consider the following game. A circle is laid out on a flat piece of ground. You stand within the circle and confine your movement to the area of the circle. Another person, preferably of the opposite sex, runs through the circle and you try to capture him (her). Why does the likelihood of capture increase as the speed of the person decreases? In what ways is this similar to the capture of a neutron by a nucleus?

31. Boron-10 (^{10}B) is commonly used as a control rod material in a nuclear reactor because it readily captures neutrons. Complete the reaction below for this process:

$$^{10}_{5}B_{5} + n \rightarrow {}^{7}_{3}Li_{4} + \underline{\quad\quad} .$$

32. If a hard sphere (such as a billiard ball) of mass M makes a head-on collision with another hard sphere (such as a bowling ball) of mass m that is initially at rest, then the kinetic energies of the mass M before and after the collision are related by

$$KE\text{ (after collision)} = \left(\frac{M - m}{M + m}\right)^2 \times KE\text{ (before collision)}.$$

What happens if the two masses are the same? A neutron and a proton have nearly equal mass; this is why water with many hydrogen atoms is used to reduce the speed of neutrons in a water-cooled reactor.

9.12 BIOLOGICAL DAMAGE DUE TO RADIATION

33. Much like light bulbs that emit visible radiation, many radioactive sources emit nuclear radiation uniformly in all directions. Knowing this, explain why getting away from a radioactive source is an effective form of protection from its radiation.

34. Strontium has many of the physical and chemical properties of calcium. Structures that can be made from calcium can also be made from strontium. Can you see why it is especially important to minimize the ingestion of radioactive strontium by young people?

35. On a yearly average, a United States citizen receives about 100 millirems of nuclear radiation from cosmic sources and minerals in buildings and the ground. A resident in Denver, Colorado, may receive twice this amount in a year's time. As the public learns more about nuclear radiation, do you think that there will be a migration away from the Denver area?

36. For identical doses of radiation, X-rays produce about 20 times less biological damage than fission fragments. Determine the relative biological effectiveness (RBE) of fission fragments.

37. How many rads of fission fragments are biologically equivalent to five rads of X-rays?

38. Some people argue that the only safe level of radioactivity emissions is zero. Using similar reasoning, what would be a safe speed for an automobile?

39. The General Electric Company states that the nearest neighbor radiation exposure from gaseous effluents from its BWR/6 reactor is less than 0.01 millirems per year. How many years of exposure are required for a person to receive 30 rems of radiation at 0.01 millirems per year?

9.13 DISPOSAL OF THE HIGH-LEVEL RADIOACTIVE WASTES

40. The disintegration rate of a radioactive sample depends on how many atoms are present and on the half-life. If the half-life is very long, then the rate will be small even if there are many atoms present.
    a) Calling N the number of atoms and T the half-life, explain why N divided by T has the form of a disintegration rate.
    b) Explain why the equation N/T agrees with the idea that very long half-life means a low disintegration rate.

c) Does it follow that radioactive wastes with very long half-lives are necessarily the most problematical?

41. Tables 9.3 and 9.7 give the half-lives and amounts of the radioactive atoms of particular concern in nuclear reactor wastes. Note that for two isotopes such as cesium-135 and cesium-137 the one with the longer half-life has the smaller amount of radioactivity. Give an explanation of this based on the idea that half-life is related to the chance that a radioactive atom will decay in some time period.

42. The total volume of solidified radioactive wastes from nuclear power plants is expected to be about 470,000 cubic feet in the year 2000. How thick is this volume if it were placed uniformly over a football field that is 100 feet wide and 300 feet long? Does this seem like a prohibitively large volume of material to deal with?

9.14 REACTOR SAFETY

43. How would you explain to a concerned citizen that a nuclear reactor cannot explode like a nuclear bomb?

44. Is the loss of cooling water in a nuclear reactor a negative feedback or a positive feedback mechanism?

45. About 1200 people are electrocuted each year in the United States. Since there are about 200,000,000 people in the United States, the individual chance of electrocution is about one in 160,000. With 100 nuclear power plants in operation the Reactor Safety Study indicated the individual chance of being killed by a nuclear accident as one in 300,000,000. What differences are there in the way these chances are assessed?

46. Living downstream from a large dam and living near a nuclear power plant pose risks. How does experience enter into figuring these risks?

47. About 7% of the thermal energy produced in the core of a nuclear reactor comes from radioactivity in the core. If a reactor is producing 3000 megawatts of thermal energy, how much of this is due to radioactivity?

48. The chain reaction terminates in a nuclear reactor when the cooling water is lost. But for a reactor producing 3000 megawatts of thermal power, about 200 megawatts remain because of radioactivity in the core.
    a) Show that in one minute, a power of 200 megawatts produces thermal energy amounting to 12 billion (12×10^9) joules.
    b) To raise the temperature of iron from 500° C to its melting point requires about 670,000 joules per kilogram of iron. How many tons of iron initially at 500° C could be melted by 12 billion joules of thermal energy? Remember, one kilogram is equivalent to 2.2 pounds and one ton = 2000 pounds.

49. The maximum steam pressure that any cylindrical vessel can withstand depends on the radius, thickness, and intrinsic strength of the container. It can be shown that these variables are related by "maximum pressure divided by intrinsic strength equals thickness divided by radius." Boiling-water reactor vessels are about 20 feet in diameter and are made from six-inch thick stainless steel having an intrinsic strength of 100,000 pounds per square inch. Show that the maximum pressure that the vessel can contain is 5000 pounds per square inch. Compare this with typical operating pressures in a boiling water reactor. (See Section 9.8.)

50. When a coin is flipped the probability that the side with a head on it will fall face up is 0.5. This is because there are two ways that the coin can end up and one event (i.e., head) is a success. If you flip three coins at once, the laws of probability predict that the probability that all three will produce a head is

$$\frac{1}{2} \times \frac{1}{2} \times \frac{1}{2} = \frac{1}{8}.$$

There are eight ways that the coins can land and one event (i.e., three heads) that is a success. You can prove this by figuring out the possible combinations of heads and tails when three coins are flipped. Do this by completing the table.

Coin 1	Coin 2	Coin 3
T	T	T
H	T	T
T	H	T
—	—	—
—	—	—
—	—	—
—	—	—
H	H	H

If some nuclear accident depends on the simultaneous failure of the closing of three valves and the probability of any single valve failing is one chance in a hundred ($\frac{1}{100}$), what is the probability that the accident will happen?

10

Solar Energy

10.1 MOTIVATION

Societal uses of energy are many and varied, and while some sources are versatile, no single one can serve all needs. Solar energy is versatile and is uniquely suited for some functions. Earl Cook* was inspired to write "life exists on earth by the grace of solar radiation" because of the unique role of solar energy in photosynthesis. Clouds form, rain falls, and rivers continually fill and empty all because solar radiation provides energy to evaporate water from the earth's surface. Winds develop because the earth's surface is heated unequally by solar radiation. Less impressive in magnitude but important to many are the warm, sun-drenched beaches with outstretched human beings bronzing their bodies with ultraviolet radiation from the sun. If solar radiation

---

* *Man, Energy, Society,* Earl Cook, W. H. Freeman, San Francisco, 1976.

Picturesque windmills have pumped water and produced electricity in rural areas in the United States for many generations. As the search for reliable energy sources continues, modern wind turbines like the one shown here may find increasingly important roles. (Photograph courtesy of Department of Energy.)

powers the rain cycle, creates the winds, and warms beaches, it would seem that buildings could be warmed, water heated, and electricity generated with energy from the sun. Indeed, all these tasks and more are possible with solar energy. But solar energy will not completely replace conventional energy sources in performing these tasks.

It is very easy to get excited about the prospects of using solar energy. After all, solar energy streams to us continually and is broadly available. We can also very easily become disillusioned, and many persons have felt their excitement wane. Solar energy is not free. True, you do not pay a solar fuel bill but a solar collector costs money, and a user must decide if the investment in a solar system is more rewarding than other investments. If the public finds that solar systems perform economically as well as technically, then solar technology will flourish. If solar systems do not measure up, then solar technology will not be used. Much thought goes into installing a solar system, and a user will invariably be disappointed if "corners are cut." In this chapter we want to explore the excitement and limitations of solar energy. Then when opportunities arise to invest in solar technology, we will be better able to weigh factors such as capital and maintenance costs, personal or national energy independence, land use, visual esthetics, and so on. Figure 10.1 maps our solar energy study route for this chapter.

10.2 GENERAL CONSIDERATIONS

A student knows well that there is a wide variation in university tuitions, and that she must pick a school that matches her financial resources and interests. More often than not, the student is able to provide only a fraction of the costs and she looks to scholarships, loans, employment, and so on for additional resources. It is much the same with homeowners seeking a household heating unit. There is a wide variation of types, and homeowners must pick one that matches their resources and requirements. Wise homeowners minimize the energy requirements and then choose a heating unit. If the

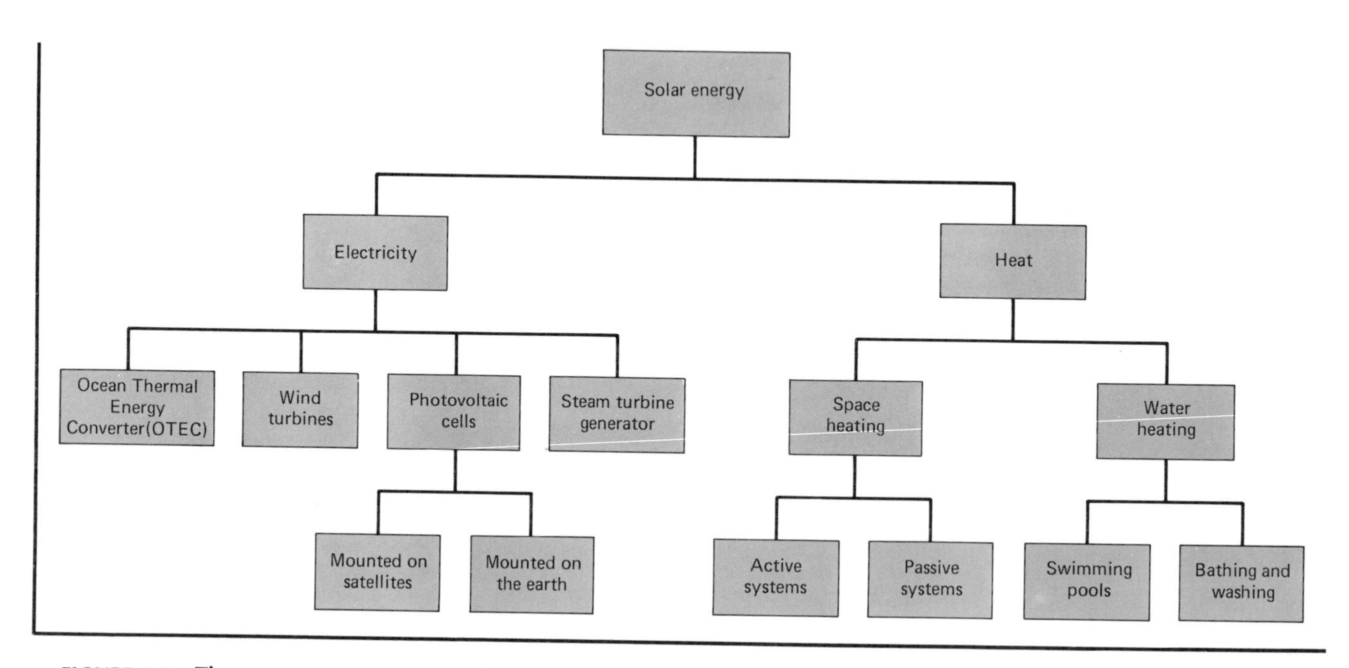

FIGURE 10.1 There are many uses of solar energy. This chart delineates most of the uses and charts our study strategy.

choice is solar, the unit will generally provide only a fraction of the requirements and homeowners look to additional energy sources such as fuel oil, natural gas, and electricity. Let us see why.

The earth rotates on an axis passing through the north and south poles and completes a full 360° rotation every 24 hours. The rotation creates the illusion of a sun that rises in the east, moves across the sky, and sets in the west. Winter daylight hours are shorter and colder than summer daylight hours, and shadows are longer in winter than in summer because the earth orbits elliptically around the sun. The northern part of the spinning earth tilts away from the sun in winter and toward the sun in summer (Fig. 10.2). A square foot of surface facing the sun just outside the earth's atmosphere intercepts 430 Btu of solar radiation each hour. The rate at which solar radiation falls on a square foot of the earth's solid surface is called the *solar insolation*. The solar insolation depends on the time of day, the day of the year, and where the square foot is located on the earth's surface. During the day the solar insolation is largest when the sun is at its highest elevation. This time of day is called solar noon. The elevation of the sun at solar noon is lowest on December 21 and highest on June 21. Thus shadows are longest on December 21 and shortest on June 21. Moving north from the equator, the solar insolation decreases because the sun's rays strike the surface less directly. Figure 10.3 shows the monthly average solar insolation at El Paso, Texas, latitude 31.8° north and Columbus, Ohio, latitude 40° north.*
Cloud cover and elevation are factors that affect the solar insolation at a particular site.

Regardless of latitude in the northern hemisphere, the availability of solar energy is smallest in winter and largest in summer. Conversely, building heating requirements are largest in winter and smallest in summer. Matching heating requirements with the heating unit is especially important in solar energy considerations.

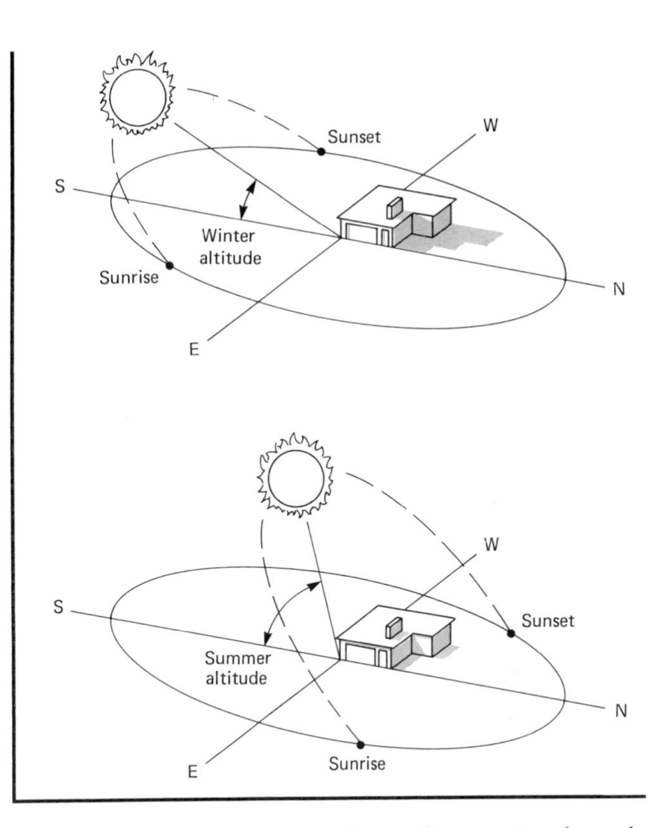

FIGURE 10.2 While rotating about a line passing through the north and south poles, the earth travels an elliptical orbit about the sun. Because the polar axis is tilted 23.5°, sunlight strikes the earth more directly in the summertime making days longer and shadows shorter than they are in wintertime.

When heat losses to the exterior of a building are balanced by heat gains from a heating unit, the temperature stays constant. A thermostat in a building is designed to turn the heating unit on when the temperature falls below a selected temperature and to turn the heating unit off when the temperature exceeds the selected temperature. The greater the heat losses, the greater must be the heat gains to maintain the selected temperature. If the heating unit has unlimited capacity and its operating costs are low, a homeowner may not worry about heat losses. But if the heating unit has limited capacity, the homeowner must reduce the heat

---

* Taken from a listing for some 50 cities in the United States and Canada in Appendix 3 of *The Passive Solar Energy Book*, Edward Mazria, Rodale Press, Emmaus, Pa., 1979. A map of the average annual solar insolation is presented on page 240.

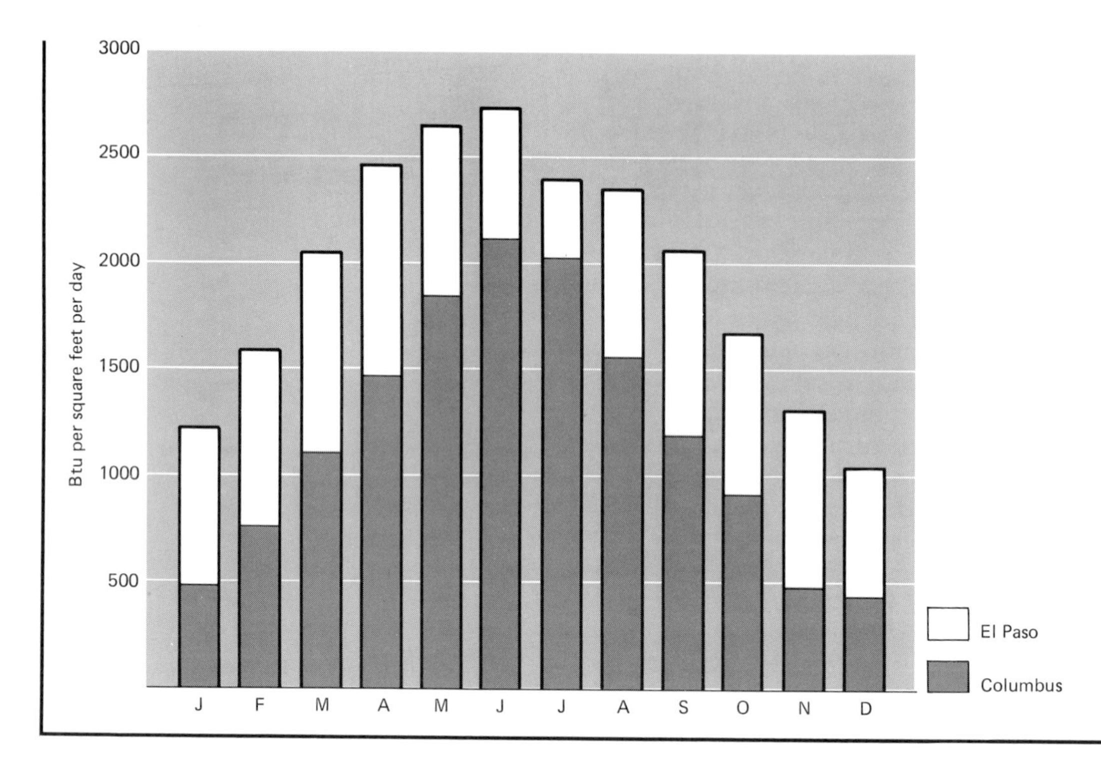

FIGURE 10.3 Illustration of the variation of solar insolation with season and with latitude. Because solar insolation varies significantly with location, detailed measurements are needed for sizing solar heating systems. Data are from Appendix 2, *Solar Heating Design,* William A. Beckman, Sanford A. Klein and John A. Duffie, Wiley, New York, 1977.

losses so that the heating unit can meet demands. As we shall see, energy provided by solar heating units is limited in most areas where building heating is required and a homeowner must make the home as energy efficient as possible.

Heat losses occur primarily by conduction (see Chapters 7 and 13) through walls, floors, ceilings, and windows, and by convection (see Chapter 7) through cracks and open doors. The care taken to make a home heat-tight is repaid in heating requirements of the home. Besides the heating unit, a home has a variety of heat sources such as people, cooking stoves, ovens, lights, and appliances. As a result, a home ordinarily requires no additional heat if the outdoor temperature stays above 65° F. When the temperature falls below 65° F, the heating unit is called on to provide heat. The longer the temperature stays below 65° F, the more heat the heating unit must provide to maintain the temperature selected on the thermostat. Thus the energy used in a home depends on the product of the interior–exterior temperature difference and the length of time the heating unit operates. Expressing the temperature difference in degrees F and the time in days yields a useful measure called heating degree days. If the average outside temperature was 42° F, we record the temperature difference as $65 - 42 = 23°$ F and the heating degree days as 23. Heating degree days are continually measured at a number of locations throughout the United States. As you would expect, the number of heating degree days increases as you proceed from south to north (Fig.

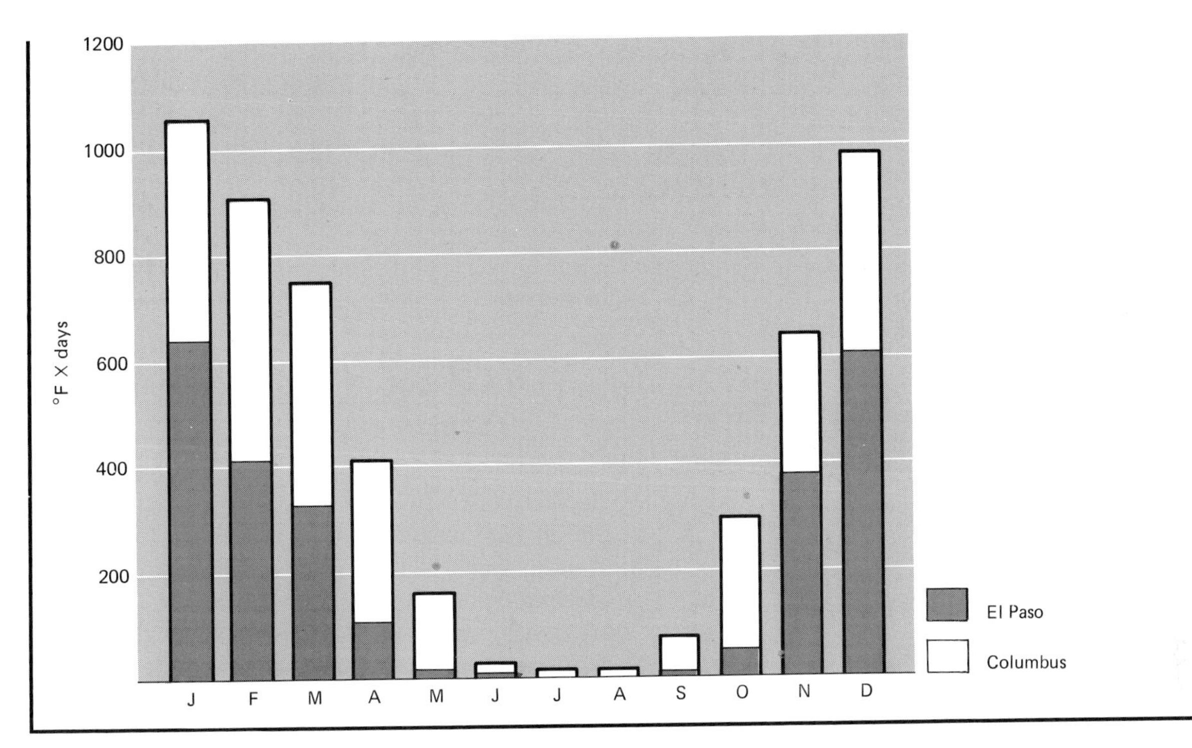

FIGURE 10.4 The average monthly heating-degree days for El Paso, Texas, and Columbus, Ohio. Comparing these data with the solar insolation data in Fig. 10.3, note that when heating requirements are highest, the solar insolation is lowest. Data are from Appendix 5, *The Passive Solar Energy Book*, Edward Mazria, Rodale Press, Emmaus, Pa., 1979.

10.4*). El Paso, Texas, averages 670 heating degree days in January while Columbus, Ohio, averages 1051 heating degree days for the same month. The number of heating degree days at the location of the home is used to estimate the heating requirements of a house.

Assessing the heating requirements of a house is straightforward. But it is tedious because you must systematically evaluate each window, wall, ceiling, and door. Someone should do this if a solar heating unit is being considered. Less accurate, but still meaningful, schemes are used to grasp the magnitude of the problems. A sufficiently accurate method for our purposes is the following. In addition to the number of heating degree days, the heat loss depends on the type of construction and the amount of floor space in the house. For a carefully built house in a suburban area, the construction factor varies from 6 to 12 Btu/degree day per square foot of floor space. For the sake of argument, let us assume the house is located in Columbus, Ohio, and has a construction factor of 9 and a floor space of 1200 square feet. As noted earlier, there are 1051 heating degree days in Columbus for the month of January. According

* Taken from a listing for the United States in Appendix 5 of *The Passive Solar Energy Book*, Edward Mazria, Rodale Press, Emmaus, Pa., 1979. Regional current values of heating degree days are presented in *Monthly Energy Review*, Superintendent of Documents, United States Government Printing Office, Washington, D.C. 20402.

to our model, the heat loss for this month is

$$\frac{9\,\text{Btu}}{\text{degree day} \times \text{ft}^2} \times 1200\,\text{ft}^2 \times \frac{1051\,\text{degree days}}{\text{month}}$$

$$= \frac{11{,}350{,}000\,\text{Btu}}{\text{month}}.$$

The average solar insolation for January in Columbus is

$$\frac{486\,\text{Btu}}{\text{ft}^2 \times \text{day}}.$$

We might reasonably expect to convert 60% of this solar insolation to useful space heating. Thus the useful heat amounts to

$$\frac{292\,\text{Btu}}{\text{ft}^2 \times \text{day}} = \frac{9050\,\text{Btu}}{\text{ft}^2 \times \text{month}}.$$

Hence the collector area needed is about

$$\frac{11{,}350{,}000\,\text{Btu/month}}{9050\,\text{Btu/month} \times \text{ft}^2} = 1250\,\text{ft}^2,$$

or roughly equivalent to all the floor area of the house. Admittedly, the calculations are approximate. But in areas in which space heating is critical, the fact is that solar insolation is limited, and it is difficult to match solar heating units to the heating requirements of well-constructed average houses. Generally, one must take advantage of every energy-saving concept in a house to minimize the heating requirements. Then a solar heating unit has some chance of providing a worthwhile fraction of the total heating requirements.

Let us now look at specific ways of utilizing solar energy for space heating. Bear in mind that we are examining the salient features of the subject. There are several textbooks treating the subject in great detail.

10.3 PASSIVE SOLAR HEATING

Two approaches to using solar energy for space heating are termed passive and active. A passive system has no mechanical pumps for circulating heat. The scheme relies on converting solar energy entering a building through south-facing windows into thermal energy. The thermal energy is distributed within the building by natural heat transfer methods of conduction, convection, and radiation. An active system uses pumps to remove thermal energy from a solar collector, usually placed on a south-facing roof and pumps to circulate heat within the building. Let us look first at passive solar heating.

In a passive system, solar radiation streams through south-facing glass windows to be absorbed by structures inside the building (Fig. 10.5). Thermal energy from the warmed structures is then distributed by natural means to other areas of the building. Thus the tasks include

1. optimizing the amount of incoming solar radiation and
2. maximizing the retention and distribution of thermal energy derived from the capture of solar energy.

A window is very effective at letting solar radiation into the building but it is very ineffective at stopping the escape of heat by conduction to the outdoors. The heat loss through a single-thickness glass window is about 20 to 25 times more than through a well-constructed and insulated wall of the same size. Thus windows through which solar radiation enters during the day must be covered with insulation at night to reduce the loss of heat to the outdoors. Building windows not on the south sides of the building should be as small as possible and double thickness with a small air space to reduce heat losses. The north-facing side being away from the sun is especially critical, and solar-heated homes often have very small, or no, windows on the north side. The north-facing side also usually bears the brunt of cold winter winds. The walls, ceilings, and floors of the building are constructed so as to minimize the escape of heat to the outdoors.* Passive solar-heated buildings are often placed partially underground to further reduce heat losses through walls and floors.

* We discuss in some detail in Chapter 13 the role of insulation in preventing conductive heat losses.

FIGURE 10.5 A single-family dwelling incorporating passive solar heating. Large, south-facing glass windows allow solar radiation to enter the living areas. Thermal energy derived from the solar radiation is circulated by natural means throughout the house. (Loren D. Wohlgemuth, Architect, Vancouver, Wash.)

Great care must be taken to prevent convective heat losses, commonly called infiltration losses, through cracks and doors opened for occupants to enter and leave. In an uninsulated house of average construction, air convection produces a complete change of air about every 30 minutes. In a quality house, a complete air change may take 80 minutes. A double-door entry whereby a person must enter a small entry room (vestibule) and close a door behind before entering the outdoors or main part of the structure minimizes heat losses when entering or leaving the building.

To prevent forced entry of cold outside air by winds, one might protect the structure with a screen of trees or some barrier serving the same purpose.

The interior structure absorbing the solar radiation may be a part of the living area, for example, a wall, a chimney, or a fireplace. This concept is termed direct (thermal) gain. The energy-absorbing unit may not be a part of the normal living area but a structure between the south-facing windows and the living area. In this case, heat is moved from the structure to the living area by natural methods. This concept is termed indirect gain.

Any material becomes a thermal energy reservoir if heat is added to it. Thus if 500 Btu of heat are added to each of two substances, the energy content of each increases by 500 Btu regardless of their physical size or makeup. At the very least, the thermal energy reservoir for a passive solar system must retain sufficient thermal energy to provide adequate space heat when the sun is not shining. To accomplish sufficient storage, the reservoir must be of reasonable size, its temperature must not be excessive, and its cost reasonable. From these considerations alone, water is the most practical medium. A cubic foot (about eight gallons) of water retains 62.4 Btu of thermal energy for each 1° F its temperature is increased above the ambient room temperature. For a temperature increase of 5° F each cubic foot of water could retain $5 \times 62.4 = 312$ Btu. On a cold January day in the central United States, a well-constructed home may require 100,000 Btu of nighttime space heating. Thus the volume of water required to store the thermal energy is

$$\frac{100{,}000\,\text{Btu}}{312\,\text{Btu/ft}^3} = 320\,\text{ft}^3.$$

In the form of a cube, each side of the required container would have a length of about seven feet, which is not prohibitively large. Many indirect gain designs employ water for the thermal energy storage medium (Fig. 10.6). The containers for the water are painted black to enhance the absorption of solar radiation. To avoid corrosion and rust, concrete and other nonmetallic materials are preferred to metallic containers. Incorporating containers of water into direct gain systems is not impossible but does require some architectural ingenuity. Direct gain systems generally employ concrete and masonry made of such things as brick, small stones, or rocks. Concrete and masonry have a heat capacity about half that of water and they are much easier to incorporate into the living areas of a building. Floors as well as specially constructed walls are used to store thermal energy. As an approximate rule, the area of the south-facing windows in a direct gain system is 20 to 25% of the floor area of the building. About 16 pounds of water or about 75 pounds of masonry are needed for each square foot of floor space. For an indirect gain system, the same amount of thermal energy storage material is needed but the south-facing window area increases to 40 to 50% of the floor area.

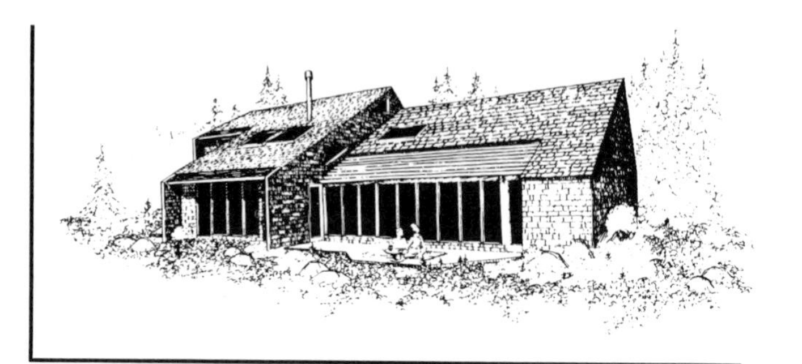

FIGURE 10.6 An example of a two-bedroom home with 1400 square feet of living space designed for the Puget Sound area in Washington. Black tubes containing 5400 pounds of water along the back wall of the greenhouse and a concrete floor in the greenhouse serve as thermal energy storage for indirect gain. An eight-inch masonry wall, masonry floors, and a masonry hearth are used for direct gain. There are 280 square feet of south-facing, double-thickness glass in front of the living area and 150 square feet of double-thickness glass for the greenhouse. Concrete and masonry in the floors and hearth total 57,000 pounds. About 56% of the heating load is provided by solar energy. Makeup heat is delivered by a wood-burning stove, a heat pump, and a forced-air electric system. Special features to minimize heat loss include minimal windows on the nonsouth sides, earth berming on the north and west sides, using the garage space for insulation on the east side, insulating curtains and rigid foam panels covering the south-facing windows at night, carefully constructed walls and floors, and using trees for wind shields. (Olson–Rowe Architects, A.I.A., Lacey, Wash.)

A rather popular indirect gain system employs a Trombe* wall. A Trombe wall is eight to eighteen inches thick and is placed about four inches behind a south-facing double-thickness glass window (Fig. 10.7). The wall is usually constructed of concrete with a rough, dark texture and with vents at the top and bottom. When the wall warms, air between the wall and window rises. Cool air from the room moves through the bottom vents to be heated in the space between the wall and window. A convection heat current is set up, warming the interior of the

* Named for Dr. Felix Trombe, a researcher at the Centre Nationale de la Recherche Scientifique in Odeillo, France, where the Trombe wall was developed in the 1960s.

building. At night, the windows are covered and the vents blocked so that the wall may radiate heat to the interior of the building. During summer the top vents may be closed and vents opened to the outside of the building. A convection current is produced that pulls warm air out of the house and moves it to the outside producing a cooling effect in the building.

A south-facing roofline extending beyond the south outside wall of a house can prevent much of the direct sunlight from entering the house in summertime. As the sun moves lower in the sky in the winter months, the extended roof no longer blocks the direct sunlight and solar radiation streams in as desired. Knowing the seasonal sun elevation at the location of the house, the extent of the overhang can be optimized. At a latitude of 40° north (about the location of Interstate Highway 70) the maximum sun elevation is 73.5° (June 21) and the minimum sun angle is 26.5° (December 21).

Greenhouses attached to the south side of a building are a popular variation of an indirect gain system (Fig. 10.8). The greenhouse gives more insulation to the south wall of the house and provides an air lock for entering and leaving the house. Some ingenuity is required to create convection currents to heat the house.

FIGURE 10.8 The black, vertical cylindrical structures are designed to store thermal energy. Sunlight entering the building is absorbed by the black surfaces. The absorbed energy raises the temperatures of the medium confined by the cylinders. At night, the thermal energy absorbed during the daytime is radiated to the interior of the house. (Photograph courtesy of Solar Components Corporation, Manchester, N.H.)

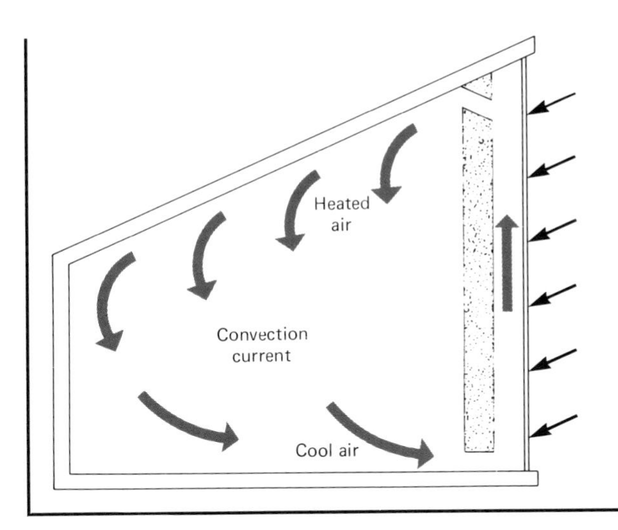

FIGURE 10.7 The Trombe wall concept. A concrete wall eight to eighteen inches thick covers south-facing windows. Convection currents formed in the space between the wall and the windows circulate heat to the interior of the house. (From *Introduction to Solar Energy,* Marion Jacobs Fisk and H. C. William Anderson, Addison-Wesley, 1982.)

10.4 ACTIVE SOLAR HEATING

An active solar heating system converts solar energy to thermal energy in collectors removed from the living area of a building. Solar collectors vary in design but are generally box-shaped about four feet wide, eight feet long, and six inches thick (Fig. 10.9). The cover plate is transparent to solar radiation and is usually made of glass. A black, metal sheet

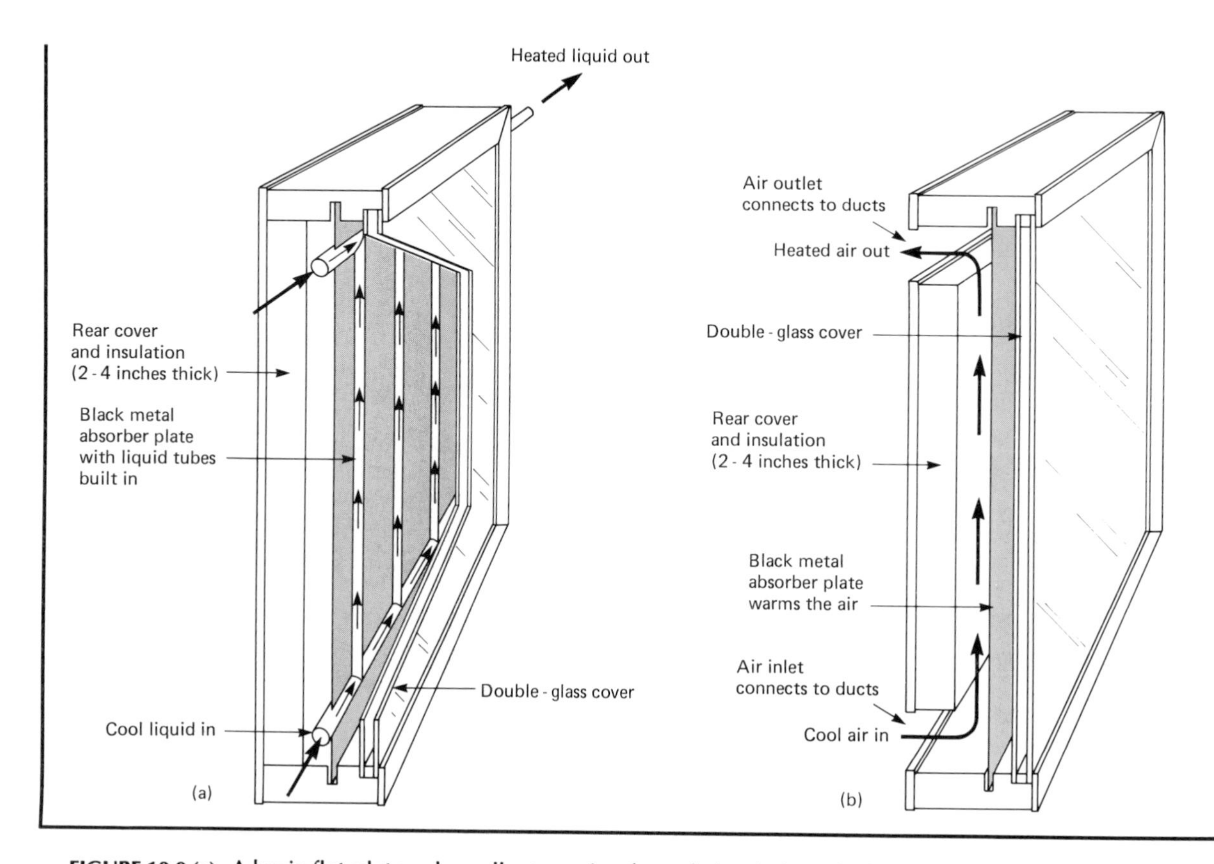

FIGURE 10.9 (a) A basic flat-plate solar collector using forced circulation of a liquid to remove captured thermal energy. The left end and part of the glass cover plate are cut away to show the interior. Solar radiation warms the black metal plate. Thermal energy is transferred to a liquid circulating through pipes bonded to the black absorber plate. The double-thickness glass cover limits the escape of infared radiation and also limits convection energy losses. Insulation on the back and sides of the collector limits energy losses by conduction. **(b)** A basic flat-plate solar collector using forced air circulation to remove captured thermal energy. The left end is removed to show the interior. The energy collection principle differs in no essential way from the flat-plate solar collector shown in **(a)**. Cool air forced into the collector is warmed when contacting the black absorber plate. The warm air is either circulated to the interior of the house or to a thermal energy reservoir usually made of rocks or small stones. (Adapted from *Introduction to Solar Technology*, Marian Jacobs Fisk and H.C. William Anderson, Addison-Wesley, 1982.)

behind the transparent cover absorbs the incoming radiation and becomes warm. Like any warm object, the absorber plate emits thermal radiation. Because the temperature of the plate is roughly 400 K, the radiation is mostly long-wavelength infrared far removed from the much shorter ultraviolet, visible, and infrared components in the incoming solar radiation. Radiation from the absorber plate cannot penetrate the cover plate, energy is trapped within the collector, and the collector becomes warm. Thermal energy is removed from the collector by air or a liquid circulating in thermal contact with the absorber plate. Two to four inches of high-quality insulation in the back of the absorber limits heat losses by conduction. The cover plate limits heat losses from convection currents in the collector and prevents wind from removing heat from the collector.

An ideal solar collector should face the sun at all times but for it to do this would require an elaborate mechanism to track the continual change of the sun's position in the sky. As a compromise, the collector can be faced toward the south at an angle relative to the horizontal which is 15° plus the geographic latitude. For example, a solar collector in Chicago would be mounted at an angle of 42° (for latitude 41.6° N) + 15° = 57°. Fifty-seven degrees is a rather steep angle, and that is why roofs supporting solar collectors appear different from roofs on conventional houses (Fig. 10.10). The collecting area for a well-designed house is roughly one-third of the floor space of the area to be heated.

The components of an active solar system using a liquid to transfer heat is shown in Fig. 10.11. Water

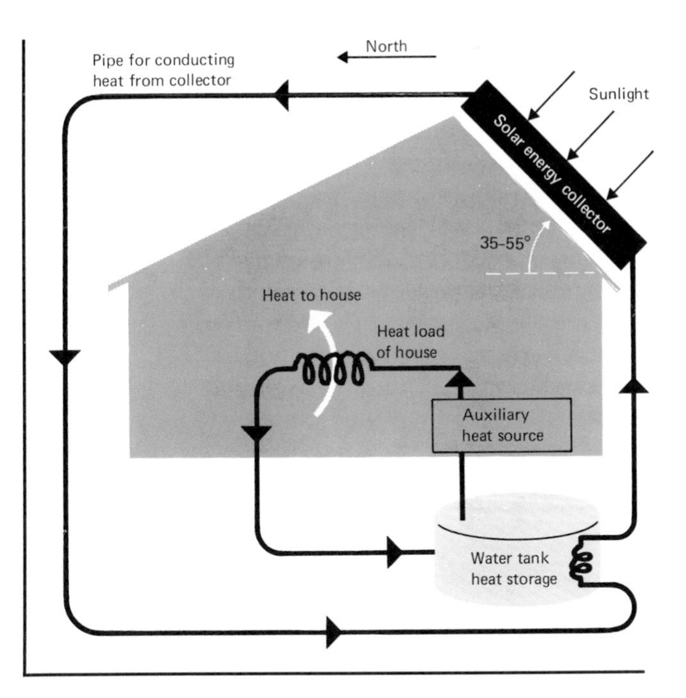

FIGURE 10.11 Schematic drawing of a solar heated house with liquid-cooled collectors and water for thermal energy storage. Heat is removed from the collectors by an appropriate fluid and transferred to the water tank. Heat is drawn from the tank as needed. An auxiliary heat source is used when the tank contains insufficient thermal energy.

FIGURE 10.10 A solar heated residence in Madison, Wisconsin. The collectors in this system heat air, and energy is stored in a pebble bed. Ideally the collector should face the sun directly to maximize the incoming solar energy. The fixed, steep angle for the collectors is a compromise because the sun cannot be continuously tracked as it passes across the sky. The small windows on the side of the house limit energy losses by conduction to the outside of the house. (Photograph courtesy of the University of Wisconsin Solar Energy Laboratory.)

is used to store thermal energy for nighttime use and for days when the sun is not shining. About three gallons of water are used for each square foot of collector area. The storage tank occupies a little over 1% of the floor area of the house.

The components of an active solar system using forced air to transfer heat are shown in Fig. 10.12. Rocks or stones are generally used for the thermal energy storage medium. The rocks or stones warm when warm air from the collectors is forced to circulate through them. The heat capacity of rocks is lower than water (Section 10.3). Consequently, about 40% more material, by weight, is required for thermal storage in an air-cooled collector system. Warm air from the collector may be circulated directly to the living areas rather than to the thermal storage area if the house is in immediate need of heat.

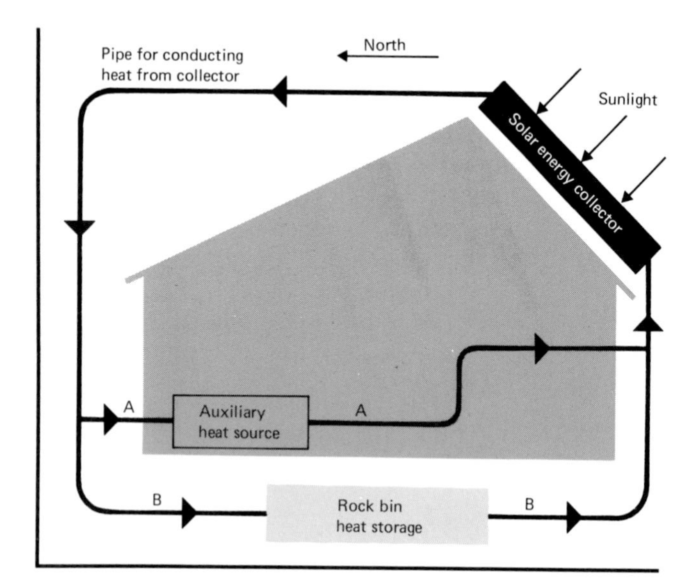

FIGURE 10.12 Schematic drawing of a solar heated house with air-cooled collectors and rocks for thermal energy storage. Heat is removed from the collectors by circulating air over them. Normally the energy is fed directly to the house (path A). If more energy is available than needed, it is fed into the rock bin storage unit (path B).

Active solar systems are simple in principle, and users are tempted to assemble their own from a plethora of available plans. However, the task is not as simple as it may seem. To take full advantage of the limited amount of solar radiation during the heating season, considerable thought and ingenuity and first-class components are required. As for passive solar houses, energy efficiency is the byword. Once the prospective user has a good grasp of the heating requirements, then the solar system can be properly sized. The variety of factors to be considered include

1. a detailed knowledge of the collector efficiency for converting solar energy to thermal energy,
2. accurate solar insolation data,
3. the effect of solar collector orientation,
4. the contributions of indirect radiation, and
5. the sizing of the thermal storage unit.

Economic performance is intimately linked with the technical performance. A solar system providing 100% of the heating requirements of a house is generally prohibitively expensive. Thus a user should pick a combination of solar and conventional heating systems. The user must decide if the economics of the combination are more favorable than for a conventional system alone. The many factors to be considered include

1. initial cost,
2. operating and maintenance costs,
3. mortgage costs,
4. property tax,
5. insurance costs,
6. state and federal tax savings and tax credits,
7. estimates of inflation, and
8. projections of conventional fuel costs.

Of the many books dedicated to the assessment of active solar systems the reader is referred to

> *Introduction to Solar Technology,* Marian Jacobs Fisk and H.C. William Anderson, Addison-Wesley, Reading, Mass., 1982; and *Solar Heating Design by the f-Chart Method,* William A. Beckman, Sanford A. Klein, and John A. Duffie, Wiley, New York, 1977.

The idea of heating a building with solar energy is obvious. Interestingly, a building can also be cooled with solar energy. Basically the solar energy is used to run a refrigerator. Refrigerators remove heat from some chamber and deposit this heat plus the energy required to do the extraction somewhere away from the chamber. In a building, the chamber to be cooled is its interior. There are different types of cooling systems. Some, like a common household refrigerator, use an electric motor to do the work. Others do not require a motor or an engine. All are based on the same general principle of a circulating gas (such as ammonia in a refrigerator) that is easily converted to a liquid under pressure (Fig. 10.13). The refrigerator motor does the work of compressing gaseous ammonia into the liquid state. Heat released in the process is shunted to the exterior. The liquid ammonia then flows to the evaporator in which it returns to its natural gaseous state. In doing

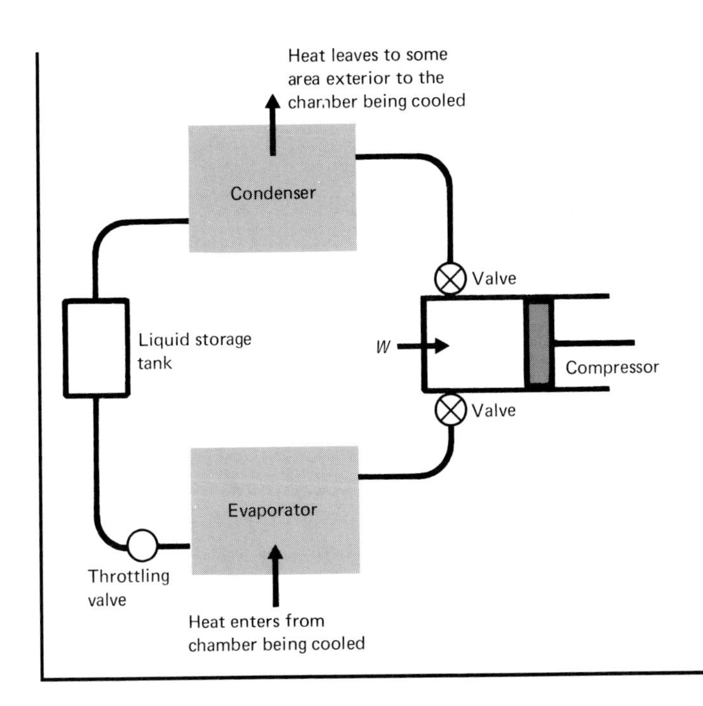

FIGURE 10.13 The cooling cycle in an ordinary refrigerator. When the liquid evaporates to a gas, it draws heat from the chamber being cooled. After being pumped from the evaporator, the gas is fed to the condenser where it loses heat in condensing back to a liquid. The liquid is fed back to the evaporator and the cycle is repeated.

this, it extracts heat from the substances it is intended to cool. The cycle is repeated over and over. In a household refrigerator, the evaporator is located in the freezer compartment and the condenser at the rear of the refrigerator. Although the evaporator is usually hidden from view, you can usually see the condenser.

While the technicalities of solar heating and cooling are well understood, it will be difficult to integrate them into our society because of the reluctance of the public to accept new technology. Tax incentives and low-interest rate loans help promote the use of solar energy. Although the widespread utilization of solar energy will take time, it will come. If 20% of the energy needs of the United States were supplied by solar energy in A.D. 2020, the arduous development path would be worthwhile.

10.5 SOLAR HOT WATER HEATING

Adapting a solar space heating system to a conventional home is generally very difficult because the house is usually not properly designed for energy efficiency and the house orientation is not optimal. Adapting (retrofitting) a solar system for providing household hot water is a much more feasible project for an existing house. Let us see why.

Every home has daily needs for hot water. Using a conveniently located water heater energized by a flame or electric heating element, water at 60° to 65° F from the local water system is warmed to 115° to 140° F. Every gallon of water warmed requires eight Btu for each Fahrenheit degree of warming (see Section 7.2). Thus 560 Btu are required to warm each gallon from 60° F to 130° F. The higher the temperature of the water entering the heater, the less energy is required to achieve the desired hot water temperature. An auxiliary source preheating the water cheaply produces a monetary saving. Even if the preheater cannot complete the heating task, the more expensive conventional oil-fired, gas-fired, or electric unit need only make up the difference. The economics and technology are very favorable for the performance of this preheating task by a solar water heater.

To estimate the requirements for a solar water

heater, we need to know

1. the daily hot water usage,
2. the desired hot water temperature,
3. solar insolation data, and
4. the collector efficiency.

Each day a household uses about 20 gallons of 130° F hot water for each person. The fraction of energy obtained from the sun is determined in large part by how much the user is willing to pay. There will be days when there is inadequate solar energy. The user who relies entirely on solar energy must have a thermal storage unit. Generally a user will opt not to use a thermal storage unit and, instead, rely on the conventional backup system on days when solar energy is inadequate. Let us assume the following conditions.

number of occupants = 4.

initial water temperature = 60° F.

daily usage rate = 20 gallons per person.

collector system efficiency = 60%.

solar insolation = 1000 Btu/ft² day, a reasonable average for the central United States.

Using these requirements we have thermal energy available

$$= \text{solar insolation} \times \text{collector system efficiency}$$

$$= \frac{1000 \text{ Btu}}{\text{ft}^2 \text{ day}} \times 0.60$$

$$= \frac{600 \text{ Btu}}{\text{ft}^2 \text{ day}}$$

and thermal energy required to heat water

$$= \frac{20 \text{ gallons}}{\text{day person}} \times \frac{8 \text{ Btu}}{\text{gallon °F}} \times 4 \text{ persons} \times 70° \text{ F}$$

$$= 44{,}800 \text{ Btu per day.}$$

Hence the collector area required is

$$\frac{44{,}800 \text{ Btu}}{\text{day}} \div \frac{600 \text{ Btu}}{\text{ft}^2 \text{ day}} = 75 \text{ square feet.}$$

If each solar panel were four feet by eight feet, then three panels would be more than the estimated size of 75 square feet.

Although values assumed in such calcuations will vary, it is clear that the size of a solar collector for heating water is not prohibitively large (Fig. 10.14). Because hot water is needed at all times of the year, the economics are generally favorable for solar hot water systems. A potential user must do or have done a careful analysis, and services are available for the analyses.*

Solar water heaters generally employ liquid-cooled, flat-plate collectors similar to those used for space heating (Fig. 10.9). In simple designs, cool water is pumped to the collector, warmed, and returned to a rather standard hot water tank having backup heaters for times when solar energy is unavailable. To prevent water from freezing in the collectors on cold nights and cold days with no sunshine, water must be drained from the collectors. This is usually done by an automatic control system. More elaborate systems use an antifreeze solution much like that used in automobiles to remove heat from the collectors. Thermal energy is transferred to the building hot water supply by a heat exchanger located in the building requiring the hot water. Most antifreeze solutions are toxic, so it is extremely important to have quality plumbing and excellent work. A potential user is well advised to examine carefully the credentials of the builder and installer of the system.

10.6 GENERATING ELECTRICITY WITH SOLAR ENERGY

The solar power impinging on a square 12 inches on a side is about 60 watts, and roughly equivalent to the electric power needed to operate a desk lamp. To achieve 60 watts of electric power from a conversion of solar power at 10% efficiency would require a solar power input of 600 watts. A square receiving

---

* For a summary of analysis methods for solar heating and cooling applications and for information on analysis services, see *Analysis Methods for Solar Heating and Cooling Applications, Passive and Active Systems,* Solar Energy Research Institute, Golden, Colo.

FIGURE 10.14 Two south-facing, flat-plate solar collectors on the roof of this home in the southwestern United States are adequate for a solar hot water system. (Photograph courtesy of American Solar King Corporation, Waco, Texas.)

area of about 38 inches (or about three feet) on a side would be required and this begins to get large as you consider more energy-demanding appliances. Still if the solar power falling on 1% of the area of the United States were converted to electric power at an efficiency of only 1%, the power produced would exceed the present electric power output of all generators in the United States. If all 50 states in the United States were of equal area, the equivalent of one half of one state would be sacrificed as a solar energy receiver. The energy numbers are attractive, and the sacrifice in land area seems small. However, the task is difficult.

A solar power system producing about 1000 megawatts of electric power requires a collecting area of a few miles on a side. Smaller systems require proportionally smaller collecting areas. A straightforward way of converting solar energy to electricity is to concentrate the solar radiation with lenses or mirrors to produce thermal energy. The heat could then be channeled into a conventional steam turbine system. This ploy is faced with the fact that solar energy is not available on a 24-hour basis whereas the demands for electric energy are. There are several possible ways to circumvent this difficulty. Conceivably, electric energy could be stored using batteries. Or some of the electric energy could be used to pump water to an elevated reservoir. Then, as energy was needed, the water could be used in a hydroelectric system. Another possibility envisions using the electric energy output to produce a fuel that could then be transported and used as needed. One such scheme dissociates water into hydrogen and oxygen (Section 10.7). The hydrogen is then used as an efficient nonpolluting fuel.

A Solar-thermal-electric Power System

A diagram of a solar-thermal-electric energy conversion scheme is shown in Fig. 10.15. The scheme differs from a fossil fuel system in two important aspects.

1. the source of heat is different, and
2. there are provisions for storing thermal energy.

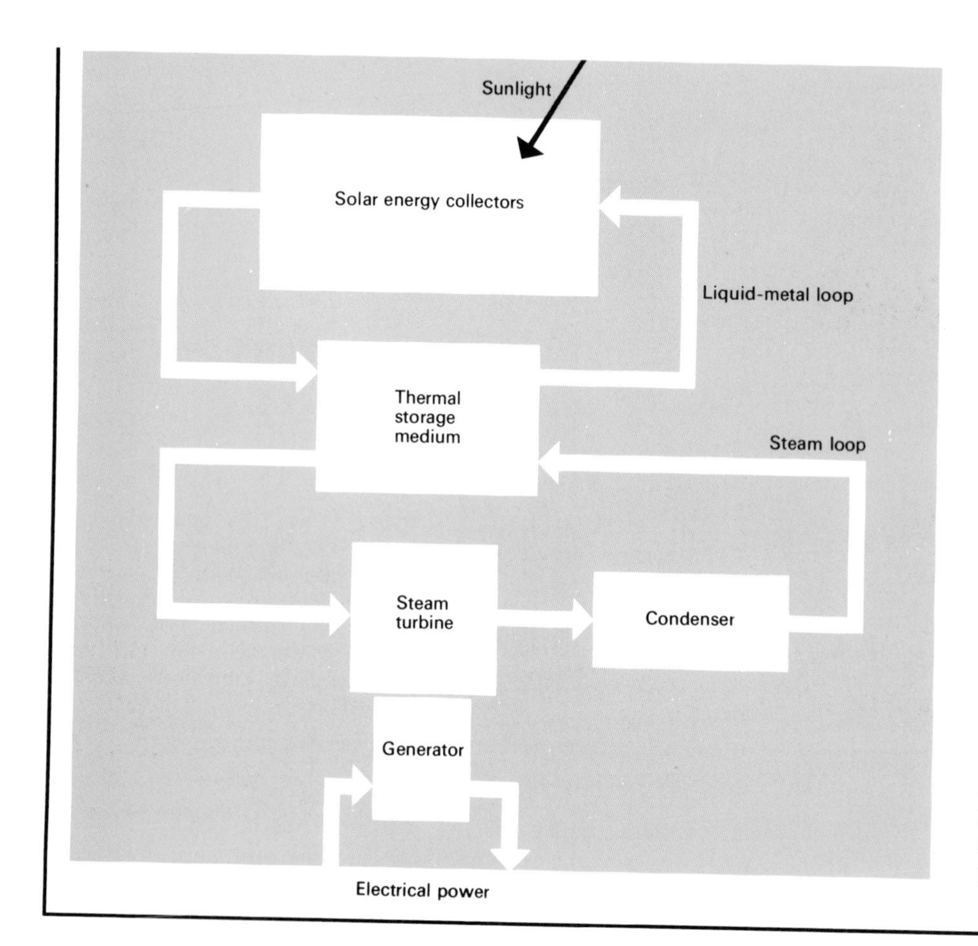

FIGURE 10.15 The major components of a solar electric plant. Except for the manner of producing thermal energy, it is very much like a conventional coal-burning or nuclear-fueled electric power plant.

Solar One is a solar-thermal-electric power plant at Daggett, California, about 12 miles east of Barstow. More than 1800 mirrors, each having a surface area of about 40 square meters, track the sun and focus solar radiation on a configuration of cylindrical tubes mounted atop a tower 86 meters high (Fig. 10.16). A computer-controlled mechanical system continually orients the mirrors toward the sun. An electrical output of about 10,000,000 watts is produced. An oil/rock thermal storage system permits the system to produce about 7,000,000 watts of electric power for four hours after sunset. Solar One began supplying electricity to existing power lines in 1981.

A scheme similar in principle to Solar One employs long, parabolic troughs with mirrored surfaces to focus solar radiation onto a pipe running the length of the collector (Fig. 10.17). Thermal energy produced in the walls of the pipe is removed by a fluid circulating in the pipe. The energy can be stored in an appropriate thermal energy reservoir or used directly to produce steam for a turbine. A facility using this concept is planned for an area near the Solar One project. About 18 acres of collecting area on 90 acres of land will convert enough solar power to produce 12,000,000 watts of electricity. These facilities are paving the way for commercial solar power plants with electric power outputs in excess of 50,000,000 watts.

Generating Electricity with Photovoltaic Cells

A photovoltaic cell, commonly called a solar cell, converts electromagnetic (solar) energy directly to electric energy. There is no steam turbine and no

FIGURE 10.16 Some of the 1818 mirrors in the Solar One project at Daggett, California. Each mirror focuses sunlight onto the black cylinder seen atop the tower in the background. Thermal energy produced in the cylinder is transferred to the boiler of a steam turbine coupled to an electric generator. (Photograph courtesy of the Department of Energy.)

FIGURE 10.17 Focusing collectors for a planned solar-thermal-electric power plant near the Solar One project. Sunlight is focused onto the pipe running the length of the collectors. Insulated pipes in the foreground are used to transfer thermal energy to a steam turbine coupled to an electric generator. (A project of Acurex Solar Corporation and Southern California Edison.)

electric generator. Miniature photovoltaic cells are used as light sensors to control the shutter speed and aperture in a camera. Arrays of photovoltaic cells provide electric power for many satellites in the space program. While the fundamental physics of photovoltaic cells is difficult, the salient features can be understood with the physics background at hand.

Metals such as copper and aluminum are referred to as conductors because electrons flow easily through them. Glass, wood, plastic, and other electrically similar materials do not readily conduct electrons. They are termed nonconductors (or insulators). There is a class of materials termed semiconductors having conductive properties intermediate to insulators and conductors. The elements germanium and silicon and compounds such as gallium arsenide, copper sulfide, and cadmium telluride are noteworthy semiconducting materials. Modern electronic innovations such as computers, hand calculators, and hi-fi sets owe their success to microscopic electrical elements made of semiconductors. Semiconductors are categorized as either n-type (negative) or p-type (positive). In an n-type semiconductor, the current mechanism involves negative charges (electrons); in a p-type semiconductor, the current mechanism effectively involves positive charges. This motion of positive charges is called hole conduction. A structurally continuous sandwich of p- and n-type materials is called a p–n junction. A photovoltaic cell is made of a p–n junction (Fig. 10.18). When solar energy is absorbed on one surface of the cell, electron-hole pairs are liberated. Electrons are attracted by an internal electric field into the n-region; holes are attracted to the p-region. A charge separation, and therefore a potential difference, evolves from the resulting separation of positive and negative charge. An electrical device connected across the cell provides a path for charge flow. As long as the cell is illuminated, it behaves like a battery producing a potential difference of about 0.4 to 0.6 volts. While the potential difference developed by a single photovoltaic cell does not depend on the size of the cell, the electric power produced will. A tiny solar cell used in the automatic exposure mechanism of a camera may have a sensitive area of about four square millimeters. Manufacturing considerations preclude building a single cell of arbitrary size. However, smaller units can be connected into arrays (see Fig. 10.19) to increase either or both the potential difference and power. The Skylab space satellite was powered with a solar cell array producing 20,000 watts of electric power. This is sufficient to provide electricity for about seven homes.

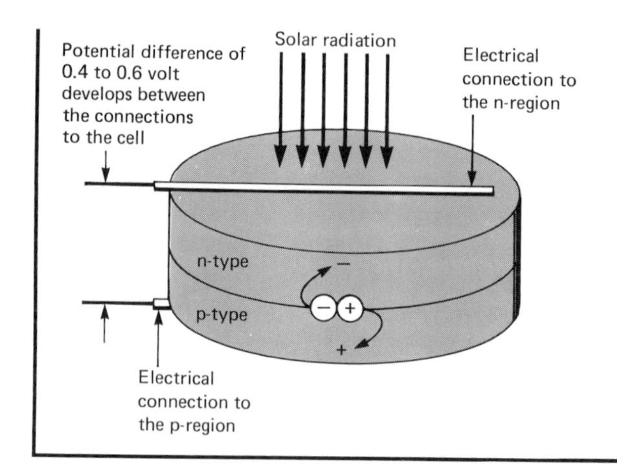

FIGURE 10.18 Schematic drawing of a solar cell. The cell is about as thick as a dime (about 1 millimeter). Depending on the electric power output, the area varies from a fraction of a square centimeter to a few square centimeters. Absorption of a photon of light liberates an electron (−) • hole (+) pair. An internal electric field across the p–n junction forces the negative charge into the n-region and the positive charge into the p-region. The accumulation of negative charges in the n-region and positive charges in the p-region gives rise to a potential difference between the n- and p-regions. The solar cell behaves like a small battery.

Photovoltaic cells in cameras are usually made from cadmium sulfide (CdS). They are relatively cheap but have a light-to-electric energy conversion efficiency of only 4–5%. Photovoltaic cells for the space program are made from silicon (Si) and have energy conversion efficiencies of 10–15%. Experimental photovoltaic cells have efficiencies as large as 18%. Being an elemental constituent of sand,

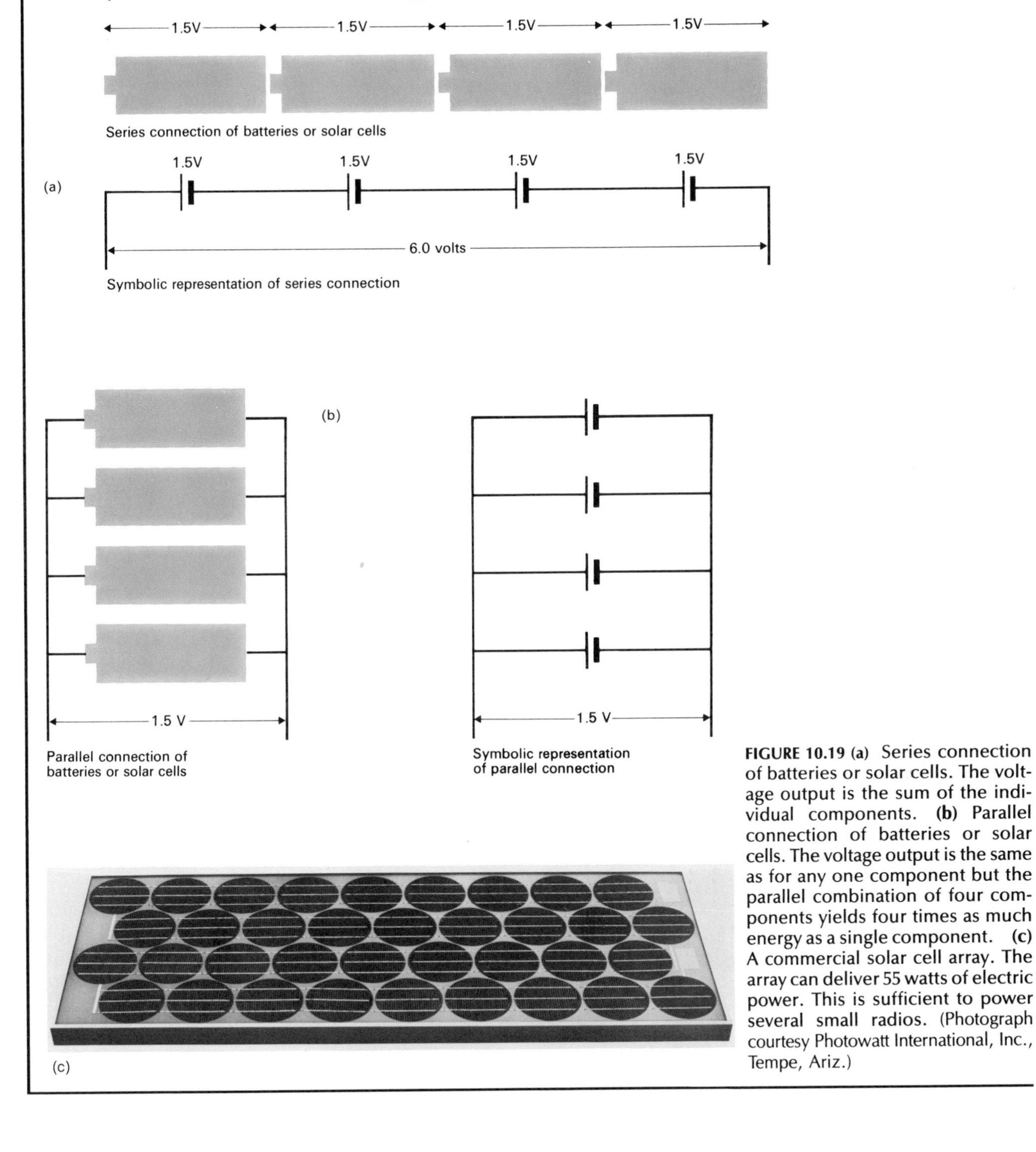

FIGURE 10.19 (a) Series connection of batteries or solar cells. The voltage output is the sum of the individual components. **(b)** Parallel connection of batteries or solar cells. The voltage output is the same as for any one component but the parallel combination of four components yields four times as much energy as a single component. **(c)** A commercial solar cell array. The array can deliver 55 watts of electric power. This is sufficient to power several small radios. (Photograph courtesy Photowatt International, Inc., Tempe, Ariz.)

silicon is extremely abundant. However, several stages of purification are required to make electronic-grade material. Coupled with a tedious fabrication process, this tends to inflate the production costs of silicon solar cells. Presently the cost of producing electric power with solar cells is not competitive with conventional methods. Persistent research and development continue to lower manufacturing costs and improve reliability, and photovoltaic cells are available commercially. Remote areas inaccessible to conventional electric power are particularly attractive for applications. An irrigation facility in Nebraska is powered with a 25,000-watt array of photovoltaic cells. Testing of photovoltaic systems in residential and commercial buildings is underway. Figure 10.20 shows an array of photovoltaic cells furnishing electricity for a home in Carlisle, Massachusetts. A photovoltaic power plant having a peak output of 1,000,000 watts to be fed into existing electric power lines is planned for Hesperia, California, about 50 miles south-southwest of the Solar One project.

A large-scale (1000-megawatts) solar cell facility would have the same land-use problem of any other solar energy collector of similar power output. A scheme that avoids this particular problem envisions 12-square-mile arrays of solar cells in orbit around the earth (Fig. 10.21). The system takes advantage of the larger concentration of solar energy outside the earth's atmosphere. It would orbit at an

FIGURE 10.20 This demonstration home in Carlisle, Massachusetts, derives electricity from an array of photovoltaic cells producing a peak power of 7300 watts. (Project of Solarex Corp., Rockville, Maryland, and Architects and Engineers: Solar Design Associates, Lincoln, Massachusetts.)

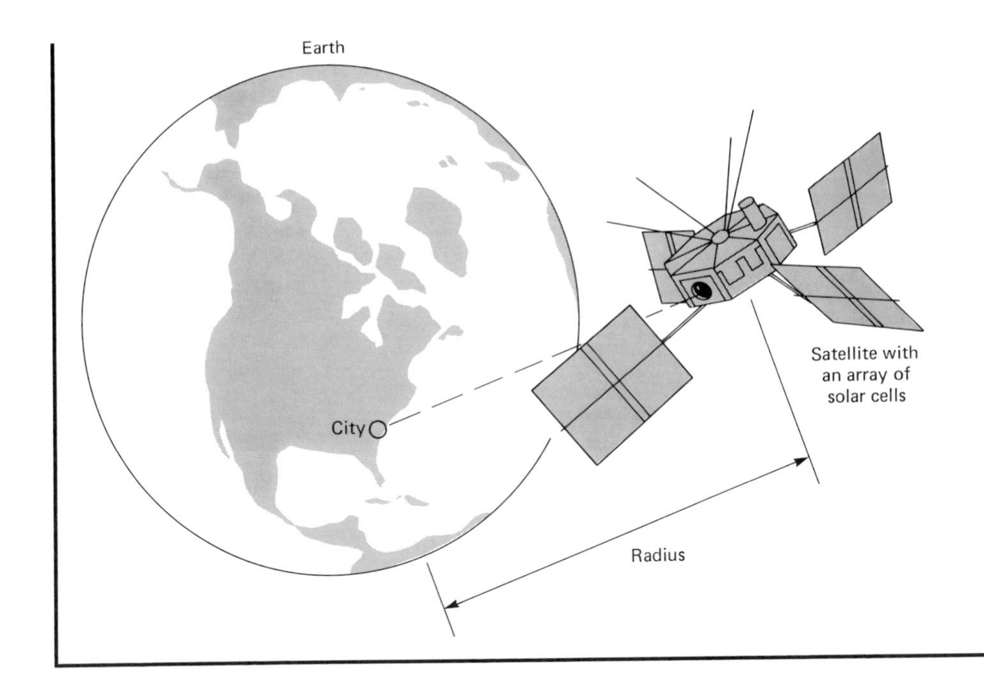

FIGURE 10.21 A satellite orbiting the earth could conceivably collect solar energy, convert it into electric energy, and transmit microwave energy to the earth. In order to serve a community on the earth, the satellite must maintain its position over the area it is serving.

altitude of about 22,000 miles from the earth's surface and would be synchronized with the earth's rotation so that it would always appear directly above some position on the earth. Being far out in space, the solar input is available on a nearly continuous basis. The electric energy generated would be transmitted to earth by electromagnetic radiation in the form of microwaves. The receiver would be an open-mesh antenna six miles in diameter. Such a system conceivably could generate and supply enough electric power to satisfy the needs of a city the size of New York City. Because the fabrication costs of the solar cell arrays, the expense of the launching, and the engineering expenses preclude the use of these space systems in the near future, the concept survives but interest in it has waned.

10.7 ENERGY FROM THE WINDS

After the rays of the sun have made their trek through the earth's atmosphere, about 1% of the energy reaching the earth is converted to atmospheric motion. Witnesses of hurricanes and tornadoes know that this energy is not distributed uniformly; however, there are regions where the winds are reasonably continuous. The most noteworthy areas are the New England and Middle Atlantic Coasts, along the Great Lakes, the Gulf Coast, and the Aleutian Islands, and through the Great Plains, Rocky Mountains, and Cascade Mountains. Wind power in these regions is estimated to be 100 million megawatts.

Wind power is not a new idea in the American scene. In the 1920s and early 1930s, wind-powered electric generators were familiar landmarks on farm buildings. Wind-driven water pumps are still widely used in the Great Plains region. Abandoned windmill towers, now often supporting TV antennas, are common sights in rural America. But the Rural Electrification Administration established by President Franklin D. Roosevelt virtually ended the rural use of wind power. Reasonably large-scale, wind-powered electric generators have also been constructed. A 1.25-megawatt wind-powered generator was operated for 16 months during World War II at "Grandpa's Knob" in central Vermont. Power generation stopped after a bearing failure in 1943. Returned to operation in 1945, one of the machine's eight-ton blades fractured and separated. The project was never revived largely because of the widespread acceptance of electric power from conventional sources. A serious interest in wind power has returned in this era of diminishing fossil fuels and air pollution problems.

A wind turbine converts the kinetic energy of wind into rotational mechanical energy. We should expect the energy imparted to the rotating turbine to increase as the wind speed increases. Additionally, larger diameter blades intercept more of the wind and therefore impart more energy to the shaft of the turbine. Therefore, we reason that the power developed by a wind turbine will increase as the propeller diameter and wind speed increase. More detailed considerations show that the maximum possible power that can be extracted from an air stream is

$$P = 0.5\, d^2 v^3 \text{ watts,*} \tag{10.1}$$

where d is the diameter of the propeller in meters and v is the speed of the wind in meters per second. With a fairly strong 8-meter-per-second (18-mile-per-hour) wind and a propeller 40 meters (131) feet in diameter, the maximum power available is 409,600 watts (409.6 kilowatts). A practical wind-turbine generator system converts about one-fourth of this to electric power. An average household requires about 3 kilowatts of electric power. Therefore, a single generator of this magnitude could service a community of about 30 homes. In remote areas, wind turbines could be very practical. But in urban areas it is clear that a large number of fairly large wind turbines would be required.

As a prelude to development of commercial wind-powered generators with one to two megawatt electrical outputs, the National Aeronautics and Space Administration (NASA) constructed a 100-kilowatt (0.1 megawatt) prototype (Fig. 10.22). The rotor blade is 125 feet in diameter and the tower is 100 feet high. It commences generating electricity in a wind of 3.6 meters per second (eight miles per hour), and achieves the 100-kilowatt output for a wind speed of 8.0 meters per second (eighteen miles per hour). Four 200-kilowatt wind turbine generators patterned after the smaller 100-kilowatt system were constructed at Clayton, New Mexico; Culebra, Puerto Rico; Block Island, Rhode Island; and Oahu, Hawaii. A substantially larger system producing 1500 kilowatts of electric power was dedicated in Boone, North Carolina, in 1979. The first complex of this

FIGURE 10.22 The 0.1-megawatt wind-powered electric generator engineered by the NASA Lewis Research Center, Cleveland, Ohio, was the first of several large, modern systems. The experimental wind turbine has a truss tower 100 feet tall and rotor blades that span 125 feet, tip to tip. The horizontal cylinder at the top contains the power transmission and controls systems. The wind turbine was tested at the NASA Plum Brook Station near Sandusky, Ohio, in late 1975. It continues to be a valuable research tool. (Photograph courtesy of NASA Lewis Research Center.)

---

* The procedure for deriving this formula is presented in Problem 35 at the end of the chapter.

generation of large wind turbine generators has been built at the Goodnoe Hills Site near Goldendale, Washington (Fig. 10.23). Each of three wind turbine generators delivers 2500 kilowatts of electric power. A still larger system that will produce 4000 kilowatts of electric power is planned for Medicine Bow, Wyoming.

Wind turbines of the NASA design, having a horizontal axis of rotation, must face the wind for optimum operation. The symmetrical Darrieus rotor (Fig. 10.24) avoids this problem. Although wind turbines of this type are not as developed as the propeller type, they do show promise and are being pursued.

Field testing the wind turbines will provide valuable practical experience and will, in large part, determine the future of large wind turbines. What to do when the wind dies down remains a challenging problem because of the difficulty of storing electric energy. Large-scale utilization of batteries and pumped-storage systems have already been mentioned. Storing energy as compressed air is another. But one of the most interesting and potentially most useful methods is the storage of hydrogen.

From the standpoint of energy, gaseous hydrogen is not a lot different from the very popular natural gas used for household heating and cooking. Its energy content is about 325 Btu per cubic foot as compared with 1030 Btu per cubic foot for natural gas. When hydrogen is burned with oxygen, the end product is ordinary water. Burning hydrogen with ordinary air also produces oxides of nitrogen. With few modifications hydrogen can be used in place of gasoline in an internal combustion engine. There is some concern about the explosive hazards associated with hydrogen but large quantities are safely and routinely shipped, burned, and used for other purposes. Because water is a compound (H_2O) formed from hydrogen and oxygen, it is possible to reverse the process and separate water into its atomic constituents. One of the most convenient ways of doing this is to pass an electric current through water (Fig. 10.25). Thus the electrical output from a wind-driven generator could be used to dissociate water into hydrogen and oxygen. The hydrogen could be stored as a compressed gas or cooled liquid and shipped to the energy market or piped to market using existing natural gas pipelines. The market

FIGURE 10.23 Each of these three wind turbine generators at the Goodnoe Hills Site near Goldendale, Washington, delivers 2500 kilowatts of electric power satisfying the requirements of about 1000 homes. A tower is 200 feet high and each turbine blade is 300 feet long. (Photograph courtesy of Bonneville Power Administration.)

could be conventional electric power plants or other industries. Wind-powered generators of megawatt capacity would be mounted on 500-foot-high floating stations and anchored offshore. The electric output would dissociate water into hydrogen and oxygen. The hydrogen would be pumped into underground storage tanks for use onshore as a fuel for conventional power plants.

The use of wind energy may seem primitive. It is, nevertheless, a significant, nonpolluting source and must be exploited in the future.

FIGURE 10.24 An experimental Darrieus-type wind turbine generator. Standing seven stories high, the system with its 17-meter-diameter rotor can produce up to 60 kilowatts of electric power in a 28-mile-per-hour wind. (Photograph courtesy of Sandia Laboratories, Albuquerque, New Mexico.)

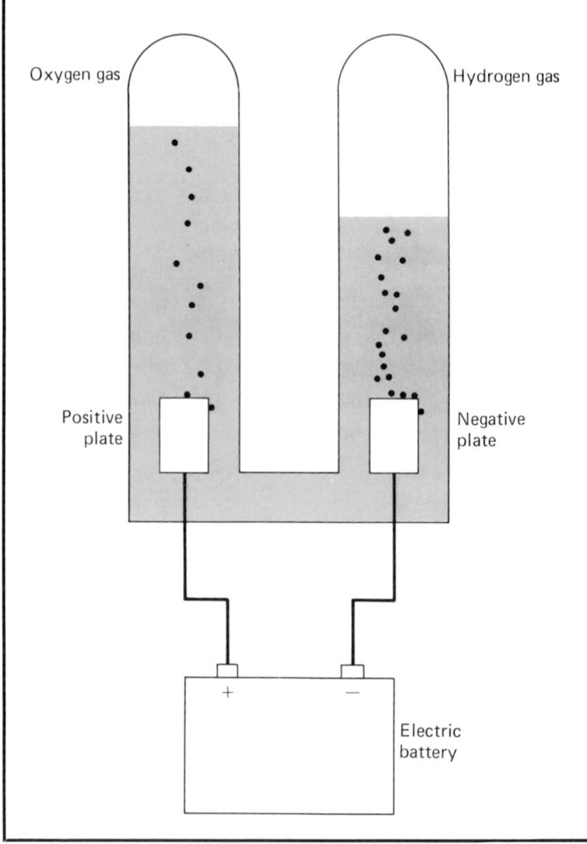

FIGURE 10.25 Water can be separated into hydrogen and oxygen if electric charge flows through it. Note that the volume of hydrogen is twice that of oxygen because each water molecule contains two atoms of hydrogen and one of oxygen.

10.8 ENERGY FROM THE OCEAN

The oceans contain an enormous amount of thermal energy but it cannot be extracted by a heat engine whose operating temperature matches the surface temperature of the ocean. However, from absorption of solar energy the surface is warmer than the ocean depths. Near the lower Atlantic Coast, in the Gulf of Mexico, and in the Pacific Ocean near Hawaii, the temperature decreases from about 25° C at the surface to about 5° C at a depth of 1000 feet.

A condition exists for extracting heat from the upper part of the ocean, converting some of the energy to useful work, and rejecting the remainder to the cooler deep region. As remote as this system appears, engineering studies indicate that there are no technical reasons precluding the construction of a commercial electric power plant deriving input thermal energy from the ocean. The system functions in principle much like a steam-turbine–electric-generator system. But because heat will be extracted at about 20° C (293 K) and rejected at about 10° C (283 K), water cannot be used as a working fluid because it could not be vaporized. A fluid vaporizing at 20° C and producing a high vapor pressure is required. Liquid ammonia is a possibility. The maximum thermal efficiency of such a system would only be a few percent because of the small temperature difference between the hot and cold sources. A conceptual design is shown in Fig. 10.26. The structure would be about 1000 feet tall and would be anchored in the ocean. Warm water would flow into the top part of the system, then past the heat exchangers that would transfer heat to a boiler that would vaporize the working fluid. The vapor would drive a turbine generator in the customary fashion. After passing through the turbine, the gas would be condensed to a liquid and returned to the boiler.

The attractive features of ocean thermal energy converters (OTEC) are

1. they are essentially pollution free, and
2. they make use of a virtually limitless resource, namely the sun, which replenishes the internal energy of the ocean surface with its radiation.

Constructing such a system in the ocean presents some unique engineering problems. Pacific island communities having limited options for generating electricity and favorable ocean conditions have shown the greatest interest in OTEC. A land-based, one-kilowatt demonstration plant has been built by Japanese interests on the island of Nauru, midway between Guam and Samoa. Larger systems are planned. In the United States, Hawaii is the

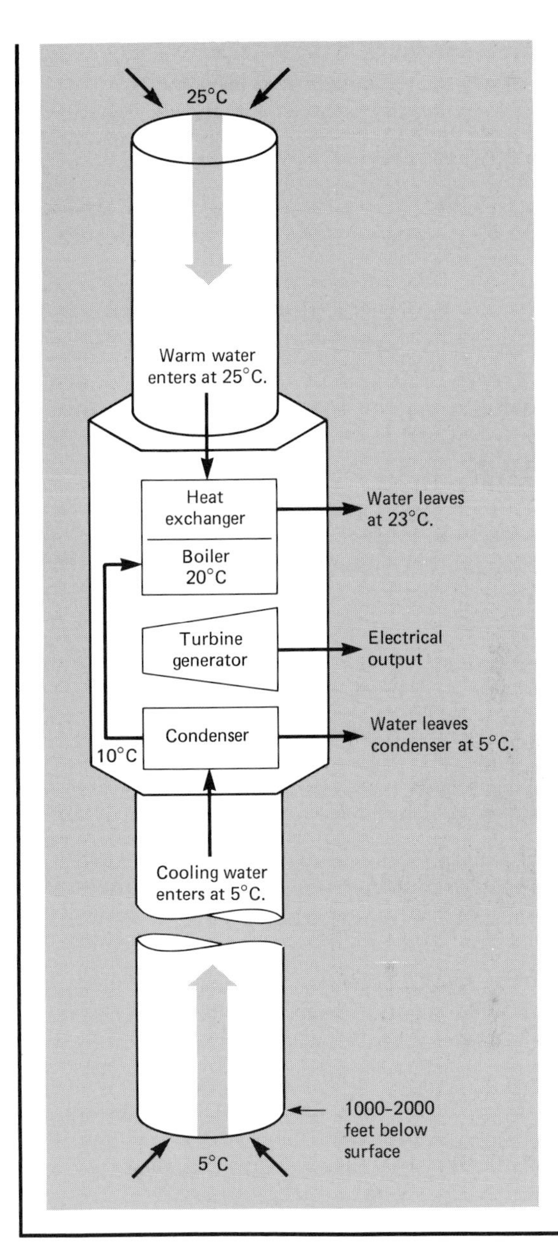

FIGURE 10.26 Conceptual design of a solar seapower system, operating between ocean temperature levels of 25° and 5° C. Water drawn from the ocean surface layer enters a heat exchanger where energy is transferred to a boiler that vaporizes an appropriate liquid. After passing through the turbine, the vapor is liquefied by a condenser cooled with water pumped from 1000 to 2000 feet below the surface.

most favorable site for OTEC development. The feasibility of OTEC was demonstrated in 1979. Whether or not commercial plants develop depends largely on federal support.

10.9 BIOMASS ENERGY

A century and a half ago, United States homes and factories were fueled primarily by wood. Even today, this solar-derived biomass energy source provides more household heating than the solar collectors which have captured the imaginations of many. If taken seriously, biomass products may provide a variety of solid, liquid, and gaseous fuels in quantities that compete with conventional types. While agricultural and lumber wastes constitute a significant ready-made biomass energy source, plants such as wood, sugarcane, and algae can be harvested for their energy content.

Burned for its heat content, wood yields about 8000 Btu per pound—roughly that of the highly attractive western coal. Additionally, wood has essentially no sulfur and burns with little ash residue. In Chapter 3 we discussed how hydrocarbon-based fossil fuels evolve from decomposition of once-living plants and animals in an oxygen-deficient environment. A similar anaerobic (without oxygen) process can be devised for biomass products to produce methane (CH_4), the dominant component of natural gas. The annual production of nearly 300-million tons of agricultural wastes (particularly corn stalks, husks, and cobs) and nearly 30 million tons of manure is a huge ready-made source for anaerobic production of methane. Because natural gas is so widely used in crop drying, the conversion of wastes to methane is appealing. While the wood products industry derives about 40% of the total energy used from bark and mill wastes, some 24 million tons of unused mill wastes and 83 million tons of discarded tree trimmings are generated annually. These wastes like nearly all biomass products could be burned for their energy content or converted to ethanol, which is an alternative to gasoline.

REFERENCES

SOLAR ENERGY FOR HEATING

Introduction to Solar Technology, Marian Jacobs Fisk and H.C. William Anderson, Addison-Wesley, Reading, Mass., 1982.

Solar Heating Design, William A. Beckman, Sanford A. Klein, and John A. Duffie, Wiley, New York, 1977.

The Solar Home Book: Heating, Cooling, and Designing with the Sun, Bruce Anderson, Cheshire Books, Harrisville, New Hampshire, 1976.

The Passive Solar Energy Book, Edward Mazria, Rodale Press, Emmaus, Pa., 1979.

Solar Energy: How to Make It Work for You, Steven J. Mueller and Peter Jones, Butterick Publishing, New York, 1980.

"Passive Design: The Solar and Conservation Potential," Christopher Flavin. In Lon C. Ruedisili and Morris W. Firebaugh (eds.), *Perspectives on Energy,* Oxford University Press, New York, 1982. From an original article in *Worldwatch,* Paper 40, November 1980.

"Solar Heating and Cooling," John A. Duffie and William A. Beckman. In Lon C. Ruedisili and Morris W. Firebaugh (eds.), *Perspectives on Energy,* Oxford University Press, New York, 1982. From an original article in *Science* **191,** 4223 (1976).

Consumer's Guide to Buying Solar Domestic Hot Water. Published by the New York State Energy Office and the New York State Energy Research and Development Authority (January 1982).

"Solar Hot Water Heaters (Ratings)," *Popular Science* **222,** 82 (June 1983).

SOLAR ENERGY FOR ELECTRIC POWER PRODUCTION

"The Technological and Economic Development of Photovoltaics," Dennis Costello and Paul Rappaport. In Lon C. Ruedisili and Morris W. Firebaugh (eds.), *Perspectives on Energy,* Oxford University Press, New York, 1982. From an original article in *Annual Review of Energy,* **5,** 335 (1980).

"Photovoltaics and Solar Energy Conversion to Electricity: Status and Prospects," *Journal of Energy* **3** (5), 263 (1979).

"Future of Photovoltaic Cells," T. Kidder, *Atlantic Monthly,* June 1980.

"Photovoltaic Systems Perspective," P. D. Sutton and G. V. Jones, *Journal of Energy* **4** (1), 7 (1980).

"Photovoltaics," Jeffrey L. Smith, *Science* **212,** 1972 (26 June 1981).

ENERGY FROM THE WINDS

"Wind Energy: Large and Small Systems Competing," *Science* **197,** 971 (2 September 1977).

The Wind Power Book, Jack Parr, Cheshire Books, Palo Alto, Calif., 1981.

"Windmills: The Resurrection of an Ancient Energy Technology," Nicholas Wade, *Science* **184,** 1055 (7 June 1974).

Wind Power and Other Energy Alternatives, David R. Inglis, The University of Michigan Press, Ann Arbor, Mich., 1980.

ENERGY FROM THE OCEAN

"Ocean Thermal Energy: The Biggest Gamble in Solar Power," *Science* **198,** 178 (14 October 1977).

"Solar Sea Power," Clarence Zener, *Physics Today* **26** (1), 48 (January 1973).

Ocean Thermal Energy Conversion: A Review, Paul C. Yuen, Hawaii Natural Energy Institute, University of Hawaii at Manoa, Honolulu, Hawaii, October 1981.

"Extracting Energy from the Oceans: A Review," Adrian F. Richards, *Marine Technology Society Journal* **10** (2), 5 (February–March 1976).

BIOMASS ENERGY

"Photosynthetic Solar Energy: Rediscovering Biomass Fuels," *Science* **197,** 745 (19 August 1977).

"Biomass Refining," Henry R. Bungay, *Science* **218,** 643 (12 November 1982).

Energy, the Biomass Options, H. R. Bungay, Wiley, New York, 1981.

REVIEW

1. In what sense is solar energy "free"?
2. What is meant by passive use of solar energy?
3. Discuss the salient features of an active solar heating system.
4. Describe the construction and principle of a solar energy collector.
5. Distinguish between water-cooled and air-cooled solar energy collectors.
6. What physical principle is exploited in a cooling system like a refrigerator or house air conditioner?
7. What are some potential large-scale uses of solar energy?
8. How is solar energy used to generate electricity?
9. What are some advantages and disadvantages of generating electricity from solar energy?
10. Why is wind energy in the same category as solar energy?
11. Present a solution to the question, "What do you do in a wind-powered system when the winds die down?"
12. What produces temperature variations near the surface of an ocean?
13. What is the attraction of ocean thermal energy for producing electricity?
14. Describe a possible method for extracting thermal energy from an ocean.
15. What is biomass energy?
16. What are some attractive features of biomass energy?
17. The rate at which solar radiation falls on a square foot of the earth's surface is called

    a) solar insulation b) solar insolation
    c) photosynthesis d) heating degree days
    e) thermosiphoning
18. Solar radiation includes not only visible light but also appreciable components of

    a) infrared and gamma radiation.
    b) beta and gamma radiation.
    c) infrared and ultraviolet radiation.
    d) alpha and ultraviolet radiation.
    e) microwave and ultraviolet radiation.
19. In order to assess the practicalities of solar energy use, one needs to know the solar input available in terms of

    a) watts. b) watts/day. c) watts/(day $\cdot$ m^2).
    d) Btu/(ft^2 $\times$ day). e) Btu.
20. The average solar input to a certain area is 200 joules per second • square meter. Given that 1 meter = 100 centimeters, the average solar input in terms of joules per second is

    a) 1.2. b) 3333. c) 12 million.
    d) 0.0003333. e) 0.02.

21. Most solar energy systems use either ____ for the energy storage system.
    a) water or moist earth.
    b) salt or rocks.
    c) rocks or water.
    d) water or styrofoam.
    e) a vacuum or lead.

22. A passive solar energy system relies on
    a) flat plate collectors mounted on a roof.
    b) photovoltaic cells.
    c) absorption of solar energy by objects in a building.
    d) absorption of ultraviolet radiation rather than light.
    e) heat pumps.

23. Large roof overhangs are normally employed on passive solar heated homes because
    a) they add greatly to the beauty of the house.
    b) they reduce the unwanted solar energy entering the house during the summertime.
    c) they act as windbreaks, thus reducing the possibility of window breakage by high winds.
    d) they allow more space to stuff insulation in the attic, thus reducing heat loss.
    e) they are required by federal law.

24. The amount of energy stored in the rocks of a solar heating system
    a) depends only on the thermal conductivity of the rocks.
    b) depends on the total mass and temperature of the rocks.
    c) varies as the square of the temperature (i.e., T^2).
    d) decreases as the temperature of the rocks increases.
    e) none of the above.

25. Solar radiation contains a significant amount of infrared radiation that is able to penetrate the glass cover plate of a solar collector. However, the infrared radiation coming from the black bottom surface of the collector does not readily escape from the collector. This is because
    a) the inside of the glass cover is silvered so as to act like a mirror.
    b) the longer wavelength infrared from the black surface cannot penetrate the glass cover plate.
    c) the infrared radiation from the black surface travels too slowly.
    d) the glass cover plate limits convection of infrared radiation.
    e) both (a) and (b) are correct answers.

26. The glass cover plate on a flat-plate solar collector minimizes energy loss due to
    a) sublimation.
    b) conduction.
    c) evaporation.
    d) reflection.
    e) radiation and convection.

27. For the optimum collection of solar energy, a solar collector should face
    a) east.
    b) west.
    c) north.
    d) south.
    e) the direction is immaterial.

28. Solar radiation amounts to about 60 watts per square foot on the earth's surface. If you can actually collect 10% of the power that falls on the reflecting surface of a solar hot dog cooker and you need 240 watts for the cooker, the minimum collector area needed is ____ square feet.
    a) 24
    b) 4
    c) 40
    d) 240
    e) 75

29. Storage of thermal energy is a problem to be reckoned with in solar energy systems. One scheme is to use a material that changes from solid to a liquid at a desired temperature. It is based on the principle that a solid
    a) liberates energy when changing to the liquid state.
    b) requires no energy to change to a liquid.
    c) absorbs energy when changing to a liquid state.
    d) releases electromagnetic energy when changing to the liquid state.
    e) expands when cooled.

30. The use of solar cells (photovoltaic cells) to generate electricity has been proposed. They are ____ and about ____ efficient in energy conversion.
    (1) expensive
    (2) inexpensive
    (3) 10–15%
    (4) 15–25%
    (5) 25–50%
    (6) greater than 50%

    a) 1; 6
    b) 1; 4
    c) 2; 3
    d) 1; 3
    e) 2; 5

31. A household hot water system generally has a reservoir of 80 gallons (660 pounds). Assuming that the local supply is at 55° F, the number of Btu of heat

required to raise the temperature of 80 gallons of water to 135° F is

a) 660. b) 52,800. c) 46,200.
d) 4600. e) 120,000.

32. Heating hot water for an average family requires about 42,000 Btu/day of thermal energy. If you want to heat this hot water with solar energy and collectors cost $40/ft^2$, then how much would your solar hot water system cost you? (Assume 1400 Btu/ft^2 falls on your roof per day and your collectors are 100% efficient.)

a) $1200 b) $500 c) $400 d) $600
e) none of the above.

33. If the collector efficiency in the previous question was a more realistic 50%, then the cost of your solar hot water system would be

a) $2400. b) $500. c) $1200.
d) $800. e) $600.

34. For the collector in the previous question, a square of between seven and eight feet on a side would be sufficient collector area.

a) true b) false

35. Energy for the winds has its origin in

a) geothermal sources of heat.
b) radioactivity in the earth.
c) gravity.
d) tidal motion.
e) the sun.

36. A windmill is based on the principle of converting

a) linear kinetic energy to rotational kinetic energy.
b) rotational kinetic energy to linear kinetic energy.
c) thermal energy to nuclear energy.
d) temperature to mass.
e) thermal energy to linear kinetic energy.

37. An electric power system that extracted energy from the surface of the ocean and rejected energy at a lower temperature would have a maximum efficiency of _ % if the surface temperature is 23° C (296 K) and the subsurface temperature is 5° C (278 K).

a) 1 b) 6 c) 12 d) 18 e) 33

38. An ocean thermal energy converter (OTEC) is based on the principle that

a) heat flows spontaneously from a cold region to a warm region.
b) heat will not flow between two regions at different temperatures.
c) heat flows spontaneously from a warm region to a cold region.
d) heat is extracted from the lower depths of the ocean.
e) thermal energy can easily be converted directly to electricity.

39. The maximum power that can be produced by a wind turbine depends on the diameter (d) of the propeller and the wind speed (v). If the wind speed doubles, the power increases eight times, but if the diameter doubles the power

a) doubles.
b) quadruples.
c) increases eight times.
d) increases ten times.
e) remains unchanged.

40. When wind blows into a wind turbine and causes it to rotate, then we would expect

a) the speed of the wind to be unchanged after passing through the turbine.
b) the speed of the wind to decrease after passing through the turbine.
c) the speed of the wind to increase after passing through the turbine.
d) the same amount of wind power to be available from the wind after passing through the turbine.
e) the wind speed to have no effect on the power developed by the turbine.

Answers

17. (b)	18. (c)	19. (d)	20. (e)
21. (c)	22. (c)	23. (b)	24. (b)
25. (b)	26. (e)	27. (d)	28. (c)
29. (c)	30. (d)	31. (b)	32. (a)
33. (a)	34. (a)	35. (e)	36. (a)
37. (b)	38. (c)	39. (b)	40. (b)

QUESTIONS AND PROBLEMS

10.2 GENERAL CONSIDERATIONS

1. Suppose that someone builds a structure that prevents solar radiation from impinging on a neighbor's solar energy collectors. Do you think that there should be legal restrictions against such structures?
2. If solar heating can be shown to work in a cold climate, what will determine whether homeowners will or will not opt for it?

3. Seventy-four million billion (74×10^{15}) Btu of energy were used in the United States in 1981.
   a) Determine the amount of energy used for space heating if 5% of the total energy used went for this purpose.
   b) Knowing that the energy content of a barrel of oil is 5.8 million Btu, show that the energy equivalent of 190 million barrels of oil could have been saved if 10% of the energy for space heating was derived from the sun.
4. The mean distance from the earth to the sun is 93 million miles. When solar radiation travels this distance it has spread out over the surface of an imaginary sphere of radius 93,000,000 miles. Measurements just outside the earth's atmosphere reveal that the radiation on each square foot of this sphere is 130 watts. Show that the radiation emitted by the sun is 400 billion-billion megawatts (400×10^{18} MW). The formula for the surface area of a sphere of radius r is $A = 4\pi r^2$. There are 5280 feet in a mile.

10.3–10.4 PASSIVE AND ACTIVE SOLAR HEATING

5. What are some ways that solar energy is routinely used in and around a home?
6. What problems arise in trying to introduce solar home heating to the public?
7. Water and rocks are commonly used to store energy in solar-heated buildings. What advantages does water have over rocks?
8. Why is a solar collector oriented in a south-facing direction?
9. The sun is lower in the sky in winter than it is in the summer. Show how a properly designed window awning allows solar radiation to enter in the winter but not in the summer.
10. Show how a tree can be a summer solar radiation shield for a house but still allow significant solar radiation to impinge on the house in winter.
11. When the outdoor temperature is 25° F, the temperature between a storm window and a single-pane glass window can easily be 60° F on a sunny day. What is responsible for this rise in temperature?
12. If you had a material that melted at 50° F, how could you use it to cool a building?
13. About 500 gallons of fuel oil are needed each year to heat a home in the central United States.
    a) Assuming a five-month heating season, how many Btu are required each day?
    b) If the solar insolation is 1200 Btu/ft^2 • day and a collecting system can utilize 50% of the input solar energy, determine the required collecting area.
14. Using the average annual solar insolation in Btu/ft^2 per day shown in the accompanying map, determine the average annual solar energy input for the area where you live. Using information from Table 2.2, express your result in kcal/m^2 • day.

15. If the tap water temperature in your area is 30° C, how many kilocalories of heat are required to increase the temperature of 20 kilograms of water to 58° C?
16. Elementary school students often demonstrate the magnitude of solar energy by constructing a solar cooker for hot dogs. It consists of a bowl-shaped structure of aluminum foil. The bowl shape allows the radiation to be focused on the hot dog.
    a) Suppose that 10% of the radiation falling on the reflector gets absorbed by the hot dog. How big a surface area would be required to achieve a power of 100 W on the hot dog?
    b) Suppose that a 1000-W household appliance cooks a hot dog in five minutes. How long would it take the solar cooker to achieve the same result?
17. The exercise below involves considerations for a solar hot water heating system. Using the example in Section 10.5 as a guide, fill in the following designated blanks:

number of occupants = 3

incoming water temperature = 30° C

hot water temperature = 57° C

usage rate = 20 gallons per person per day

collector efficiency = 65%

average energy available = 1200 Btu/ft^2 • day

thermal energy available for heating water = ____

thermal energy required to heat the water = ____

collecting area required = ____

installed cost of collectors at \$30/ft^2 = ____

18. Glauber's salt (sodium sulfate decahydrate) having a melting point of 90° F is being studied as a heat-storing medium for homes and industries. Energy is required to melt the salt but this energy can be recovered when the molten salt changes back to a solid. It takes 0.3 Btu to melt one cm^3 of Glauber's salt. When the outdoor temperature is 0° F, a well-insulated home needs about one million Btu of heat for a two-day period. How many cubic centimeters of Glauber's salt would be required to store one million Btu of thermal energy? Compare this volume with a household refrigerator measuring 100 cm × 200 cm × 50 cm.

19. Suppose that you wanted to build a device that would convert solar radiation to electric power. If the radiation amounts to 0.1 W/cm^2 and you could convert the solar energy to electric energy with an efficiency of 1%, how big a square would be required to produce 1000 W of electric power?

20. a) If you were to place a 12-inch-square piece of paper on the ground, on the average it would receive about 65 watts of solar radiation. Approximating the United States as a rectangle 3000 miles long and 1000 miles wide, how much radiation falls on 1% of the land area of our country?
    b) If this radiant power is converted to electric power with an efficiency of 1%, show that it would exceed the 1981 electric generating capacity of about 630,000 megawatts (6.3×10^{11} watts).

10.6 GENERATING ELECTRICITY WITH SOLAR ENERGY

21. What are some environmental questions that might be raised when solar energy is considered as a large-scale energy source for producing electricity?

22. Compare the priorities of using land for solar farms as against highways and roads.

23. In early 1978, electricity could be derived from solar cells at a cost of \$11 per watt of electric power. What would the cost of the solar cells be for a home requiring 2000 watts of electric power?

24. Suppose that the batteries in a small electric car had the energy equivalent of four gallons of gasoline (about 500,000 Btu) and that a recharging system using photovoltaic cells (Section 10.6) is used to recharge the batteries. If the batteries are to be recharged in a day's time and the average solar energy available is 1600 Btu/ft^2 • day, how big a collecting area would be required if the recharging system has an efficiency of 10%? Does the size required seem practical to you?

25. The data in the map on page 240 are expressed in Btu/ft^2 • day. A Btu is a unit of energy and day is a unit of time. Thus Btu/day is a unit of power.
    a) Knowing that one Btu = 1055 joules and one day = 86,400 seconds, show that one Btu/day = 0.0122 watts.
    b) A small car is about five feet wide and ten feet long. In an area where the average solar insolation is 2000 Btu/ft^2 • day, about how many watts of solar power impinge on a small car?
    c) If this solar power is used to run an electric motor, what is the largest possible power output that can be obtained from this motor?
    d) If this car could function with a ten horsepower motor, could you ever get enough solar power to operate it? (One horsepower = 746 watts.)

26. The Skylab space satellite was powered with an array of solar cells producing 20,000 watts of electric power. If the energy conversion efficiency of the solar cells is 15% and the solar power available is 1400 watts per square meter, how big a collecting area is required? Compare this area with a classroom seven meters wide and twelve meters long.

10.7 ENERGY FROM THE WINDS

27. Wind turbines for powering large electric generators are located a few hundred feet off the ground. What environmental problems result from having to do this?

28. What energy transformations result when energy is extracted from wind to produce electricity?

29. Why is energy required to separate a water molecule (H_2O) into hydrogen and oxygen?

Assume the power in watts available to a wind-powered electric generator is $P = 0.5d^2v^3$ where d is the diameter of the propeller in meters and v is the wind speed in meters per second. The following problems involve manipulation of this expression.

30. How much power is available to a wind-powered electric generator having a diameter of 40 meters if the wind speed is eight meters/second?
31. Suppose that a rural home could manage with a wind-powered electric generator producing 250 watts. If the system had an efficiency of 25% and the wind speed was five meters/second, how large a diameter must the propeller have?
32. If the efficiency for generating electricity does not depend on wind speed, how will the power output of a given generator change for a doubling of the wind speed?
33. If the efficiency for generating electricity does not depend on propeller diameter, how will the power output change for a doubling of the propeller diameter?
34. The first experimental wind turbine engineered by NASA used blades 38 meters long. The maximum wind speed where it was tested is about eight meters per second.
    a) Determine the available wind power for these conditions.
    b) The electrical output for these conditions is 100,000 watts. What is the efficiency of the system for converting wind power to electric power?
35. Most of the expressions that we have used for quantitative analyses involve terms that do not involve a numerical exponent. For example, the potential energy of water atop a dam is $E = w \cdot H$. On the other hand, the power associated with wind flowing into the blades of a wind turbine varies as the cube of the speed; i.e., P is proportional to v^3. This is very important because it means that if the speed is doubled the power increases eight times. It is fairly easy to see why the power varies as the cube of the speed. Imagine a cylinder of air of cross-sectional area A and length L moving by the turbine blades with speed v.

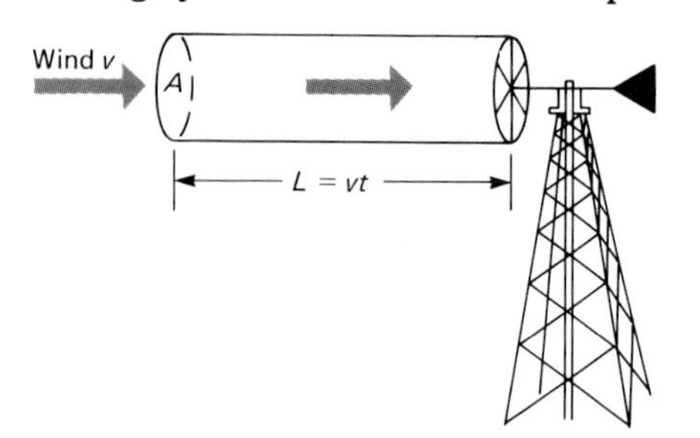

The mass of this amount of air is $M = \rho v = \rho AL$ where ρ is the density of the air (in kilograms per cubic meter, for example). The kinetic energy of this mass is just $E = \frac{1}{2} Mv^2 = \frac{1}{2} \rho ALv^2$. If this cylinder flows by in a time t, the power is

$$P = \frac{E}{t} = \frac{1}{2}\rho A\left(\frac{L}{t}\right) \cdot v^2.$$

However, L/t is just the speed. So $P = \frac{1}{2}\rho Av^3$ showing that power is proportional to the cube of the speed.

a) Does all the wind power entering the wind turbine get transferred to the turbine blades?
b) How does the concept of efficiency apply to this situation?
c) Certainly the power developed should increase as the length of the turbine blades increases. Why should you expect the power to depend on the square of the blade length, i.e., P is proportional to d^2?

10.8 ENERGY FROM THE OCEAN

36. Why would you not expect electric power plants utilizing heat from the ocean to be built offshore of the North Atlantic and North Pacific coasts?
37. What environmental problems might result from electric power plants submerged in the ocean?
38. About 30% of the electric power output from an ocean thermal energy electric power plant is needed to power the pumps that circulate the water in the system. Why does such a system require such a large pumping capacity? What sort of engineering problems does this requirement present?
39. Because the temperature difference of the source and sink for an ocean thermal energy electric power plant is very small, the energy efficiency at best is very small. If such a system is to produce electricity on a competitive scale, why would you expect the rate at which water is taken in to be very large?
40. In Chapter 7 we showed that the maximum thermal efficiency for a heat engine extracting heat at a temperature designated T_{hot} and rejecting heat at a temperature designated T_{cold} is

$$\epsilon_{max} = \left(\frac{T_{hot} - T_{cold}}{T_{hot}}\right) \times 100\%.$$

Show that the maximum thermal efficiency for a solar sea power system operating betwen temperatures of 20° C (293 K) and 10° C (283 K) is 3.4%.

41. If the turbine of an electric power plant deriving input thermal energy from the ocean has an energy-conversion efficiency of 2%, at what rate will it have to extract thermal energy in order to have a power output of 1000 megawatts?

42. Hydroelectric units at the Grand Coulee Dam discharge water at a rate of 1,870,000 pounds per second. Because of the intrinsic, low energy-conversion efficiency, an ocean thermal energy electric power plant generating 100 megawatts (remember, a large electric power plant generates 1000 megawatts) will have to process water at a rate of 100,000 gallons per second. Compare this rate with the discharge rate at the Grand Coulee Dam. (One gallon of water weighs 8.3 pounds.)

11

Other Energy Systems

11.1 MOTIVATION

In this chapter we discuss some alternative energy systems. Some are old, some are visionary. Hydroelectricity has been viable for many decades and continues to make an important impact on our energy budget. Hydropower will continue to grow slightly, but it will have less of an overall effect on our energy appropriations in the future. Energy from the tides has been a dream for half a century, but it is likely that none of us will see a single kilowatt of electric power generated from it in this country. Some systems are fascinating. Can you imagine your reading lamp being energized with electric power from a power plant whose input energy comes from garbage and refuse! The creation of some of these facilities may seem remote, but none is beyond the realm of pos-

Shown here is an electric power plant that derives its input energy from steam emanating from the earth's interior. Located at The Geysers in northern California, this facility generates 24 megawatts of electric power. (Photograph courtesy of Pacific Gas and Electric Company.)

sibility. If priorities change or unexpected snags develop in the advanced electric power technologies, these systems may well come to the forefront of the energy picture.

11.2 HYDROELECTRICITY

The monumental dams on the Columbia, Colorado, and Tennessee rivers are multipurpose. The water they impound provides recreation facilities, municipal water supplies, and irrigation water. The dams also help control flooding. Additionally, and very importantly, each dam shuttles water to a turbine coupled to an electric generator.

A hydroelectric system is a way of supplying power to a generator other than by a steam turbine. The energy language is much the same. Mechanical energy in a steam turbine is produced by steam flowing from an area of high thermal potential energy to an area of lower thermal potential energy. Energy is transferred to the turbine in the process. The cycle is completed by condensing the steam and reheating the water to make steam for another cycle. We know that water at the top of a dam is in a higher state of gravitational potential energy than water at the bottom of the dam. As water flows to the bottom of the dam, energy is extracted by a water turbine to drive the electric generator. The water cycle is eventually completed when water below the dam evaporates, clouds form, and rain refills the reservoir above the dam.

A volume of water of weight w has potential energy wH at the top of a dam of height H (Section 2.8). At the bottom of the dam, the water's potential energy is converted to kinetic energy in accordance with the work–energy principle (Section 2.7). Dividing wH by the time for falling the distance H gives the power associated with the moving water.

power = energy ÷ time,

$$P = \frac{wH}{t} \tag{11.1}$$

A large fraction of this waterpower is transferred to the water turbine. Note that w/t is just the rate of flow of the water. To achieve units of watts for power, we usually express the weight (w) in newtons, the distance (H) in meters, and the time (t) in seconds. To illustrate, each hydroelectric unit in the third power plant at the Grand Coulee Dam (Fig. 11.1) discharges water at a rate of 8,320,000 newtons per second from a height of 87 meters. The waterpower of 720 megawatts entering a turbine is converted to electric power at an efficiency of 85%, giving a net output of 600 megawatts for each unit. Thus the electric power produced by a single hydroelectric unit is comparable to a typical 1000-megawatt coal-burning or nuclear unit. In total, the third power plant produces 3900 megawatts of electric power. Coupled with an electric power output of 2300 megawatts from two older units, yields a total of 6200 megawatts for the Grand Coulee system. If plans are fulfilled, the Grand Coulee Dam will someday generate 10,000 megawatts of electric power.

Hydroelectric generating capacity grew from 3700 megawatts in 1921 to 77,100 megawatts in 1981. Forty-one percent of this generating capacity is located in the Pacific Northwest. Although hydroelectric energy production continues to grow, the percentage contribution to the total electric energy production has declined from 17% in 1960 to 11% in 1981. The decline is a consequence of having already developed most of the best hydroelectric sites and resistance from environment-conscious groups opposed to the development of available sites. About two-thirds of the remaining potential generating capacity is located in five states—Washington, Alaska, California, Oregon, and Idaho. About 30% of this potential is located in Alaska, far removed from large users of electric energy.

Large hydroelectric power plants are symbolic of central generating facilities that grew in size with the growth in electric power use. Smaller hydroelectric systems producing five megawatts or less were very common until the 1940s but the favorable economics for the large systems forced the closure of smaller units. Estimates of electric power obtainable from new small hydroelectric systems are as large as 27,000 megawatts. An equal amount could conceivably be produced from rehabilitated systems. With the passage of the National Energy Act in 1978, there has been a resurgence in the devel-

FIGURE 11.1 The Grand Coulee Dam on the Columbia River in Washington is 4173 feet long and 550 feet high. The electric power output from the generators at its base amounts to over 6000 megawatts. Roosevelt Lake, impounded by the dam and used as a recreational facility, extends for 151 miles upriver. (Photograph courtesy of the United States Department of Agriculture, Soil Conservation Service.)

opment of small hydroelectric systems. The Public Utilities Regulatory Policies Act (PURPA) directs existing electric utilities to purchase electric energy from small producers at a rate consistent with conventional costs. Applications for permission to build small hydro projects increased from 18 in 1978 to 1800 in 1981. A number of abandoned small hydroelectric facilities are being resurrected and new units planned. Most of the resurrected plants are situated east of the Mississippi River while the new units are for the most part located in the Pacific Northwest.

We stressed in Chapter 4 that a conventional electric power plant does not store electric energy. It produces electric power on demand. If users turn off all their energy-consuming devices, the power plant stops generating electricity. Under normal operating conditions, the demand varies with the time of day and with the time of year (Section 4.9). Because a power plant operates most efficiently if its output is constant, a power company often utilizes auxiliary units, such as gas-turbine electric units burning natural gas, to handle the peak demand. One way of storing energy for daily peak periods and making use of the base load generators during slack times is to use the electrical output for motor-driven pumps that force water to an elevated reservoir (Fig. 11.2). When the daily demand for electricity peaks, the water is drained through a hydroelectric system to produce electricity for the peak periods. This type of hydroelectric project is called a pumped-storage system. Although only about two-thirds of the electric energy used to run the pumps can be recovered, this system can still be economical. For example, nuclear power produced at night when demand is low may be purchased relatively inexpensively. Then the electricity from the hydroelectric units may be sold at a higher price during the afternoon when air conditioner demands rise.

Pumped-storage systems require both a source of water and terrain favorable for a reservoir. Some systems operate in conjunction with a conventional

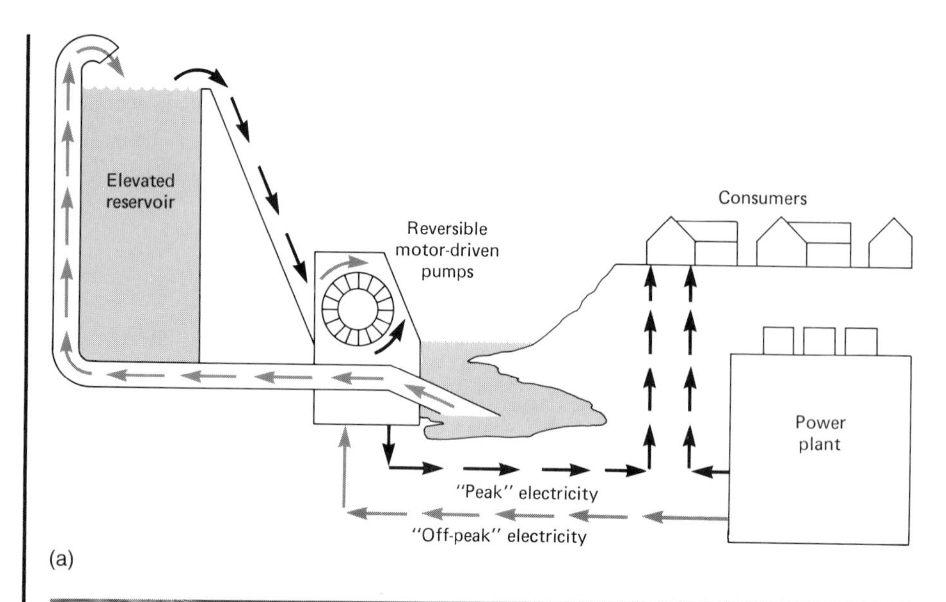

(a)

(b)

FIGURE 11.2 (a) The pumped-storage principle. Consumer demands for electricity are provided by a combination of the large power plant and the hydroelectric unit using water from the elevated reservoir. When daily consumer demands for electricity are low, the power plant provides electricity to the hydroelectric unit for pumping water to the elevated reservoir. **(b)** This aerial view shows the Philadelphia Electric's Muddy Run pumped-storage project. The powerhouse is shown at the Susquehanna River's edge, in foreground. The horseshoe-shaped construction above the powerhouse is the intake canal for the system. This system utilizes eight 110-megawatt reversible pump-turbine combination units that function alternately as motordriven pumps and turbine generators. (Photograph courtesy of the Philadelphia Electric Company.)

hydroelectric power plant by using pumps to push water back up to the reservoir behind the dam. This is done to some extent at the Grand Coulee Dam. A single-purpose, pumped-storage system uses water turbines to power electric generators when electricity is desired and the same turbines running in reverse to pump water when the reservoir is refilled. Because pumped-storage systems are designed to handle daily peak electricity demands, they are most practical in areas in which peak demands are large relative to base load. About three-fourths of the pumped-storage capacity is located east of the Mississippi River. The generating capacity of pumped-storage systems grew from 400 megawatts in 1964 to over 10,000 megawatts in 1982. There were 29 pumped-storage systems of varying design in 1979 with 39 more having a total generating capacity of 40,000 megawatts either proposed or in the construction phase. The largest system located at Ludington, Michigan, produces 2000 megawatts of electric power from water pumped from Lake Michigan. Both the Northfield Mountain Plant on the Connecticut River in Massachusetts and the Blenheim-Gilboa Plant on Scholarie Creek in upstate New York produce 1000 megawatts of electric power. Table 11.1 tabulates the single-purpose, pumped-storage systems having generating capacities of at least 240 megawatts.

11.3 GEOTHERMAL ENERGY

Thermal energy within the earth is termed geothermal energy. It is not a new energy source and it is not going to be a complete answer to energy demands. However, the drive to vary our energy resource base has attracted new attention to geothermal energy.

We live on a crust of earth varying in thickness from two to thirty miles. Beneath the crust is a semimolten rock layer called the mantle. Beneath the mantle is a molten core of iron and nickel. The earth's crust is warmed from within by the hot interior and by radioactive decay products, in particular emissions from uranium-238, thorium-232, and

TABLE 11.1 Pumped-storage systems having a generating capacity in excess of 240 megawatts. Taken from *Hydroelectric Plant Construction Cost and Annual Production Expenses—1978,* Energy Information Administration, United States Department of Energy.

Project	Water source	State	Capacity (megawatts)
San Luis	San Luis Creek	California	424
Cabin Creek	South Clear Creek	Colorado	300
Bear Swamp	Deerfield River	Massachusetts	600
Northfield Mountain	Connecticut River	Massachusetts	846
Ludington	Lake Michigan	Michigan	1980
Taum Sauk	East Fork Black River	Missouri	408
Yards Creek	Yards Creek	New Jersey	387
Blenheim-Gilboa	Scholarie Creek	New York	1000
Lewiston	Niagara River	New York	240
Salina	Salina Creek	Oklahoma	260
Muddy Run	Susquehanna River	Pennsylvania	800
Seneca (Kinzua)	Allegheny River	Pennsylvania	396
Jocassee	Keowee River	South Carolina	612
Fairfield	Broad River	South Carolina	518

potassium-40. The temperature of the crust increases about 2° C for each 100-meter penetration.

The thermal energy stored under the continental United States to a depth of six miles is equivalent to the energy obtainable from the combustion of trillions of tons of coal. But, like solar energy, geothermal energy is very dispersed and therefore difficult to exploit. However, there are areas in which hot molten rock (magma) is forced up through structural defects and either surfaces in the form of volcanoes or is trapped close to the surface. These masses trapped close to the surface sometimes produce concentrated sources of hot water or steam that can be tapped for a variety of energy uses.

As long ago as 1894, geothermal energy provided district heating in Boise, Idaho, and hot water was pumped to homes in Klamath Falls, Oregon, in 1900. But the most significant geothermal venture in the United States began in 1960 at The Geysers in northern California. There exists in this region sources of rare dry steam that can be tapped by drilling technology akin to that employed for removing oil from the earth's crust (Fig. 11.3). In 1960, the Pacific Gas and Electric Company (P. G. and E.) piped geothermal steam to a steam turbine coupled to an electric generator producing 84 megawatts of electric power (Fig. 11.4). Since then additional geothermal plants have been built bringing the total electric generating capacity to 1020 megawatts in 1982. Desirable dry steam sources like those at The Geysers are extremely limited. More likely, but still representing only 10% of the geothermal resource base, are sources yielding hot water rather than steam. Converting the thermal energy of hot water to electricity is a difficult technology. Minerals and salts in geothermal hot water complicate matters. Nevertheless, technical barriers have been broached to the extent that two 10-megawatt electric power plants using geothermal hot water have operated successfully. They are at East Mesa, California (1979), and at Brawley, California (1980). Each kilowatt of electric generating capacity was assessed at $1600. While still expensive, the per kilowatt cost is some ten times lower than for Solar One (Section 10.6) and the energy is available around the clock. Plants capable of producing 50 megawatts of electric power are under construction.

Some 70% of the geothermal resource base is in subsurface rock heated by low-lying magma. Extracting the thermal energy involves drilling to the rock, injecting water into the entrance, and recovering the steam formed when the water contacts the hot dry rock. Successful hot dry rock experiments in 1979 led to the production of electric power from a hot dry rock resource at Fenton Hill, New Mexico.

11.4 ENERGY FROM BURNING TRASH

We are accustomed to clever packaging for merchandise. Countless hours are spent determining the carton that optimizes sales. More breakfast cereal is bought for its colorful packages with catchy sayings, enclosed trinkets, and outrageous claims than for its nutritional value. But the interest in packaging declines once the product is purchased even though the disposal of the containers and trash, in general, is a most perplexing societal problem. Solid waste is generated at an average rate of five pounds per person per day in the United States. A city of a million people produces 2500 tons of solid waste each day. It takes 500 trucks each carrying five tons to haul these wastes. To compound matters, trash production is growing at a rate of 4% per year. Still, in a technological era that witnessed a landing on the moon, this trash is likely to be dumped into a hole in the ground and covered with dirt. A growing shortage of land disposal sites is stimulating interest in alternate disposal schemes. One such scheme is to burn the trash for input energy to an electric power plant. Some recovery of marketable products may be done in the process.

This idea is relatively new in the United States, but it is quite old in Europe. Over 50 years ago, Paris began burning garbage to produce electric energy and today enough electricity is generated to supply the needs of 50,000 people. Since 1968, a 7.3-megawatt trash-burning power plant has been operating in Milan, Italy. Additional plants are

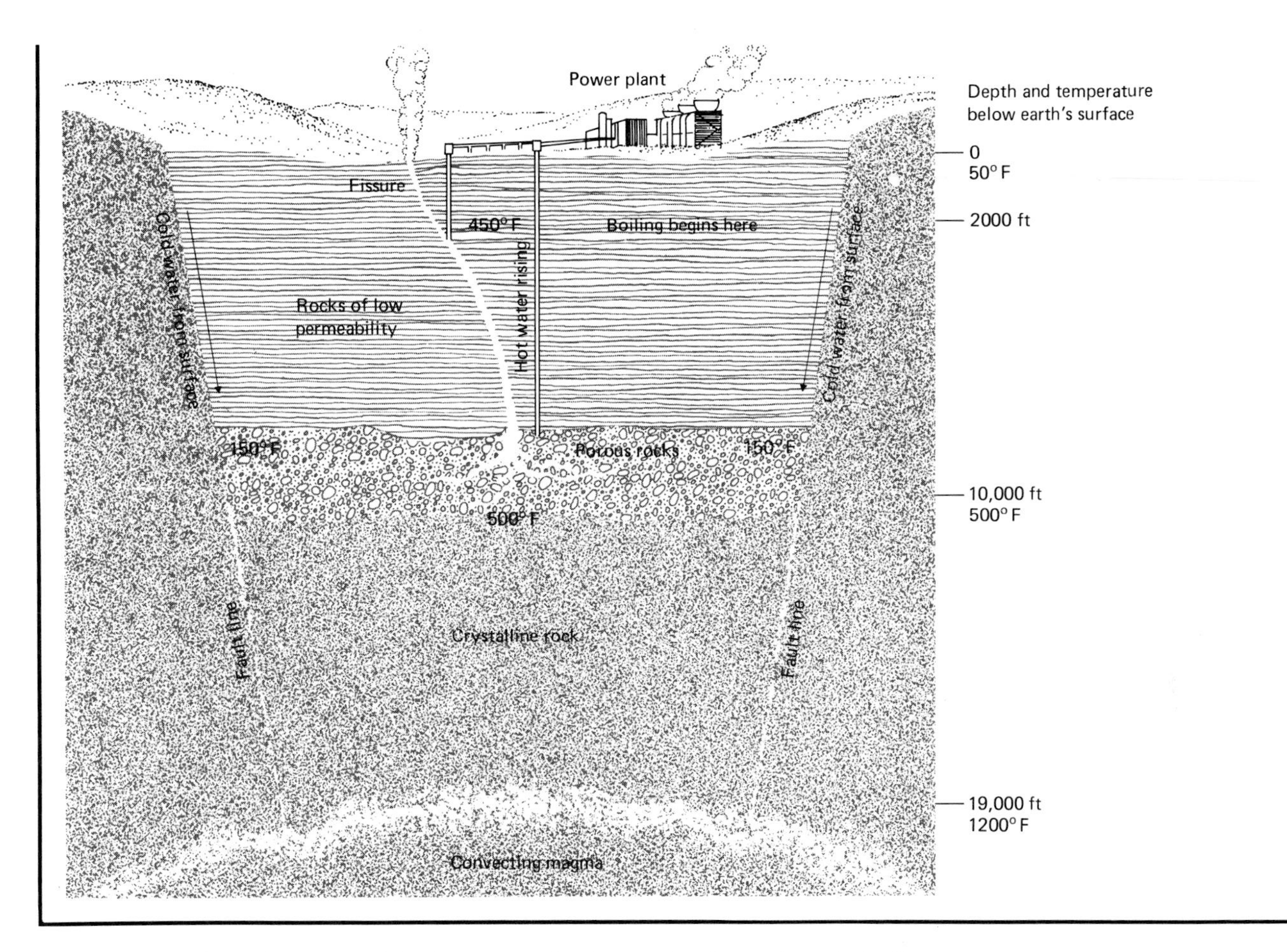

FIGURE 11.3 A geological model of a geyser. Surface water flows down into the earth through faults where it is trapped in a porous layer that is bounded by crystalline rock on the lower side and a low-porosity layer on the upper side. The water is heated by the magma, pressure builds up, and the water rises through a fissure. When the pressure is released, the water begins to boil and vaporize, and steam and hot water emerge at the earth's surface. If a natural vent for the water is not available, it is sometimes possible to drill a line to the reservoir. (Adapted from original material from *Fortune* by Max Gschwind; June 1969.)

planned. In West Germany, steam heat is produced from the burning of refuse produced by nearly eleven million people.

Combustible paper products account for about 53% of the total weight of solid wastes. Another 7% is in other combustible carbon-based products. When burned, each pound of solid waste produces approximately 5300 Btu. Although this is low when compared with an energy content of 13,000 Btu per pound for high-quality coal, it is still significant. Furthermore, the sulfur content of solid waste is about 0.1%, putting it in a class with the best low-sulfur coal. These two features have brought attention to solid wastes as an energy source of great potential.

FIGURE 11.4 This geothermal electric power plant in Sonoma County, California, produces 54 megawatts of electric power. The total capacity of all geothermal plants in this area amounted to 1020 megawatts in 1982. This capacity is expected to increase to 1700 megawatts in 1990. (Photograph courtesy of Pacific Gas and Electric Company.)

The energy potential of solid wastes is impressive. If 150 million tons of the solid waste produced in 1981 had been burned, then it would have produced 1500 trillion Btu of heat. If converted to electric energy at an efficiency of 32%, the burning would have produced 140 billion kWh, which amounts to about 6% of the electric energy produced by conventional means in 1981.

Pioneering efforts for deriving energy from solid waste were made in St. Louis. Beginning in 1972, a prototype system operated by the Union Electric Company burned a mixture of trash and coal to generate electricity. The plant ultimately processed about 650 tons a day of the city's output of trash. In February 1977, the Union Electric Company announced cancellation of its solid waste utilization system that included plans for a capacity of 7,500 tons a day. The reasons cited were rising costs and a Missouri law that made it difficult to finance the construction of electric power plants. Several cities are producing either burnable fuel or steam that can be used for heating purposes or input to a steam turbine (Table 11.2).

Whether burning solid wastes solely for the energy content is the best solution for a mounting problem remains to be seen. Perhaps recycling the wastes is more appropriate. However, in an era of concerns for energy resources, the burning of solid wastes for the energy content is a viable alternative (Fig. 11.5).

TABLE 11.2 The November 1981 issue of the trade journal *Waste Management* lists 69 materials and energy-recovery facilities in the United States. Some are in the construction phase, others are temporarily closed. This table lists trash-to-energy systems that are operational and have a capacity of at least 100 tons per day.

Location	Capacity (tons/day)	Output
Ames, Iowa	200	fuel
Auburn, Me.	200	steam
Baltimore County, Md.	1200	fuel
Braintree, Mass.	250	steam
Chicago, Ill. (Northwest Incinerator)	1600	steam
Durham, N.H.	108	steam
Dyersburg, Tenn.	100	steam
Hampton, Va.	200	steam
Harrisburg, Pa.	720	steam
Madison, Wisc.	400	fuel
Nashville, Tenn.	530	steam
New York City, N.Y.	1000	steam
Norfolk, Va.	360	steam
North Little Rock, Ark.	100	steam
Oceanside, N.Y.	750	steam
Pittsfield, Mass.	240	steam
Salem City, Va.	100	steam
Saugus, Mass.	1200	steam
Tacoma, Wash.	500	fuel
Waukesha, Wisc.	175	steam

FIGURE 11.5 The Columbus (Ohio) Refuse and Coal Fired Municipal Electric Plant. This facility has a capacity to burn 3000 tons a day of a mixture of 90% shredded refuse and 10% coal. Ninety megawatts of electric power are produced by three turbine-generators. (Photograph courtesy of City of Columbus.)

11.5 COAL GASIFICATION AND LIQUEFACTION

Environmental problems from using coal inspired the statement, "There are two things wrong with coal today. We can't mine it and we can't burn it."* There is enough coal in the United States to satisfy much of our energy needs for many years—*if* it could be mined and burned. Natural gas can be burned in an environmentally acceptable way, but it is in extremely short supply. Converting coal to a synthetic gas takes advantage of the best features of these two fuels. The most pressing environmental problem from burning coal involves sulfur oxides (Chapter 5). Techniques exist for removing the sulfur oxides after burning, but they are expensive, and their reliability questionable. Coal gasification removes most of the sulfur content of the coal during the gasification process.

Coal gasification is not a new concept. Coal gas was widely used for lighting and cooking until natural gas came into prominence after World War II. However, the conversion of energy from coal gas to electricity on a commercial scale has yet to be demonstrated although the principle is relatively simple.

When coal is burned to liberate energy, enough oxygen is provided to optimize the burning. Ideally, only carbon dioxide (CO_2) and heat are produced. If there is insufficient oxygen, combustion is incomplete and gases other than carbon dioxide are produced. Coal gasification capitalizes on incomplete combustion conditions. A limited amount of oxygen (or air) is used to instigate burning for the purpose of producing heat. The basic heat-producing reactions from carbon and hydrogen in the coal are

$$C + O_2 \rightarrow CO_2$$

and

$$2H_2 + O_2 \rightarrow 2H_2O.$$

When a certain temperature is reached, steam is injected into the system promoting reactions with carbon and carbon dioxide.

$$C + H_2O \rightarrow CO + H_2$$

and

$$C + CO_2 \rightarrow 2CO.$$

Both carbon monoxide (CO) and hydrogen (H_2) are flammable. Methane (CH_4), the molecular form of natural gas, is formed to some extent by the reactions

$$C + 2H_2 \rightarrow CH_4$$

and

$$\text{coal} + \text{heat} \rightarrow CH_4 + \text{hydrocarbons}.$$

The resulting gas is a mixture of carbon monoxide and hydrogen, with lesser amounts of methane. At the very least, the energy content is about 120–160 Btu per cubic foot. At further expense, distillation increases the energy content to 500–600 Btu per cubic foot. These energies are low compared with an energy content of 1030 Btu per cubic foot in natural gas. However, gasification converts the sulfur content of the coal to hydrogen sulfide (H_2S) that can be removed by proven methods. The resulting low-sulfur content gas can then be burned to produce heat in conventional electric power plants. Locating the gasification system and the power plant near the coal-producing sites minimizes transportation of the raw material but does not avoid the problems associated with both strip and shaft mining. There is, however, interest in gasifying coal while it is still in the ground using, basically, the same principle as in gasification plants. Tests have produced a combustible gas of variable energy content, but recovery has been poor. Nevertheless, the possibility of extracting the energy content of coal without ravaging the landscape is enough of a virtue to prompt a continued exploration of this technology.

Current research is proceeding with the development of large-scale, high-energy content coal gasification systems for producing a gas comparable in quality to natural gas and with coal liquefaction technology. The coal liquefaction process involves

* S. David Freeman, Director of the 1974 Ford Foundation Energy Policy Project.

the reaction of hydrogen with coal to produce a liquid hydrocarbon. Industrial processes for doing this have been known for some time. A large part of the gasoline for Germany's World War II effort was produced from coal liquefaction. Although it is attractive to manufacture a product that can replace gasoline, which is refined from a dwindling natural resource, the process at present is expensive both in terms of money and energy usage. Demonstration plants for coal liquefaction and high-energy content gasification have been built. A successful technology would greatly alleviate the plight caused by dwindling natural gas and oil reserves.

11.6 MAGNETOHYDRODYNAMICS (MHD)

The efficiency of any single energy conversion process is always less than 100%. If several sequential operations are required to produce a desired form of energy, the overall efficiency is less than the smallest efficiency in the chain. Even if every individual conversion stage were nearly perfect, except one, then the overall efficiency would still be low. Thus any method of producing electric energy using a steam turbine is intrinsically inefficient because of the limited thermal efficiency of the turbine (Section 7.2). Why, then, isn't a scheme devised that avoids the steam turbine? Despite considerable thought and physical effort, few schemes have emerged as being commercially practical in this century. One method bears the name magnetohydrodynamics, abbreviated MHD.

Charges in motion produce an electric current. A flow of gas or a stream of liquid could produce an electric current if the constituents possessed a net charge. A gas turbine used to power a jet airplane produces a stream of high-speed gas but little electric current because there is practically no net charge associated with the atoms and molecules of the gas. If the moving gas could be made to contain free electric charges, then it could produce an electric current. This condition is achievable by "seeding" the gas with atoms or molecules that readily lose their electrons in a hot gas environment. Elemental potassium and cesium or appropriate compounds including one or the other of these elements are used for "seeds." Producing an electric current in a gas is one thing; diverting it into the electrical wires of a house is another. The transfer in an MHD system is accomplished by directing the gas through a magnetic field (Fig. 11.6). Because of the force experienced by the moving charges in a magnetic field, the charges are deflected laterally into metallic electrodes connected to the wires of an electric utility system. This is the same physical principle as in the magnetic confinement of a plasma (see Section 12.7).

Using the hot exhaust gases to vaporize water for a conventional steam turbine, an MHD system should have an overall energy conversion efficiency of between 45 and 60%. This is a substantial improvement over that attainable with a steam turbine system. A variety of fuels can be used to heat

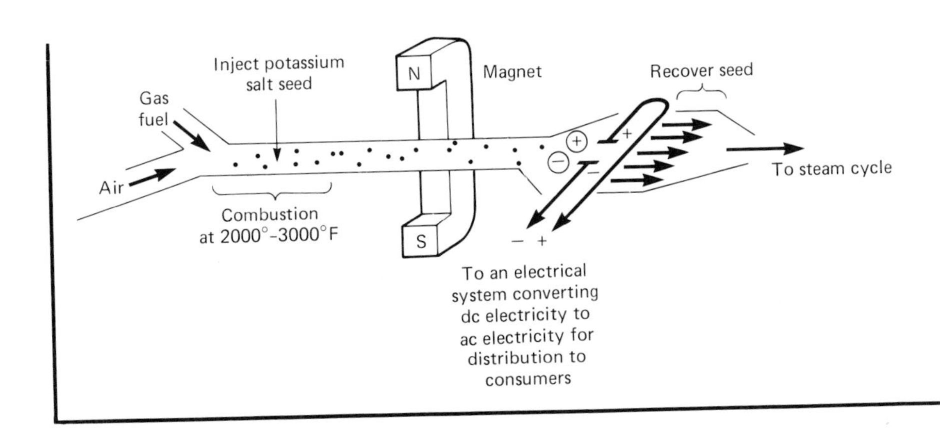

FIGURE 11.6 Principle of the magnetohydrodynamic electric generating process. The electrically charged particles in the moving gas are deflected by the magnetic field into electrodes connected to an external electrical circuit.

the gas to high temperature (2600° C), but coal is especially attractive because it is abundant. As in any coal-burning process, this scheme would produce particulates and sulfur oxides. The particulate problem is minimized because the required high temperature permits more complete burning of the coal. Economics demands recovery of the seed used to enhance the charge in the gas. During this recovery operation, the sulfur oxides can also be collected. Environmentally, MHD systems are attractive in terms of both air and thermal pollution.

The technical aspects of an MHD system are far from trivial. Consider the following two examples. The electrodes picking up the charges deflected from the gas stream must withstand temperatures as high as 2000° C. If they are cooled to prevent melting, the cooling decreases the amount of net charge in the gas. Huge electric currents are required in the massive coils of wire producing the charge-deflecting magnetic field. It has been difficult to make an MHD system whose output power is greater than the input electric power required for the magnets. Recent advances in magnet technology using wires having zero electrical resistance at temperatures near that of liquid helium (4.2 K) may alleviate this problem.

In 1965, an American-built MHD unit produced 30 megawatts of electric power for a short period of time (Fig. 11.7). Despite encouraging results, interest waned until federal legislation in 1974 called for a demonstration of commercial feasibility. Commercial production of electric power by MHD principles has already been demonstrated in Russia where a 25-megawatt system provides electricity for the Moscow area. A 500-megawatt generator is planned. MHD offers a rare opportunity for substantially increasing the efficiency for converting heat to electricity. Whether MHD emerges as a viable energy source in the United States remains to be seen.

11.7 ENERGY FROM BURNING HYDROGEN

Hydrogen is an alternative to natural gas. But unlike natural gas that can be burned as soon as it is removed from the earth, hydrogen must be recovered

FIGURE 11.7 This experimental MHD generator has produced 2300 kilowatts of electric power. The channel is installed within the core of a large magnet. Gas, produced by burning coal or oil, is made conductive by bringing it to a temperature of 5000°F and adding small amounts of potassium carbonate. When this gas flows through the channel, it interacts with the magnetic field to produce electric power. (Photograph courtesy of AVCO Everett Research Laboratory, Inc.)

from chemical reactions and these require energy. This energy is not completely recovered when the hydrogen is burned. So the cost per Btu for hydrogen in an era of plentiful natural gas will always be more expensive. Additionally, the public knows well that hydrogen is a highly flammable, sometimes explosive, gas. Fear of catastrophes, however remote, may sway public acceptance of hydrogen as a fuel. But as natural gas supplies diminish, hydrogen could be a replacement transported in the same pipeline manner.

About half the cost of electricity is due to transmission and distribution. Savings can be made if electric power complexes are built near the consumption areas but then public objections arise as do concerns for pollution. Hydrogen on the other hand is easy and economical to transport in underground pipelines. For those energy applications that could be handled by hydrogen as well as electricity, it makes some sense to produce the hydrogen from the electric output of a plant well removed from the consumption area and then pipe the hydrogen in. This process would also allow a power plant to operate at constant power, which is the most efficient way to operate. This would be particularly applicable to solar-powered electric systems because they will probably have to be in remote areas. To a somewhat lesser extent it applies to nuclear plants, because for safety reasons it is especially desirable to locate them in remote areas.

Hydrogen has uses other than for heating. To mention a few, it is an important ingredient in the production of fertilizers and foodstuffs. The production of synthetic fuels from organic material and trash will require large quantities of hydrogen. Although there are some hazards associated with the handling of hydrogen, they are not insurmountable. Because of the many uses of hydrogen and the clean way that it burns, it probably will play an increasingly important role in future energy scenes.

11.8 TIDAL ENERGY

Few things capture the imagination of poets and writers more than the incessant motion of the oceans and the tantalizing ebb and flow of the tides. The thought of harnessing the enormous energy content of both the oceans and tides has pervaded the minds of human beings for centuries. To some extent we have not been denied the use of this energy. Water-powered mills operating from tidal motion were used in New England in the eighteenth century. A tidal-powered sewage pump functioned in Hamburg, Germany, until 1880, and a water pump installed in 1580 under the London Bridge operated successfully for two and a half centuries. These systems were eventually replaced by the more efficient and convenient electric motors. The search for alternatives to coal-burning and nuclear power plants has promoted some interest in tidal power plants even though they are difficult to construct and are marginally economical.

The tides have their origin in the gravitational force exerted on the earth by the moon and the sun. Gravitational forces hold the earth in its orbit about the sun, and the moon in its orbit about the earth. The earth and moon make complete revolutions every 365 and 28 days, respectively. Thus there are times when the moon comes between the earth and sun. When this happens, the gravitational force exerted upon the earth by the moon and sun is maximum. The fluid ocean bulges out under this force and produces a high-tide condition (Fig. 11.8a). Interestingly a high tide also occurs on the opposite side of the earth. This is because the gravitational force between two masses decreases as the separation increases. Because the water on the side of the earth opposite the moon is farther away than the earth's solid core, the core is pulled toward the moon and away from the water causing it to bulge on the side of the earth opposite the moon. The tide falls as the earth rotates on its axis. Because the earth makes a half rotation on its axis in 12 hours, a high-tide condition will return to the initial position when the earth rotates to a position away from the moon. Thus two high (and two low) tides occur each day. Maximum high tides also occur when the moon is on the opposite side of the earth from the sun (Fig. 11.8b). These tides that occur for the aligned positions of the earth, moon and sun are called spring tides. Tides also occur when the moon is not on a line with the earth and sun, but the tides are less

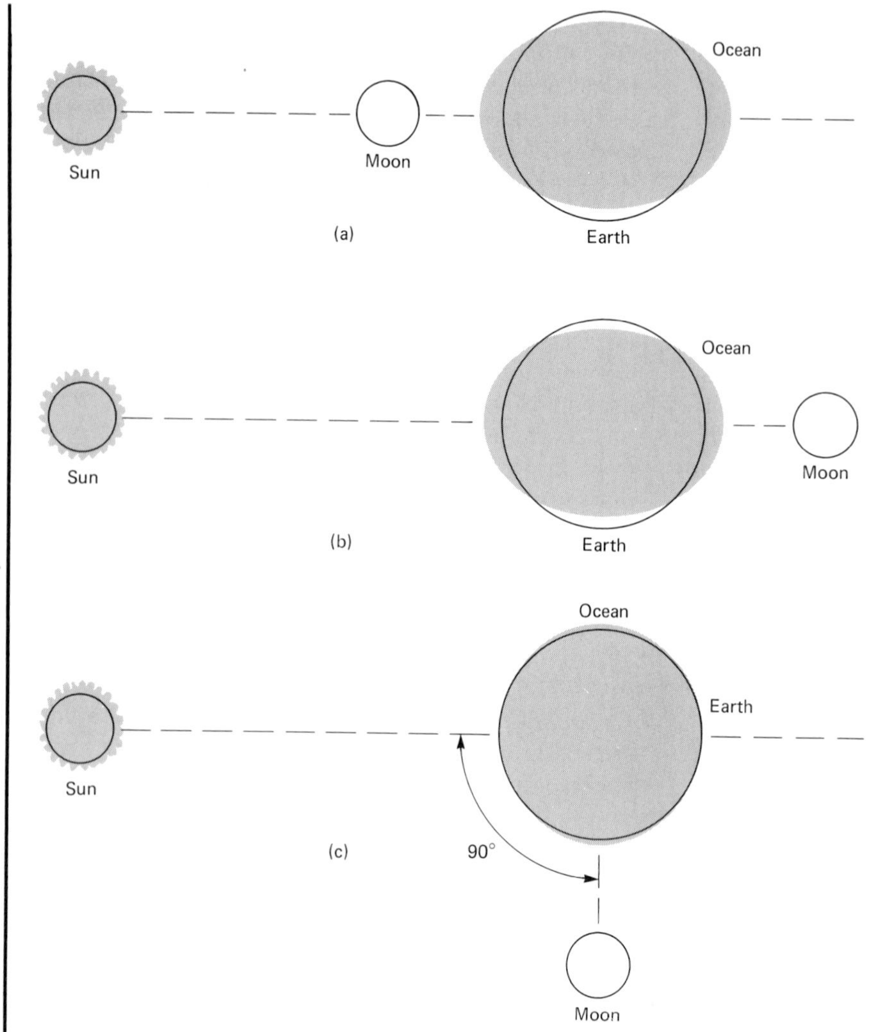

FIGURE 11.8(a) When the moon is between the earth and the sun, the oceans bulge toward the earth and away from the earth. The bulging causes a spring-tide condition. **(b)** Spring tides are also produced when the earth comes between the sun and the moon. **(c)** Neap tides are produced when the angle between the lines connecting the earth and the sun and the earth and the moon is 90°.

pronounced. The minimum effect occurs when a line connecting the earth and moon is at right angles (90°) with a line connecting the earth and the sun (Fig. 11.8c). These tides are called neap tides. Both spring and neap tides occur every 14 days because the earth, moon, and sun line up twice in a 28-day lunar month.

In a tidal electric power plant, water flows into a basin during high-tide. As the water flows in, it is directed into the blades of a water turbine that turns an electric generator producing electric power. At the peak of the tide, the water gates are closed and the water is trapped in the basin. When the tide recedes, the gates are opened and water again flows through the water turbine. Power can be produced on the return of the water to the ocean by reversing the blades on the turbine.

To evaluate the power developed from a hydroelectric system we used the relation

$$P = \left(\frac{w}{t}\right)H \tag{11.2}$$

where w/t is the water flow rate (newtons/sec) and H is the height from which it falls. Basically the same relation applies to tidal systems except that the equation has to be divided by two because the height changes as the water flows out of the tidal basin. This height is called the tidal range when discussing tidal energy systems. For tidal systems there is much less flexibility in the parameters. The time (t) is fixed by the period of the tides. If the water is used as it flows both into and out of the basin, the period is six hours. The tidal range (H) is fixed by the prevailing tidal conditions. In some contemplated tidal energy sites the average tidal range varies between 18 and 35 feet. Because both the large time and small range are working to limit the power, one must look for natural areas where the tidal basins are very large so that the total weight (w) of the water can be made large. Suppose that a basin at a given site is nine square miles (if a square, it would be three miles on a side) and that the average tidal range is 18 feet. The tidal power based on a six-hour period from this site would be 159 megawatts. Thus the power is considerable.

The first commercial tidal power plant was built in France where the Rance River empties into the Atlantic Ocean on the Brittany Coast (Fig. 11.9). Completed in 1966, it generates 240 megawatts of electric power. It produces electricity both on the entrance of water into and exit of water out of an 8.5 square mile basin having an average tidal range of 27.6 feet. The experimental Kislayaquba plant in Russia produces 400 megawatts of electric power. Though the outputs of these systems are not comparable to a 1000-megawatt contemporary nuclear or coal-burning power plant, these tidal power plants provide valuable operating experience for this unique type of system.

United States tidal power plants have been in various stages of planning for nearly 50 years, but a single kilowatt of power has yet to be produced. The most celebrated planned project would have used Passamaquoddy Bay (which links Maine and New Brunswick, Canada) as a 100-square-mile tidal basin. The impounded water would have flowed through 100 ten-megawatt turbogenerators into Cobscook Bay in Maine (Fig. 11.10). Building seven miles of dams and 160 water gates indicates the complexity of the project. If built at current prices, the

FIGURE 11.9 A tidal-electric power plant located at the mouth of the Rance River in France. The water basin is shown in the upper portion of the photograph. Six water-turbine–electric-generator systems are housed in the left-most part of the retainment structure. Boats may enter the tidal basin through the opening on the right-most part of the retainment structure. The complex is capable of generating 240 megawatts of electric power, which is about one-fourth the output of a large coal-burning or nuclear power plant. (Photograph courtesy of French Embassy.)

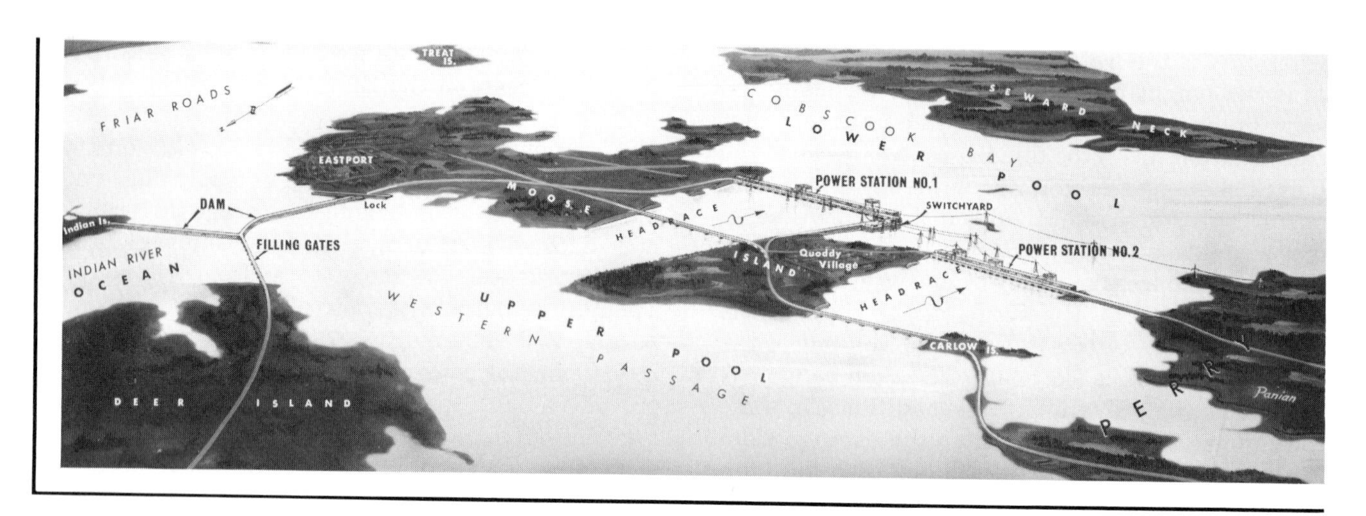

FIGURE 11.10 An oblique map of the once-conceived Passamaquoddy tidal power system. Water flows into the upper pool during high tides. At the peak of the tide, the filling gates are closed and the water is trapped in the upper pool. When the tide recedes, the water is channeled into the power stations by the headraces and empties into Cobscook Bay. (Map courtesy of Department of the Army Corps of Engineers.)

project would probably cost over a billion dollars. Such a huge cost per kilowatt of power puts it in a class with peaking units. Although the project is attractive from an environmental standpoint and makes use of a source of power that goes unused everyday, it is unlikely that it or any United States tidal power system will be built in the near future because of the capital cost.

REFERENCES

HYDROELECTRIC POWER

Energy: Electric Power and Man, Timothy J. Healy, Boyd and Fraser, San Francisco, Calif., 1974.

GEOTHERMAL ENERGY

"Geothermal Power," Joseph Barnea, *Scientific American* **226,** 1, 70 (1972).

"The Story of Geothermal Energy," brochure, Union Oil Company, Los Angeles, California.

"Geothermal Energy," L. J. P. Muffler. In Lon C. Ruedisili and Morris W. Firebaugh (eds.), *Perspectives on Energy,* Oxford University Press, New York, 1982.

Geothermal Energy, Paul Kruger and Carel Otte (eds.), Stanford University Press, Stanford, Calif., 1973.

"Hot Dry Rock Geothermal Energy," Grant Heiken, Hugh Murphy, Gregory Nunz, Robert Potter, and Charles Grigsby, *American Scientist* **69,** 400 (July–August 1981).

"Extracting Geothermal Energy Can Be Hard," Richard A. Kerr, *Science* **218,** 668 (12 November 1982).

ELECTRIC ENERGY FROM BURNING TRASH

"New Progress for Big Recovery Plants," *Waste Age* (April 1982).

"Power from Trash," William C. Kasper, *Environment* **16,** 2, 34 (March 1974).

"Tapping Resources in Municipal Solid Waste," S. L. Blum, *Science* **191,** 669 (20 February 1976).

COAL GASIFICATION AND LIQUEFACTION

"The Gasification of Coal," Harry Perry, *Scientific American* **230,** 3, 19 (March 1974).

"The Chemistry and Technology of Synthetic Fuels," Alan Schriesheim and Isidor Kirshenbaum, *American Scientist* **69,** 536 (September–October 1981).

"Clean Power from Dirty Fuels," Arthur M. Squires, *Scientific American* **227,** 4, 26 (October 1972).

"Oil and Gas from Coal," Neal P. Cochran, *Scientific American* **234,** 24 (May 1976).

MAGNETOHYDRODYNAMICS (MHD)

Energy: From Nature to Man, William C. Reynolds, McGraw-Hill, New York, 1974.

"MHD Power Generation," A. Kantrowitz and R. J. Rosa. In M. D. Fiske and W. W. Havens, Jr. (eds.), *Physics and the Energy Problem—1974,* American Institute of Physics, New York, 1974.

ENERGY FROM BURNING HYDROGEN

"The Hydrogen Economy," Derek P. Gregory, *Scientific American* **228,** 1, 13 (January 1973).

"Energy and the Future," Allen L. Hammond, William D. Metz, Thomas H. Maugh II, American Association for the Advancement of Science, Washington, D.C., 1974.

TIDAL ENERGY

"Power from Ocean Waves," J. N. Newman, *Technology Review* **86,** 50 (July 1983).

"Harnessing the Tides," James A. Fau, *Technology Review* **86,** 51 (July 1983).

"Harnessing Power from the Tides: State of the Art," Paul R. Ryan, *Oceanus* **22,** 64 (Winter 1979–1980).

"Whatever Happened to Tidal Power?", Roger H. Charlier, *Oceanus* **14,** 30 (November–December 1981).

REVIEW

1. List the energy conversions in a hydroelectric power system.
2. What determines the power available in water flowing over a dam?
3. Describe the anticipated impact of hydroelectricity in the future energy scene.
4. List some environmental concerns emanating from hydroelectric power plants.
5. What is a pumped-storage hydroelectric system?
6. What is geothermal energy?
7. Describe how electricity is produced from geothermal energy.
8. What are some problems with the use of geothermal energy?
9. For what reasons has trash emerged as a potential energy source?
10. To what extent is "energy from trash" realistic?
11. Can trash be used for any energy purpose other than producing electric energy?
12. Describe the coal gasification process.
13. Why is there considerable interest in gasifying coal for use in electric power plants?
14. What does MHD mean?
15. How would an MHD electric power plant differ from a nuclear power plant?
16. When can we expect to see MHD power plants in the United States? in the world?
17. What is meant by the "hydrogen economy"?
18. Name some advantages and disadvantages of burning hydrogen.
19. Name some industrial uses of hydrogen.
20. Describe how tides are produced.
21. How might tidal energy be used for producing electricity?
22. Realistically, when can we expect to have our homes lighted with tidal-produced electricity?
23. Calling R the rate of flow of water through a hydroelectric system and H the difference in height of water level, pick the correct equation for the power available from the moving water.

    a) $P = RH$ b) $P = R/H$ c) $P = H/R$
    d) $P = 1/HR$ e) $P = R^2H$

24. If the rate of flow of water in a hydroelectric power plant is doubled, then we should expect the power output to
    a) double.
    b) halve.
    c) remain unchanged.
    d) increase but reach a limit before the power doubles.
    e) be unaffected unless the height of the dam also changes.
25. Of the percentages listed below, pick the one that most closely represents the hydroelectric contribution to the total United States electric energy production in 1983.

    a) 0.1% b) 1% c) 5% d) 11% e) 45%

26. Hydroelectric systems have been and still are an important source of electric power in the United States. In the future we can expect
    a) the existing systems to be phased out and replaced by nuclear power plants.
    b) numerous new systems on the Ohio and Mississippi rivers.

c) massive financial investments in new hydroelectric systems.
d) some growth in hydroelectric systems but their overall contribution to electric power production will decline.
e) that no new hydroelectric systems will be constructed.

27. A pumped-storage system used by some electric power companies stores
   a) electric energy by pumping electric charge into batteries.
   b) gravitational potential energy by pumping water into an elevated reservoir.
   c) rotational kinetic energy in a massive, rotating wheel.
   d) electric power in giant electric transformers.
   e) electromagnetic energy in nuclear pumped lasers.

28. Power from a dammed river is given by the equation $P = W/T \cdot H$. Power from a tidal energy system is given by $P = \frac{1}{2} W/T \cdot H$. The number $\frac{1}{2}$ arises because
   a) in a tidal system H is not constant.
   b) the tidal system is only 50% efficient.
   c) of the density of salt water.
   d) of the role of the high and low tides.
   e) hydroelectric dams are always twice the height of tidal ranges.

29. Power is the rate of doing work or, alternatively, the rate of converting energy. As a formula, $P = E/t$ where E is energy and t is time. To get significant power out of a tidal energy system, the energy of the water must be significant because the time
   a) is relatively large.
   b) is related to the speed of waves on the ocean.
   c) is relatively small.
   d) is 24 hours, the length of a day.
   e) is 365 days, the length of a year.

30. The nation where tidal energy was first developed, and where a sizeable facility is in operation, is
   a) the United States. b) England.
   c) Canada. d) France.
   e) New Zealand.

31. In the earth's crust the temperature increases about 2° C for each 100 meters depth below the surface. If the surface temperature is 30° C, the temperature is about 100° C at a depth of
   a) 5 km. b) 500 m. c) 35 km.
   d) 100 km. e) 3.5 km.

32. A conglomerate of geothermal electric power plants at The Geysers, California, produces power comparable to a coal-burning power plant. The site was chosen because
   a) high-level technology has permitted drilling to extreme depths.
   b) surface water can percolate down to be heated by radioactive hot spots near the surface.
   c) geological faults extend down to the mantle of the earth.
   d) high-sulfur bearing springs rendered the land unsuitable for farming.
   e) it produces dry steam, allowing easier and earlier development.

33. A geothermal electric power plant has a thermodynamic efficiency
   a) about the same as a coal-burning power plant.
   b) about the same as a nuclear power plant.
   c) about the same as an oil fired power plant.
   d) all the above because all involve conversion of thermal energy to electric energy.
   e) none of the above. Geothermal power plants have an efficiency of about 99%.

34. The use of trash as a fuel is attractive because
   a) trash has a reasonable heat content when burned—about half that of coal.
   b) trash is a low-sulfur fuel.
   c) trash is readily available.
   d) trash solves a waste disposal problem.
   e) all of the above.

35. An electric current involves moving charges. The larger the rate of movement, the larger the electric current. The movement of air in a jet engine involves little electric current because
   a) the speed of the atoms is very small.
   b) the atoms of the gas are, for the most part, electrically neutral.
   c) there are no electrons and electric currents always involve electrons.
   d) only positive charges are involved.
   e) there is no magnetic field present.

36. A magnetohydrodynamic (MHD) generator works on the principle that
   a) a hot flowing gas always has a large electric current associated with it.
   b) the energy in coal can be converted *directly* to electricity.
   c) moving charges in the presence of a magnetic field will experience a force.

d) sulfur dioxide can be converted to electricity.
e) the energy in high-temperature steam can be converted to rotational energy by using a steam turbine.

37. In magnetohydrodynamic generators (MHD), the hydrodynamic fluids most often used are
    a) cesium and potassium.
    b) boron and cadmium.
    c) deuterium and tritium.
    d) ammonia and water.
    e) hydrogen and helium.

38. One should expect that carbon monoxide (CO) could be burned to produce energy because
    a) carbon monoxide is a hydrocarbon.
    b) carbon monoxide is a result of incomplete combustion of carbon and therefore should undergo further combustion.
    c) carbon monoxide is closely related to methane gas which is easy to burn.
    d) carbon monoxide is toxic.
    e) carbon monoxide can be hydrogenated.

39. Coal is one of the most abundant energy resources in the United States. By converting coal to a combustible gas,
    a) an actual gain in total energy content is possible.
    b) the sulfur dioxide problem associated with burning coal is minimized.
    c) the problems associated with photochemical smog would be alleviated greatly.
    d) the massive coal deposits in the southeastern United States can be utilized.
    e) there is produced a fuel that generates no pollutants when burned.

40. When coal is gasified, ordinary air is used as a source of oxygen for burning the coal. What gaseous pollutant also produced by automobiles would you expect to be formed?
    a) particulates
    b) sulfur oxides
    c) carbon dioxide
    d) sulfuric acid
    e) nitrogen oxides

Answers

23. (a)	24. (a)	25. (d)	26. (d)
27. (b)	28. (a)	29. (a)	30. (d)
31. (e)	32. (e)	33. (d)	34. (e)
35. (b)	36. (c)	37. (a)	38. (b)
39. (b)	40. (e)		

QUESTIONS AND PROBLEMS

11.2 HYDROELECTRICITY

1. For reasons of transportation and water supply, many cities are located near rivers. Why haven't most cities derived their electricity needs from hydroelectric plants on these rivers?
2. Even though a river flowing through a city may be able to support only a small hydroelectric project, why might a city still want to have it available?
3. The Bonneville hydroelectric plant near the mouth of the Columbia River utilizes a small reservoir of water and a difference in water level of 59 feet. The Grand Coulee hydroelectric plant located several hundred miles up the Columbia River utilizes a huge reservoir of water and a difference in water level of some 285 feet. Why are both facilities able to produce such large power outputs with such different water facilities?
4. What complications arise when river flow is restricted by the presence of dams?
5. Does the electric generator used on a bicycle for operating a horn or headlight produce electricity all the time that the bike is in motion? Is the generator able to store electric energy for future use?
6. Why are pumped-storage hydroelectric systems not practical in all geographic locations?
7. Water in a pumped-storage hydroelectric system must flow to a lower geographic level. In a region where a natural high level is unavailable for a reservoir, how could a pumped-storage system be built below ground level?
8. A "cube" of water one foot on a side weighs 62.4 pounds. If water flows over a dam at a rate of 200 cubic feet/second, what is the flow rate in pounds/second?
9. A city of 15,000 people wants to build a small hydroelectric power plant for emergency use of electricity. If the dam is to be eight meters high and the flow rate 200 cubic feet/second, how much hydropower is available for conversion to electric power? (Remember, one cubic foot of water weighs 62.4 pounds and one pound = 4.45 newtons.)
10. It is desired to build a pumped-storage system that has 100,000 kWh of gravitational potential energy. The vertical height between the reservoir and turbine is 80 meters.
    a) Using Eq. (11.1), calculate the weight in newtons of the amount of water needed. One cubic

meter of water weighs 9800 newtons.

b) If the water depth is five meters, how big an area will be required for the reservoir?

11. The formula for determining power produced by falling water is

$$P = \frac{wH}{t}.$$

If weight, height, and time are expressed in pounds, feet, and seconds, respectively, then the power is in foot-pounds per second. This can be converted to watts by multiplying the result by 1.35. Show that 720 megawatts of power are produced by a water flow of 1,870,000 pounds per second falling through a height of 285 feet.

11.3 GEOTHERMAL ENERGY

12. Why isn't the geothermal energy in The Geysers region of northern California used for household heating rather than for input energy for electric power plants?

13. Why is the availability of water an important consideration for a geothermal electric power plant?

14. A geothermal electric power system converts thermal energy to electric energy. Why is a geothermal system less efficient than a contemporary coal-burning system?

15. The temperature of the core of the earth is not accurately known, but the general consensus is that it is less than 10,000° F. The variation of temperature with depth into the earth's crust is fairly well known; 1° F increase for every 100 feet down into the crust. If this variation were to continue, what would you expect for the temperature of the earth's core? Assume the distance to the earth's surface from the edge of the core is 2000 miles. This illustrates how an extrapolation can go astray.

16. The weight of six miles of the earth's crust below the United States is about 100 billion billion pounds (10^{20} pounds). If the crust were cooled 1° F, then each pound would release about one Btu of heat. Therefore, cooling six miles of the crust 1° F would release 100 billion billion (10^{20}) Btu of energy. Assuming the heat obtained from one ton of coal is 25 million Btu, show that the energy derived from cooling six miles of the earth's crust below the United States by 1° F is equivalent to the heat which can be produced from about 4 trillion (4×10^{12}) tons of coal. This should convince you of the enormous energy content of the earth's crust.

11.4 ENERGY FROM BURNING TRASH

17. Based on your own experience, does the generation of five pounds of trash per person per day seem unrealistic? You might like to make your own estimate.

18. Open burning was once a popular way of disposing of trash. Why do you think this method is now forbidden? Would the problems of open burning also be present to some extent in power plants burning trash for energy?

19. Historically, cities have looked to suburban areas for land-fill waste disposal sites. How has the population settlement pattern complicated the waste disposal problems of cities?

20. Although relatively few United States cities (see Table 11.2) have commitments to trash-energy systems, the technological success of trash-energy systems has been demonstrated. What, then, are the barriers preventing many more cities from constructing trash-energy systems?

21. What percentage of the daily winter heat requirement of 500,000 Btu could possibly be obtained from burning the trash of a family of four producing five pounds of trash per occupant?

22. Solid waste is generated at an average rate of five pounds per person per day. Show that a city of one million people produces 2500 tons of solid wastes each day.

23. One scheme for converting municipal wastes to a burnable gas produces per ton of waste 24,000 cubic feet of gas having an energy content of 300 Btu per cubic foot. Compare the gas energy content per pound of waste with an energy content of 13,000 Btu per pound of coal.

24. You will need the following data for this problem:

    1 pound of trash = 5300 Btu of energy,
    1 kilowatt-hour = 3413 Btu.

    a) Show that the energy content of 136 million tons of trash is 1400 trillion Btu.
    b) Show that 1400 trillion Btu are equivalent to 410 billion kilowatt-hours.
    c) Assuming a conversion efficiency of 32%, show that 1400 trillion Btu of energy will produce 130 billion kilowatt-hours of electric energy.

25. Several of the trash-energy systems listed in Table

11.2 have trash-handling capacities of about 1000 tons per day. About how many people will such a plant accommodate each day?

11.5 COAL GASIFICATION AND LIQUEFACTION

26. The increasing reliance on foreign oil from which gasoline is refined is a perplexing American problem. What other alternatives do we have for producing gasoline or gasoline substitutes?
27. Carbon monoxide (CO) is a burnable gas that can be derived from coal. If carbon monoxide is used as a replacement for natural gas, what special precautions would be in order?
28. Coal can be used to produce substitutes for either natural gas or gasoline. From a knowledge of the relative complexities of the molecules in the gas and gasoline, why would you expect liquefaction to be the more difficult of the two processes?
29. What is the end product in the chemical reaction describing the burning of carbon monoxide (CO)?
30. Before the advent of natural and synthetic gases for heating and cooking, a relatively low-grade fuel called coal oil was common. What would you guess is the origin of this product?
31. If air is used in the initial burning process for coal gasification, what gaseous pollutant will be produced that is also associated with the internal combustion engine?
32. A certain pilot coal liquefaction plant produces three barrels of a low-sulfur, utility fuel for each ton of coal processed. If the heat content of the liquid is 5,800,000 Btu per barrel, how many Btu of oil heat is obtained for each pound of coal processed? Compare this result with the original coal heat content of about 13,000 Btu per pound.
33. A proposed coal gasifier will process 1.5 tons of coal per hour to produce one million cubic feet of gas having an energy content of 160 Btu per cubic foot. Assuming that the heat energy content of the coal is 13,000 Btu per pound, show that the heat energy content of 1.5 tons of coal is 39 million Btu and the energy content of the gas produced is 160 million Btu. Where does the extra energy for the gas come from?

11.6 MAGNETOHYDRODYNAMICS (MHD)

34. What environmental problems are associated with the MHD method of generating electricity?
35. The exhaust gases from a MHD generator are at a temperature of about 3000 K. Why does this allow for the possibility of a thermodynamic efficiency substantially higher than that in a coal-burning or nuclear-fueled electric power plant?
36. The United States has developed an unrivaled technology for producing large magnetic fields. Why are countries, Russia, for example, with an interest in MHD systems also interested in United States magnet technology?

11.7 ENERGY FROM BURNING HYDROGEN

37. Why is it possible to recycle hydrogen after it is burned but not possible to recycle natural gas after it is burned?

11.8 TIDAL ENERGY

38. Why is the concentration of energy in a tidal power plant small compared to that in a nuclear power plant?
39. The energy available from a tidal basin is related to the mass (M) of the water contained and the tidal range (H) by $E = \frac{1}{2}MgH$ where g is the acceleration due to gravity (9.8 meters/second^2). The amount of energy available per kilogram of water contained is then $E/M = \frac{1}{2}gH$.
    a) Determine the energy per kilogram for a tidal range of ten meters.
    b) Compare this with coal having an energy content of 13,000 Btu per pound. From this comparison, comment on the general impracticality of tidal energy systems. (A kilogram is equivalent to 2.2 pounds and a Btu = 1055 joules.)
40. a) A "cube" of water one foot on a side weighs 62.4 pounds. Knowing this, show that a basin three miles wide, three miles long, and eighteen feet thick weighs 282 billion pounds. (There are 5280 feet in a mile.)
    b) If this water is drained out of the basin in six hours, show that the average rate of drainage is 13 million pounds per second.
    c) Using the expression

$$P = \frac{1.35}{2}\left(\frac{w}{t}\right) \cdot H$$

for determining tidal power, show that the power developed is 159 megawatts.

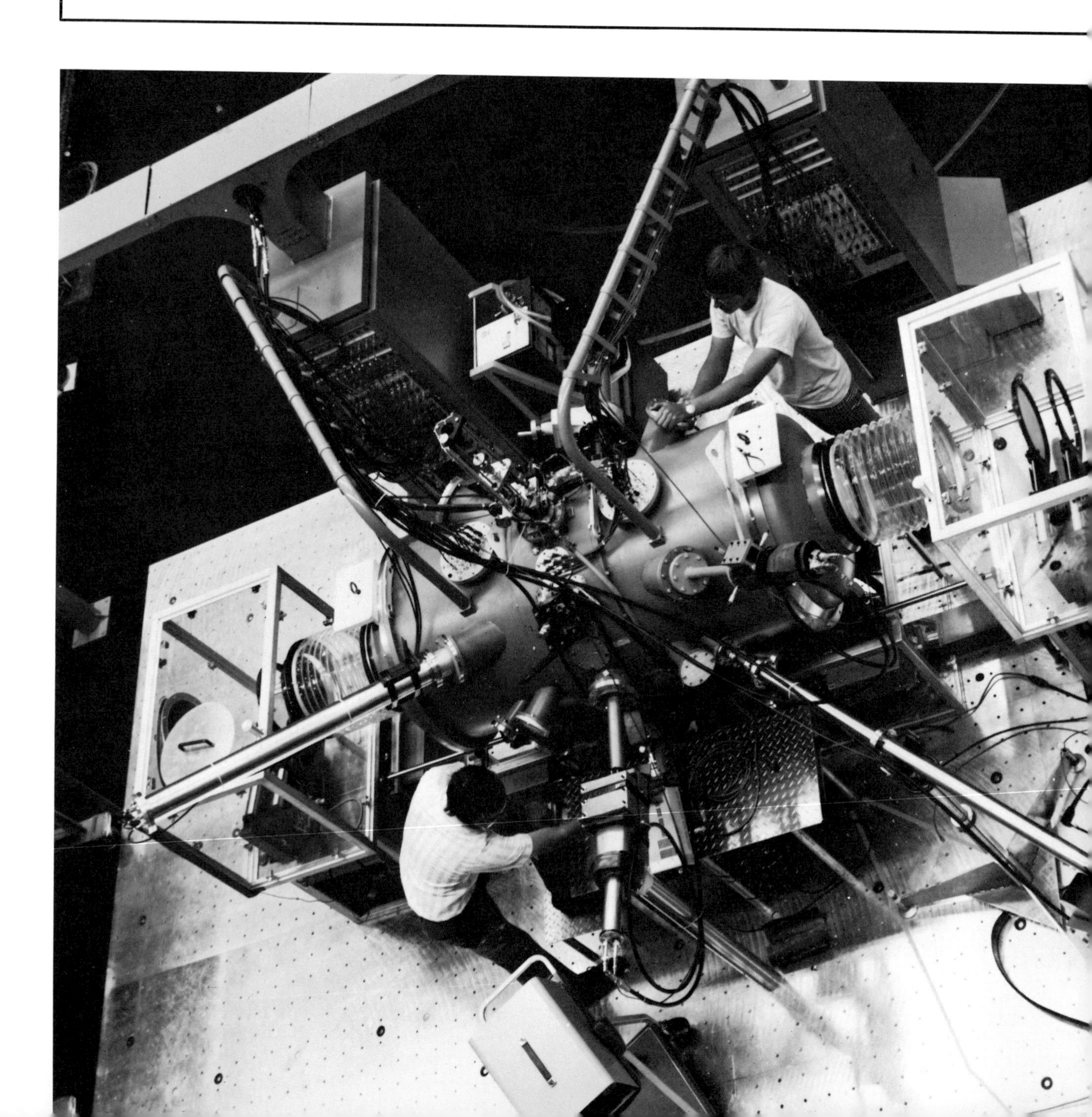

12

Nuclear Breeder and Nuclear Fusion Reactors

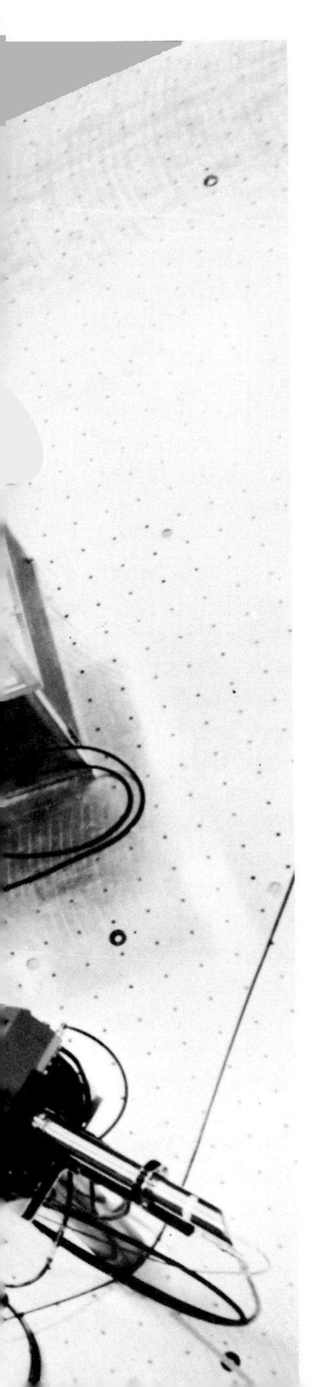

12.1 MOTIVATION

Several methods for generating electricity on a commercial scale have been discussed, but none has been touted as a complete answer to long-range energy problems. Hydroelectricity is limited by the availability of useful dam sites. Coal is abundant, but there are health and environmental problems from mining, there are air pollution problems from burning it, there is concern for the greenhouse effect brought on by carbon dioxide from burning it, and there is the realization that there are better uses for it than burning. Cheap uranium-235 fuel is limited. If less abundant, more costly ores are exploited, electricity will be more expensive and the same damage to the land results as with strip mining of coal.

Research on laser-induced nuclear fusion is performed with the apparatus shown here. Hopefully, these tedious, expensive endeavors will reward human beings with an abundant, socially acceptable energy supply. (Photograph courtesy of Lawrence Livermore Laboratory.)

The nuclear breeder and nuclear fusion reactors are widely heralded possibilities for large-scale, long-term energy production. Although controversy shrouds the use of nuclear breeder reactors, they do make better use of natural uranium resources, and several foreign nations have commitments to nuclear breeder reactor technology. It remains to be seen whether the United States also commits itself to the nuclear breeder reactor technology. Conceivably, nuclear fusion energy technology can avoid many of the concerns put forth for the nuclear breeder reactor and may be the long-term energy solution. But even though hopes are high for nuclear fusion energy, as yet its technology is unproven. In order to respond intelligently to questions surrounding these exciting, but controversial, technologies, it is important to understand their basics. Obtaining this understanding is the goal of this chapter.

12.2 THE NUCLEAR BREEDER REACTOR

To review, $^{233}_{92}U$, $^{235}_{92}U$, and $^{239}_{94}Pu$ are the only useful nuclear fission reactor fuels. Only $^{235}_{92}U$ occurs naturally and it comprises only 0.7% of natural uranium which is predominantly $^{238}_{92}U$. The isotopes $^{233}_{92}U$ and $^{239}_{94}Pu$ must be produced through nuclear reactions. A breeder reactor is fueled initially with $^{235}_{92}U$. In addition to liberating energy by the normal nuclear fission process as in a conventional nuclear reactor, it produces $^{233}_{92}U$ or $^{239}_{94}Pu$ from nuclear transformations of $^{232}_{90}Th$ (thorium) or $^{238}_{92}U$. Producing the new fuel requires $^{232}_{90}Th$ or $^{238}_{92}U$ that is used up in the production of $^{233}_{92}U$ or $^{239}_{94}Pu$ and must be replaced as the process demands.

The fuel production principle is straightforward. When a nucleus such as ^{235}U or ^{239}Pu is fissioned with a neutron, there is no way to predict the composition of the reaction products. Some products are more likely to be formed than others. Some reactions produce no neutrons and others produce several. Depending on the nucleus, the average number of neutrons produced in each fission reaction is between two and three. One of these neutrons is required to sustain the fission chain reaction process. The remainder can be used for some other purpose. The purpose in a breeder reactor is not only to produce heat with fission but also to generate fissionable nuclei. The sequence begins by arranging for a neutron to be captured by an appropriate nucleus. The nucleus formed by neutron capture is radioactive. After two successive disintegrations, the desired stable (or very long-lived) end product is a fissionable nucleus. The nucleus capturing the neutron is called a fertile nucleus and the process leading to a fissionable nucleus is called breeding. The isotope $^{238}_{92}U$ is a suitable fertile nucleus. When $^{238}_{92}U$ captures a neutron, the unstable isotope $^{239}_{92}U$ is formed. $^{239}_{92}U$ disintegrates by emitting a negative beta particle. Because the proton number increases by one, the species formed is an isotope of neptunium ($^{239}_{93}Np_{146}$). This isotope is also unstable and disintegrates by emitting another negative beta particle. The end product is plutonium-239 ($^{239}_{94}Pu_{145}$) which is a suitable nucleus for a reactor fuel. The sequence of events leading to the formation of $^{239}_{94}Pu$ is shown in Fig. 12.1. A similar breeding process can be devised using thorium-232 as the fertile isotope to produce fissionable uranium-233.

Novel schemes for producing energy are often impractical because of low production rates. Although the mechanism by which $^{239}_{94}Pu$ is made from neutron bombardment of $^{238}_{92}U$ is attractively simple, it is important to know just how fast a fuel can be produced. One way of characterizing the production rate is to determine how long it takes to produce twice as many plutonium atoms as there were fissionable atoms in the initial fuel of the reactor. This time is called the doubling time.

A nuclear breeder reactor is not a perpetual energy source. It requires both fissionable nuclei (for energy) and fertile nuclei (for making new fuel). Both resources are depleted as time passes. If either the source of fissionable nuclei or the source of fertile nuclei is used up, then the nuclear breeder reactor cannot function. Both a nuclear breeder reactor and a contemporary water-cooled reactor are designed to produce energy from nuclear fission chain reactions. Thus both reactors must have a strategic arrangement of fuel elements, a method of removing thermal energy from the reactor, and a scheme for controlling the energy output. A nuclear breeder

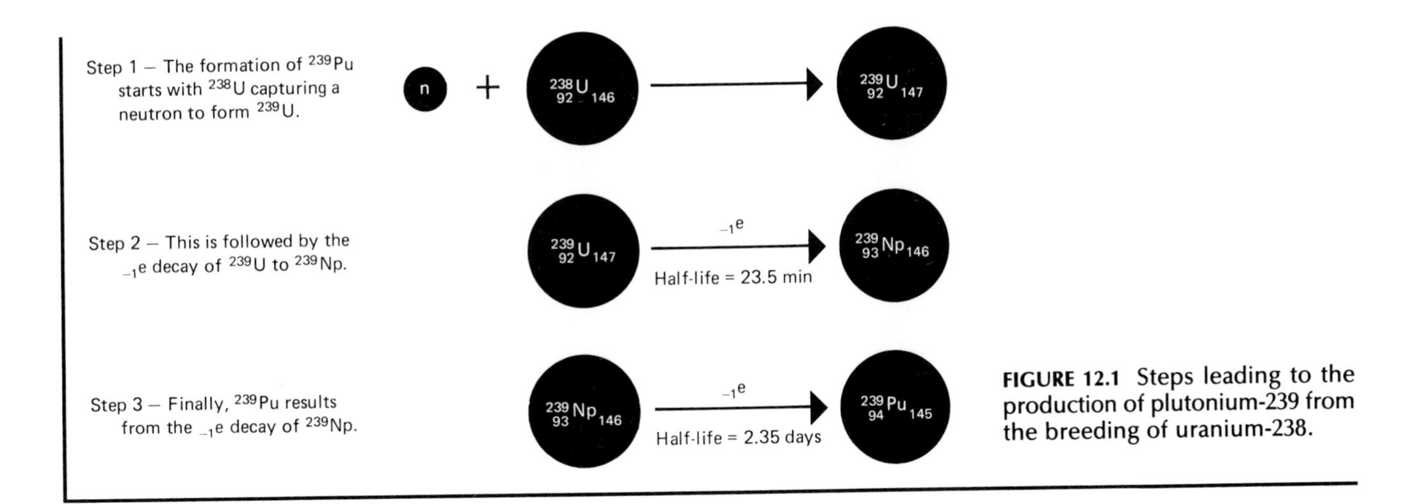

FIGURE 12.1 Steps leading to the production of plutonium-239 from the breeding of uranium-238.

reactor, unlike a water-cooled reactor, is designed to optimize the production of a fissionable fuel. Thus the fertile nuclei, $^{238}_{92}U$ or $^{232}_{90}Th$, must absorb neutrons from nuclear fission reactions in order to start the breeding process. At some point in time, the material containing $^{238}_{92}U$ or $^{232}_{90}Th$ is reprocessed and the fissionable fuel recovered. In a period equal to the doubling time, the recovered fertile material yields two fissionable atoms for every initial fuel atom loaded.

The doubling time for a very efficient nuclear breeder reactor is about six years. This means that a very efficient breeder reactor produces enough fuel to refuel itself and another similar reactor after six years of operation. The doubling time for first generation breeder reactors is expected to be 20 to 25 years. Clearly, nuclear breeder reactors could substantially extend the lifetime of uranium resources.

12.3 THE LIQUID METAL FAST BREEDER REACTOR

Remember (Section 9.6), the chance of nuclei fissioning when bombarded with neutrons depends sensitively on the energy of the neutrons. Neutron-induced fission of $^{235}_{92}U$ works best with slow neutrons. The same chance aspect prevails in the attempt to have $^{238}_{92}U$ capture neutrons to make $^{239}_{92}U$. Neutron capture (*not* fission) by $^{238}_{92}U$ is significant only for energetic (or fast) neutrons.* Hence fast neutrons are required to optimize breeding. Because both breeding and fission reactions are desired, the fission reactions must be initiated by fast neutrons even though it is more difficult. Water cannot be used as a coolant because it slows (or moderates) the speed of the neutrons. The alternatives to water for a coolant are limited. Molten sodium and molten salt are possibilities. Liquid sodium is used as the coolant in most breeder reactors. These reactors are designated LMFBR which means liquid metal (for cooling) fast (for energetic neutrons) breeder reactor.

Sodium has several desirable features. Even though it is a metal, its melting point is only 210° F. Thus it melts at the temperature of boiling water. It does not boil until it reaches 1640° F. Because the maximum temperature of sodium circulating in the core of a reactor will be around 1150° F, there is a substantial temperature margin before boiling occurs. While sodium is in the liquid state, the pressure it exerts on the walls of a container is insensitive to temperature even when the temperature reaches 1150° F. In contrast, the pressure in a pressurized water reactor rises to about 2250 pounds per square inch when the water temperature is 600° F.

* A neutron with energy greater than 10,000 electron volts (10 keV) is called a fast neutron.

The only pressure on the liquid sodium is that required to circulate it around the reactor core. Thus concerns are minimal for pressure-induced leaks and failures in pipes and valves.

Liquid sodium is an excellent heat conductor. While removing heat from a reactor, it does not significantly capture or moderate neutrons needed to sustain the chain reaction and to instigate the breeding reactions. Fission fragments that do escape from the fuel rods and leak into the sodium coolant are readily tied up chemically making it difficult for them to escape to the environment. It seems that sodium is a reactor designer's dream, but there are disadvantages. Liquid sodium oxidizes and burns if exposed to air and reacts violently producing hydrogen if exposed to water. Continual bombardment of the sodium coolant with neutrons produces the radioactive sodium isotopes ^{22}Na and ^{24}Na, which can accumulate to a hazardous extent. So, despite the fact that few problems are encountered in the pressure considerations involved with containing liquid sodium, extensive precautions still must be taken to ensure that the sodium does not escape.

Heat is removed from the reactor core by liquid sodium, but steam is required for the turbine coupled to the electric generator. This is accomplished by transferring heat from the sodium-cooling circuit to a water-cooling circuit. As a precaution against radioactivity leaking into the water circuit, an intermediate liquid sodium cooling circuit is used (Fig. 12.2).

The feedback method (Section 9.14) employed in the control of the water-cooled nuclear reactor is also used in the fast breeder reactor. When the power output increases above some desired level, signals are fed back to the controls and action is taken to decrease the power. For example, control rods are inserted to limit the availability of neutrons for fission reactions. Routine control of a fast breeder reactor and a water-cooled reactor are somewhat different. One reason is that the time between the birth of a neutron and when it is captured or produces fission is substantially less in a fast breeder reactor. Another reason is that there are fewer delayed neutrons in the fast breeder reactor if ^{239}Pu is used for the fuel. Routine control is manageable unless a malfunction occurs that produces a feedback tending to increase the output. A void (or bubble) in the liquid sodium is such a condition. To a small but important extent, sodium in the cool-

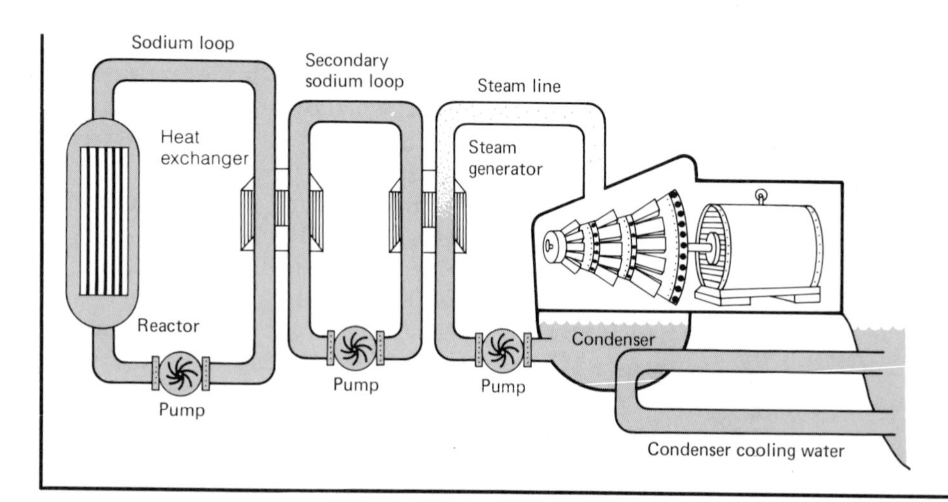

FIGURE 12.2 The heat-transfer circuits in a fast breeder reactor. Thermal energy generated in the reactor core by nuclear fission reactions is transferred to a heat exchanger. This closed loop retains any radioactive material acquired in the heat transfer. A second closed liquid sodium loop transfers heat to a steam generator that vaporizes water for the steam turbine.

ant captures neutrons. If a void occurs, there are more neutrons available for fission because fewer neutrons are captured. The extra neutrons cause the power output of the reactor to increase. Increased output produces more heating and possibly more voids that may increase the power still further. The process takes place so quickly that conventional controls cannot operate. A control is needed that operates just as fast as the void mechanism, but which produces a diminishing effect on the output power. Such a control utilizes a principle named the Doppler effect.*

Imagine a situation where you are riding in a car listening to the sound from the horn in a parked car. As you pass the car, the pitch (frequency) of the sound changes suddenly. This phenomenon is caused by the Doppler effect. It is due to the fact that the relative speed between you and the crests of the sound waves changes abruptly when you pass the stationary car.† Molecules in a solid are not free to migrate but they do vibrate. The frequency and speed of their vibrations depend on temperature. The greater the temperature, the greater the frequency and speed. Thus the relative speed between an impinging neutron and a nucleus attemting to capture it depends on temperature. Because the chance of neutron capture by a nucleus in the fuel element depends sensitively on the relative speed, the chance also depends on temperature. As the temperature increases the chance of capture increases. More captures mean fewer neutrons for fission. When the output of a reactor suddenly increases and the temperature of the core rises, this effect instantaneously comes into play and tends to reduce the output. Valuable time is gained so that conventional controls can be brought into play.

The fast neutron principle in a breeder reactor necessitates a more highly enriched fuel than required by a water-cooled reactor. Whereas the fuel in light-water-cooled reactors is about 3% fissionable nuclei, the percentage is 15–30% for a nuclear breeder reactor. In addition, the energy-producing core may occupy a volume of only a few cubic meters (or cubic yards, if you want). These features lead to a concern called secondary criticality; a concern which is insignificant in light-water reactors. Criticality in a light-water reactor is attained after the fission neutrons have been sufficiently reduced in energy. If the coolant which also serves as the moderator is accidentally removed, then the mechanism for achieving criticality is removed (negative feedback) and the chain reactions stop. In a nuclear breeder reactor the liquid sodium functions only as the coolant and if it is suddenly removed, the reactor is still critical until other sensors detect that something is wrong. If a meltdown of the core occurs, there is the remote possibility of the fuel assuming a different geometric arrangement and developing a critical mass that can produce an uncontrolled chain reaction. This is the secondary criticality. Consequently, the design considerations of fast breeder reactors are more stringent. Because water and sodium are highly reactive, an emergency cooling system obviously cannot employ water. Current designs take advantage of the exceptional characteristics of sodium for conducting heat. In one design the reactor is sealed in a pool of sodium. In another the heat is transferred out of the reactor into a vat of sodium.

12.4 STATUS OF NUCLEAR BREEDER REACTOR PROGRAMS

Russia, France, and England have successfully operated nuclear breeder reactor electric power plants having electric power outputs ranging from 235 to 350 megawatts (Table 12.1). Second-generation systems are being built by all three countries. Japan and West Germany have systems in the construction stage.

An experimental nuclear breeder reactor has been operating at Idaho Falls, Idaho, since 1963, and a commercial nuclear breeder reactor electric power plant was built in Monroe County, Michigan,

---

* Named for Christian Johann Doppler, Austrian physicist (1803–1853).

† Relative means the speed of the wave crests as "seen" by you in the car. If two cars are traveling side by side on an interstate highway at identical speeds of 50 miles per hour each, then the speed of one car relative to the other is zero. On the other hand, if two cars approach each other doing 50 miles per hour, then their relative speed is 100 miles per hour.

TABLE 12.1 Survey of active nuclear breeder reactor electric power plants.

Location	Name of system	Electric power output (megawatts)	Approximate operational date
Idaho Falls, Idaho	EBR-II	16.5	1963
Russia	BN-350	350	1973
France	Phenix	250	1974
England	PFR	250	1974
Russia	BN-600	600	1980
West Germany	SNR-300	330	1985
Japan	Monju	300	1984
France	Super-Phenix	1200	1984

in 1963. The latter facility was named for Enrico Fermi. It was permanently terminated in 1966 after a blockage in a sodium cooling line produced a partial meltdown of its core. Plans were announced in February 1972 for a major nuclear breeder electric power plant to be built on the Clinch River in eastern Tennessee (Fig. 12.3). Intended for completion between 1978 and 1980, it was to have an electric power output between 300 and 500 megawatts and was to be financed by both federal and private sources. Interest in the project has varied with presidential administrations and while it may be completed eventually, the date is highly uncertain.

Although the liquid metal fast breeder technology is the most advanced, there are other alternatives using the breeder concept that have merit and are receiving nominal support. Two examples are the thermal breeder using fuel bred from ${}^{232}_{90}Th$ and the high-temperature gas-cooled fast breeder reactor (GCFBR).

The thermal breeder derives its name from its use of slow (or thermal) neutrons. Fissionable ${}^{233}_{92}U$ would be bred from ${}^{232}_{90}Th$ using slow neutrons (see Problem 7). Recognizing that a standard pressurized water nuclear reactor could be converted to a nuclear breeder reactor operating on the conversion of ${}^{232}_{90}Th$ to ${}^{233}_{92}U$, plans were begun in 1965 for converting a reactor at Shippingport, Pennsylvania.* Now completed, its breeding and energy-producing characteristics are being studied. Compared to the breeding of ${}^{239}_{94}Pu$ from ${}^{238}_{92}U$, the conversion of ${}^{232}_{90}Th$ to ${}^{233}_{92}U$ is less efficient. In fact, the doubling time is beyond practicality. Nevertheless, the ${}^{232}Th$ conversion process should not be minimized because not only is it a way of making better use of limited naturally occurring fissionable material but it also avoids the ${}^{238}_{92}U$–${}^{239}_{94}Pu$ breeding cycle.

The GCFBR utilizes gaseous helium as the heat transfer medium. This scheme is already used commercially in a type of ${}^{235}U$-burning reactor (Section 9.9). Helium has the advantage of being chemically inert. It is also very difficult to form radioactive products through neutron bombardment of helium. The main disadvantage is that the helium has to be compressed to pressures about 100 times atmospheric pressure, a process entailing extremely reliable pressure vessels and auxiliary equipment. However, experience with helium in conventional reactors indicates that these problems are solvable.

12.5 SPECIAL PROBLEMS WITH HANDLING PLUTONIUM

It is not the function of a light-water reactor to produce ${}^{239}Pu$ through breeding. Nevertheless, it does produce some because of the ${}^{238}U$ content of the fuel. This plutonium is valuable, and when the fuel rods are reprocessed it is extracted. Extracting plutonium is not simple because the element is extremely dangerous. Not only does it produce deleterious chemical effects on the body, but it is radioactive and decays by emitting alpha particles. Internal to

* The first commercial nuclear electric generating station became operational in Shippingport, Pennsylvania, in 1957. It utilized a pressurized water reactor that operated successfully for 17 years.

FIGURE 12.3 Conceptual drawing of the proposed nuclear breeder reactor electric power plant to be built on the Clinch River in eastern Tennessee. Although similar facilities are a reality in several European countries, enthusiasm for the American project has wavered with presidential administrations. (Photograph courtesy of Westinghouse Electric Corporation.)

the human body, alpha particles are among the most damaging nuclear particles (Table 9.4). If plutonium is taken into the body in soluble form, it concentrates in the bones and the liver and tends to remain. If taken into the lungs as small particles, for example, it can produce intense local damage and possibly induce cancer. For these reasons, plutonium processing is done by remote control methods. Considering that the maximum allowable plutonium burden on the body is set at 0.6 of a billionth of a kilogram, the potential hazard is obvious even without nuclear breeder reactors. Inherent dangers exist both at the reprocessing plant and in the transportation of spent fuel rods and the processed plutonium. Prudent policy can minimize the chance of accidents, but accidents may still occur because of the human element involved.

If the nuclear breeder reactor proves successful and if we proceed to an energy economy based upon it, then the problem with plutonium will become magnified tremendously. For example, the plutonium radioactivity in a nuclear breeder reactor will amount to about one million curies and the stores of plutonium are anticipated to be around 720,000 kilograms in the year 2000.

While the dangers associated with the handling of plutonium are serious enough, even more serious is the possibility that plutonium might be stolen and used by terrorist groups. A convincing argument can be made for the impossibility of making a bomb from the $^{235}_{92}U$ fuel used for light-water reactors. The concentration is only a few percent and, to make a bomb, nearly 100% pure $^{235}_{92}U$ is needed. No one, save the federal government, has the enrichment facilities and processing technology to make $^{235}_{92}U$ bomb material. However, the plutonium in a fuel rod is highly enriched and only five kilograms of $^{239}_{94}Pu$ are needed to make a bomb like the one that destroyed Nagasaki in World War II. Granted, considerable danger would be involved in processing the plutonium and considerable technical knowledge would be required in order to construct a large bomb, but the possibility cannot be ruled out. Even if a bomb were not made from stolen plutonium, the thieves could demand high ransom for its return. Whether these problems can be solved remains to be seen. But certainly other energy options, though not so highly promising as the nuclear breeder reactor, must be pursued vigorously if in so doing we can avoid the potential problems associated with a plutonium economy.

12.6 THE NUCLEAR FUSION REACTOR

Energy is liberated in any nuclear reaction in which the total mass of the reacting nuclei exceeds the total mass of the reaction products. This follows from the principle of conservation of energy and the fact that mass is equivalent to energy ($E = mc^2$). Splitting a heavy nucleus into fragments is but one of many energy-releasing nuclear reactions. It is possible to sort of reverse the fission process and combine the very lightest of nuclei together to form other nuclei whose combined mass is less than the total mass of the reacting nuclei. This reaction is called nuclear fusion. Fusion normally involves the isotopes of hydrogen and helium. The interaction of two protons to form a deuteron, a positive beta particle, and a neutrino takes place routinely in the deep interior of the sun and is illustrative of the nuclear fusion process. Pictorially,

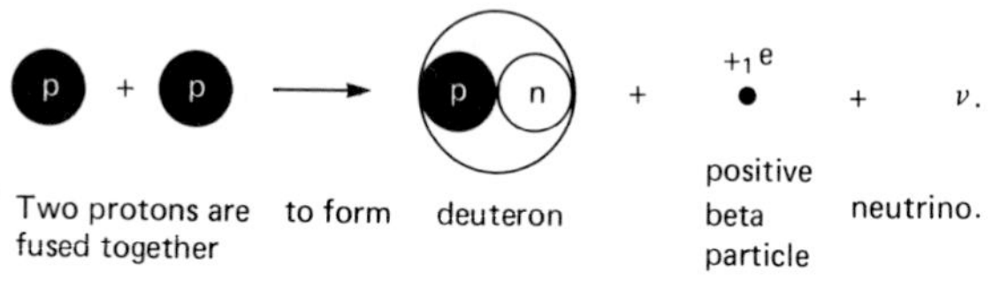

In symbols,

$$^1_1H_0 + {}^1_1H_0 \rightarrow {}^2_1H_1 + {}_{+1}e + \nu.$$

The sum of the masses of a deuteron, beta particle, and neutrino is 0.001 amu less than the combined masses of two protons. This mass is converted into 0.93 MeV of energy as a result of the fusion reaction. The energy released in each reaction is small compared with the energy released in a nuclear fission reaction, but it is still several thousand times more than the energy released in a chemical reaction.

It is simple to write the equation for the proton–proton fusion reaction and to determine the energy released, but it is no easy task to actually achieve the fusion process. Because each interacting proton possesses a positive charge, a mutual repulsive electric force tends to separate them. Energy must be supplied to overcome this electric repulsion, and it must be sufficient to get the protons to within the interaction distance (about 0.0000000000001 centimeters) of the nuclear force. Even when this is achieved, fusion is very improbable because of the variety of possible nuclear reactions. In the laboratory, the necessary energy can be supplied by a proton accelerator such as a cyclotron. In the sun, and in any practical nuclear fusion energy source, the energy is supplied thermally requiring a temperature of millions of degrees. This ignition temperature and a high density of protons needed for large numbers of collisions are available only in the core of the sun. Because of the low reaction probability and large ignition temperature, the proton–proton reaction as a practical manufactured energy source is extremely remote.

Some other fusion reactions have the same energy-producing character as the proton–proton reaction and have greater reaction probability. The deuteron–deuteron and deuteron–triton reactions are of particular interest (Fig. 12.4). Deuteron–deuteron reactions are extremely attractive because deuterium is not radioactive and occurs naturally. One of every 7000 naturally occurring hydrogen isotopes is deuterium, and there is essentially an unlimited supply in the oceans beause of the enormous amount of hydrogen in the water. Tritium does not occur naturally and is radioactive, but the deuteron–triton reaction is more appealing than the deuteron–deuteron reaction because the energy release is substantially larger and the ignition temperature is lower (about 40,000,000° C compared with about 400,000,000° C). Producing an ignition temperature of millions of degrees for a sufficient length of time and containing the "nuclear fire" may seem

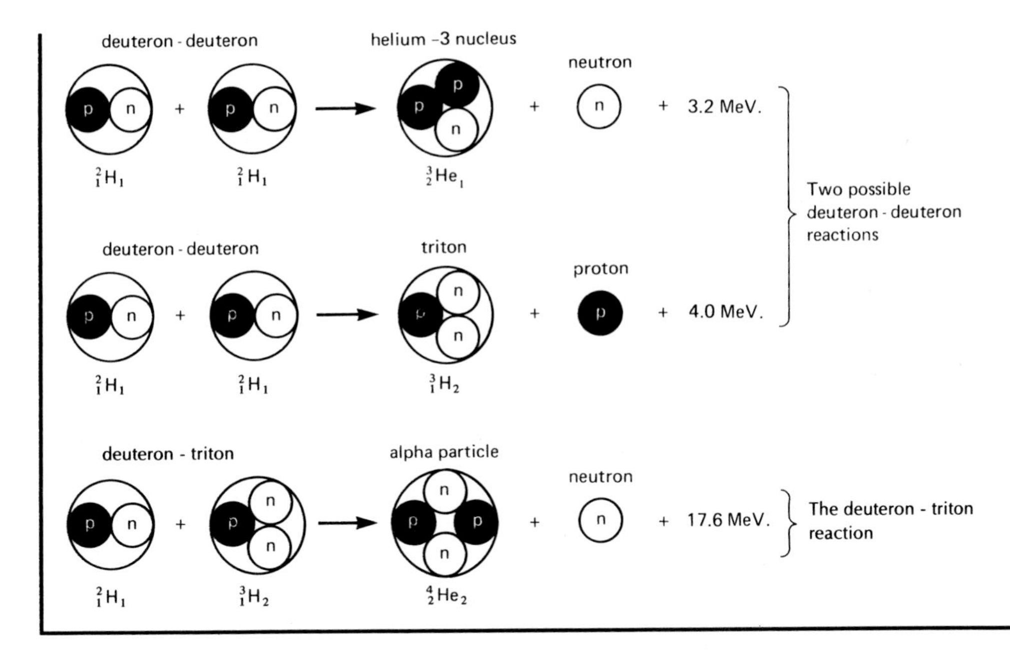

FIGURE 12.4 Pictorial representations of the deuteron-deuteron and deuteron-triton nuclear fusion reactions.

technologically inconceivable. Nevertheless, the practical application of fusion energy appears to be within our grasp and it offers a nearly unlimited source of energy without the bulk of the radioactivity problems associated with nuclear fission (Fig. 12.5).

12.7 PRACTICAL CONSIDERATIONS FOR NUCLEAR FUSION REACTORS

An ordinary flame produces a temperature of about 1000° C. The energy of an atom or molecule associated with this temperature is about 0.1 electron volts. This is sufficient to excite some atoms to higher energy states, but not enough to remove an electron from an atom. Conversely, the thermal energy associated with a nuclear fusion fire is much more than enough to completely remove all electrons from the atoms of the gas. In this condition the gas is termed a plasma. Once kindled, the natural tendency for a fusion fire is to expand—thereby tending to extinguish itself. So the reactions must be contained if energy is to be liberated at a controlled rate. The time of containment of the fusion reactions is called the confinement time.

Any burning process requires a sufficiently high ignition temperature. Ignition ensures that the fire starts, but does not guarantee sustaining the fire. Once a fire is self-sustaining, one hopes to derive more energy than was required to kindle the fire. For example, a lighted match provides both an ignition temperature and a certain amount of energy for lighting a gas stove. Once the gas is lit, more energy is derived from it than was provided by the match. A nuclear fusion fire, once ignited with an appropriate ignition temperature, becomes self-sustaining when energy production exceeds energy loss. Mechanisms such as conduction and convection that remove energy from any burning process are also important in nuclear fusion. Another energy loss unique to nuclear fusion is radiation from the rapidly moving charged particles in the plasma. The goal of nuclear fusion research is to minimize these energy losses to achieve a net gain in energy.

FIGURE 12.5 A controlled nuclear fusion research device called the Princeton Tokamak Fusion Test Reactor. This reactor is scheduled to begin operation in late 1985. It could be the first nuclear fusion reactor to produce as much energy as is required for its operation. (Photograph courtesy of Princeton University Plasma Physics Laboratory.)

The ignition temperature, the confinement time, and the density of the ions* in the plasma are the crucial characteristics determining whether or not nuclear fusion is a practical energy source. Energy balance considerations reveal that the ion density multiplied by the confinement time must be at least 100 trillion (10^{14}) seconds/cm^3 for systems utilizing the deuterium–tritium reaction. This is called the Lawson criterion.† If the plasma used has an ion density of 1000 trillion (10^{15}) per cm^3, then it must be contained for at least 0.1 second. A minimum ignition temperature of 40,000,000° C is required for net energy gain, and an ignition temperature between 80,000,000 and 100,000,000° C is needed in a practical nuclear fusion reactor. All research on fusion energy sources is centered on maximizing this product of ion density and confinement time and creating the required ignition temperature. The containment of a 40,000,000° C plasma is a challenging problem because the plasma must not contact the walls of a containing structure. The concern is not that the hot plasma will melt the structure. The ion density, and, therefore, the energy density of the plasma is just too low. But if the plasma does strike the walls, it loses energy, which tends to extinguish the self-sustaining fusion process. The force that a pipe would exert on the plasma to balance the force exerted by the plasma on the pipe is simulated with a magnetic force. These systems are labeled magnetic bottles (Fig. 12.6). All sorts of exotic configurations have been devised to try to achieve this containment. Most are based on the physical principle that a charge experiences a force when it moves in a magnetic field (Section 4.7). The idea is to have moving charged particles (ions) interact with an imposed magnetic field (or fields) in such a way that a containing force is exerted on the ions in the plasma. This keeps the plasma completely away from any confining structure that would extract energy from it.

---

* An atom devoid of one or more electrons is called an ion. A plasma consists of electrons and ions.

† Named for British physicist J. D. Lawson.

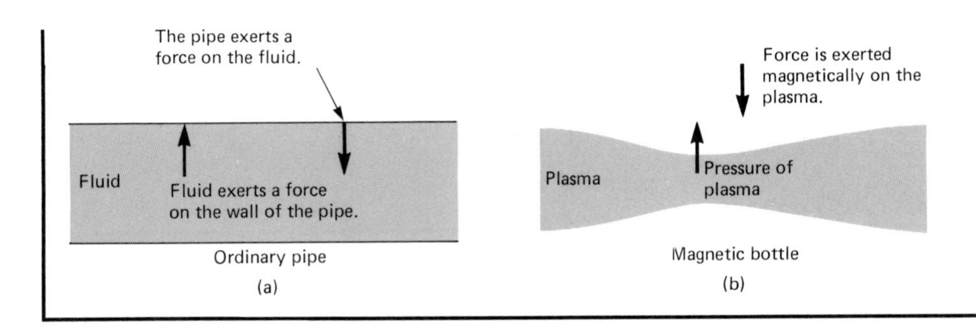

FIGURE 12.6 (a) In an ordinary pipe, the walls exert a containing force on a gas under pressure. This force is the reaction to the force exerted on the walls by the gas. **(b)** In a magnetic bottle, a magnetic force contains the charged ions of the plasma.

Confinement can be understood more quantitatively in the following manner. When the net force on an object is always perpendicular to its instantaneous direction of travel, the object travels in a circular path (Fig. 12.7). Examples include the circular orbit of a satellite about the earth (Section 4.2) and the circular movement of an object attached to a string. If a charged particle moves in a direction perpendicular to a magnetic field, the particle experiences a force perpendicular to both the magnetic field and its direction of travel (Fig. 12.8). Hence it moves in a circular path. The larger the magnetic field, the greater the force and the smaller the radius of the orbit. The charged particle moves in a circle around the magnetic field lines. If the direction of travel is not perpendicular to the magnetic field, then the force is still perpendicular to the direction of travel and the magnetic field, and the particle now moves in a helical path (Fig. 12.9). It is this principle that is used to regulate both the density and temperature of ions in a plasma.

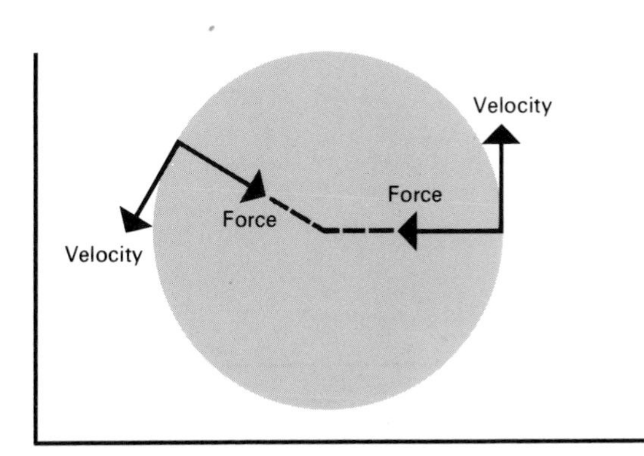

FIGURE 12.7 An object travels in a circular path when the net force is perpendicular to its velocity.

While the principle is straightforward, instabilities develop for which there is no explanation in elementary terms. These instabilities lead to leakage of ions in the plasma from the magnetic field and preventing achievement of the necessary confinement time.

12.8 TYPES OF NUCLEAR FUSION REACTORS

Two types of magnetic confinement show special promise. The principles of the devices being developed in these programs are referred to as Tokamak and magnetic mirror. Schematic diagrams of the concepts and principles are shown in Figs. 12.10 and 12.11.

Research and development of nuclear fusion concepts have inched the confinement parameters upward. The Princeton Large Torus (PLT) machine using the Tokamak concept achieved a temperature of 80,000,000 degrees in 1980. The Alcator C machine at the Massachusetts Institute of Technology, also using the Tokamak principle, managed to achieve the Lawson criterion in 1975. The energy breakeven point seems at hand. Two pending experiments are critical. Tests of the magnetic mirror approach in a reactor designated MFTF–B are to be concluded in 1984 at the Lawrence Livermore Laboratory in California. Scientists working with the TFTR (Tokamak Fusion Test Reactor) at Princeton University will attempt to achieve an ignition temperature of around 100,000,000 degrees and exceed the Lawson criterion in 1985. The outcome of these and other ventures could pave the way for commercial nuclear fusion reactors.

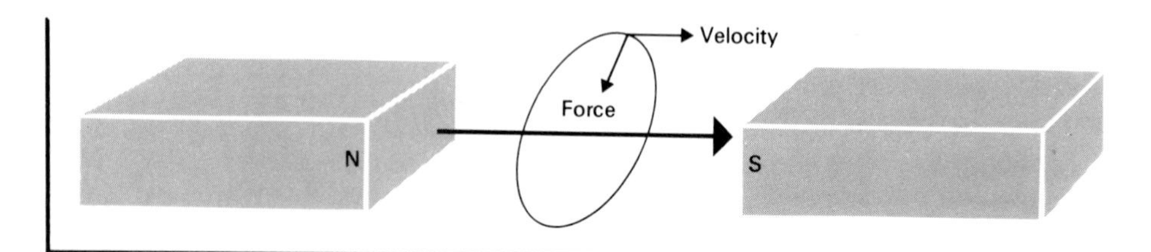

FIGURE 12.8 The magnetic field is directed from the N pole to the S pole. The charge moves at right angles to the direction of the magnetic field and revolves in a circle. This is the principle employed in cyclotrons used to accelerate nuclei such as protons and alpha particles to very high speeds.

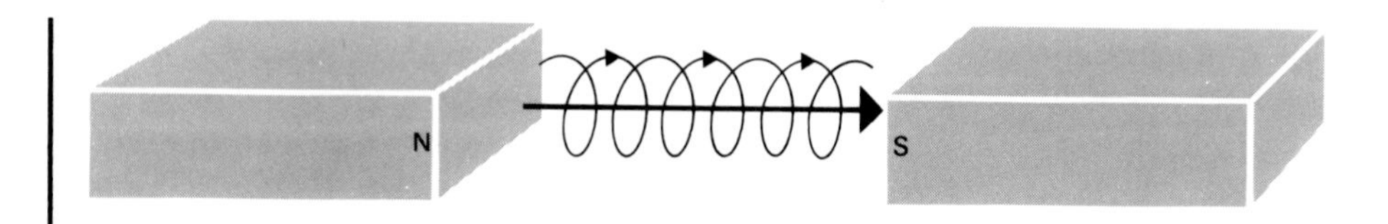

FIGURE 12.9 If the direction of travel of the charged particle is not perpendicular to the magnetic field direction, then the particle moves in a helical path. To visualize this motion, consider the following analogy. If a person rotates a mass on the end of a string, then an observer nearby sees the mass moving in a circle. If the person starts walking with the rotating mass, then the observer sees the mass moving in a spiral path.

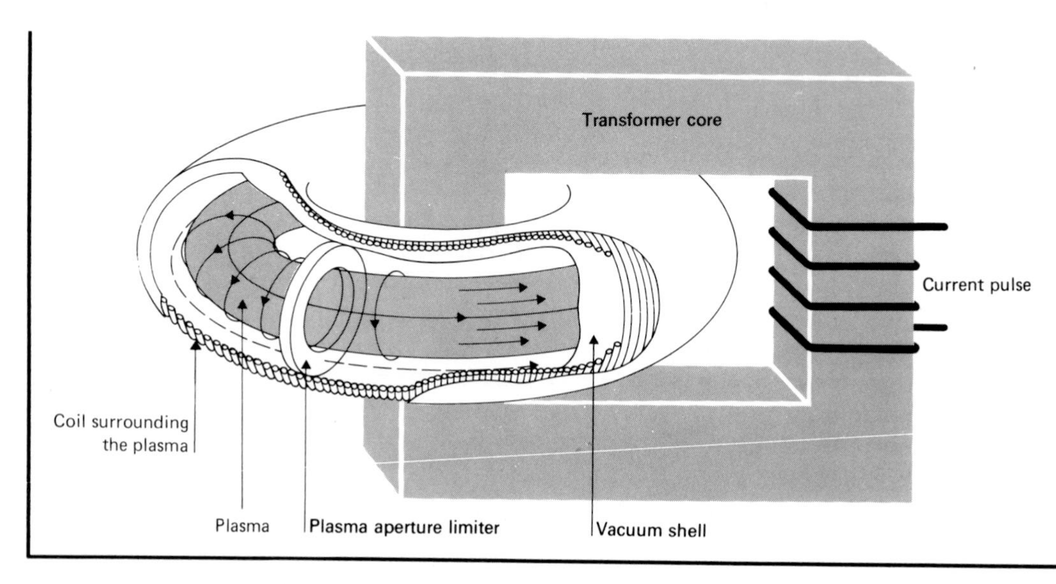

FIGURE 12.10 The Tokamak confinement geometry. The plasma constitutes the secondary winding of a transformer (Section 4.9). When a current is sent through the primary winding, a large current is produced in the plasma. This current heats the plasma producing the required ignition temperature. Current in a coil surrounding the plasma produces a magnetic field that contains the plasma. Several Tokamak systems are being studied in the United States.

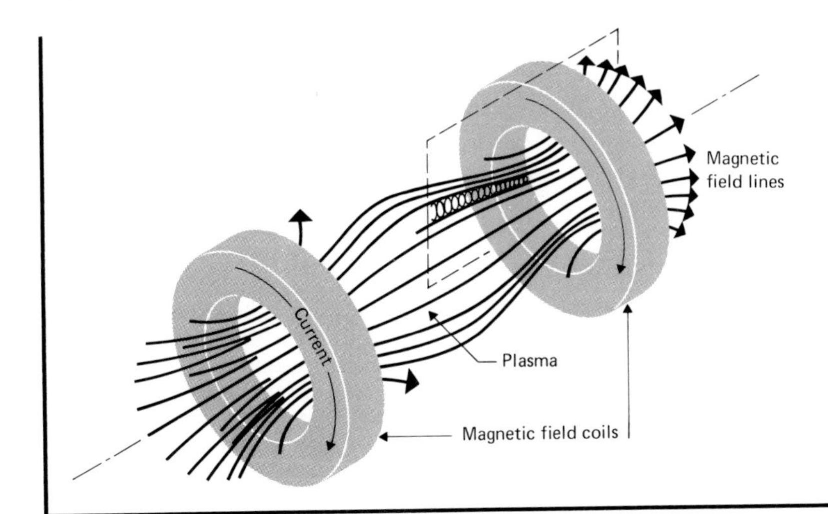

FIGURE 12.11 A magnetic mirror confinement scheme. Magnetic fields are created by circulating currents in coils at the ends of the mirror. Ions spiral along the magnetic field lines and, ideally, are reflected from the ends where the field is more concentrated. In practice, some particles leak through the ends. Research on the magnetic mirror system is conducted at the Lawrence Radiation Laboratory, Berkeley, California.

Figure 12.12 illustrates the principle of a nuclear fusion reactor employing deuterium–tritium fuel to produce thermal energy for producing steam for a turbine coupled to an electric generator. The energy system has three main components.

1. a mechanism for producing a magnetic field to contain the plasma,
2. the central plasma region, and
3. a lithium blanket surrounding the plasma.

Nuclear fusion reactions producing neutrons and alpha particles (Fig. 12.4) take place in the plasma. The energy of interest is possessed primarily by neutrons spewing out from the central core. Thus provisions must be made to convert the energy of neutrons into heat. This is the role of the third component—the lithium blanket. Unlike fission fragments, neutrons have no net electric charge and are impossible to slow down by electric forces. Within the liquid lithium blanket, neutrons induce nuclear reactions with lithium nuclei producing helium and tritium as reaction products. The reaction products are charged and easily stopped, and warm the lithium blanket. The heated liquid lithium is removed and replaced continuously. In the process of removing the heat, tritium is recovered and recirculated as fuel to the reactor.

The Magnetic Fusion Engineering Act of 1980 mandates the construction of a prototype nuclear fusion reactor by 1990. This prototype, dubbed Fusion Engineering Device (FED), is to be built at a Center for Fusion Engineering. It is intended to incorporate all known technology and provide answers to engineering questions relating to

1. materials for the walls and blanket,
2. efficient production of the magnetic fields,
3. tritium processing and handling,
4. coping with radioactivity induced by neutrons, and
5. reactor controls.

It is true that a nuclear fusion reactor will not produce radioactive fission fragments and actinides (Section 9.10). But a nuclear fusion reactor is not devoid of radioactivity. Constant bombardment of walls of the structure with neutrons induces radioactivity. The first wall outside the plasma is especially critical because it weakens structurally from intense neutron bombardment. Consequently the first wall will have to be replaced regularly, possibly every five years. Tritium has a half-life of only twelve years and low radiotoxicity, yet it must be handled carefully because the quantities involved are large and difficult to contain in their gaseous form.

There appear to be no technological barriers to the development of a practical nuclear fusion reac-

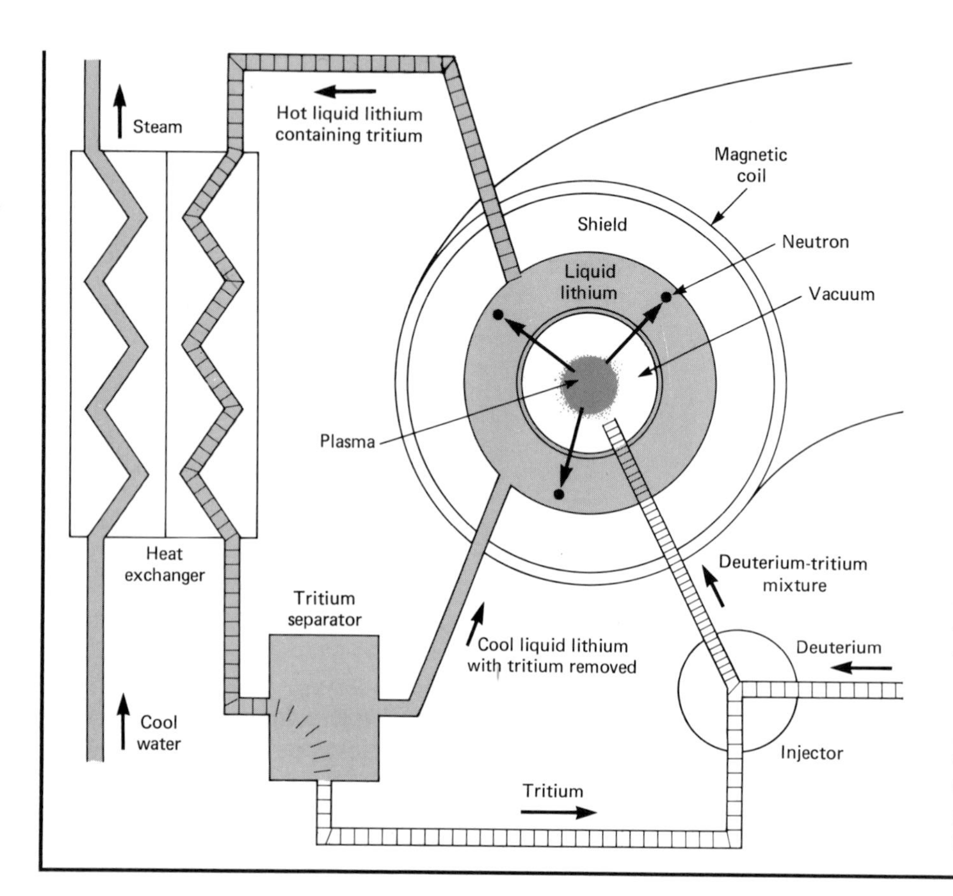

FIGURE 12.12 A nuclear fusion scheme using the Tokamak concept. The view is inside the "doughnut-shaped" chamber. Deuterium-tritium fusion reactions are instigated in the central core. Neutrons produced by the reactions are absorbed in the lithium blanket surrounding the central core. Thermal energy generated in the lithium blanket is transferred to a heat exchanger that vaporizes water for a steam turbine. Tritium formed from neutron-induced reactions in the lithium blanket is recirculated back into the reactor for fuel for the fusion reactions. Containment of the plasma is provided by magnetic-field windings surrounding the entire structure.

tor. To propose a timetable is risky because of uncertainties in funding and energy economics. It seems unlikely, though, that a commercial system will be developed before the turn of this century.

12.9 ENVIRONMENTAL CONSIDERATIONS AND OUTLOOK

The nuclear fusion reactor avoids one of the major drawbacks of the nuclear fission reactor—the production of large quantities of radioactive wastes in the form of fission fragments and actinides. The alpha particles produced from the interaction of neutrons and lithium are stable against nuclear decay. By capturing electrons, alpha particles become helium, which is recovered, condensed to a liquid, and used as a low temperature coolant (4.2 K) in a variety of industries. Tritium reclaimed from the lithium coolant for new fuel is not recycled immediately. Hold-up times of about a day are required. As a result, about 100 million curies of tritium may be held up in the tritium recovery process. This is comparable to the amount of radioactivity in the fission products of a breeder reactor. Tritium emits only a low-energy negative beta particle so there is little problem in shielding it. But tritium, like hydrogen, is difficult to contain and will diffuse to some extent through the walls of a container. Because tritium has the chemical characteristics of hydrogen, it can combine with oxygen to form water. In this state, it poses some biological hazard. The tritium radioactivity problems are not without solution, but nonetheless constitute a major consideration in fusion reactors utilizing deuterium-tritium or deuterium-deuterium fuel.

The nuclear fusion reactor avoids the problems of storage of radioactive fission fragments and actinides, requires no critical mass of fuel that might lead to a nuclear explosion, extends the lifetime of fuel supplies to millions of years, makes no demands on a type of fuel that may have other national priorities, and will probably be more thermodynamically efficient. Like the nuclear breeder reactor, it emits no particulate or chemical compounds to the atmosphere. Looking well into the future, the nuclear fusion reactor offers the possibility of direct energy conversion into electric energy.* If this is achieved, the efficiency could rise to, perhaps, 90% and the thermal pollution problem using steam turbines would be minimal. The high temperature technology gained in the development could be used in a multitude of ways.

Tokamak concepts are an exciting prospect, and research is forging ahead, but there is no guarantee that the Tokamak principle will be the ultimate nuclear fusion scheme. What is termed inertial confinement of nuclear fusion reactions is an equally exciting prospect. Let us look at this venture into controlled nuclear fusion.

12.10 INERTIAL CONFINEMENT OF NUCLEAR FUSION REACTIONS

When you jump toward the shore from a canoe, the canoe moves away from the shore. When a uranium nucleus emits an alpha particle, the thorium nucleus formed moves opposite to the travel direction of the alpha particle. The resulting motions are in accord with Newton's third law of motion. The force of the recoiling nucleus on the alpha particle causes it to move in one direction and an equal but oppositely directed force of the alpha particle on the residual nucleus causes it to move diametrically opposite. If you could arrange for a surface layer of particles on a sphere to explode radially outward, then the inner core would experience a reaction force from each escaping particle. The core would not move because

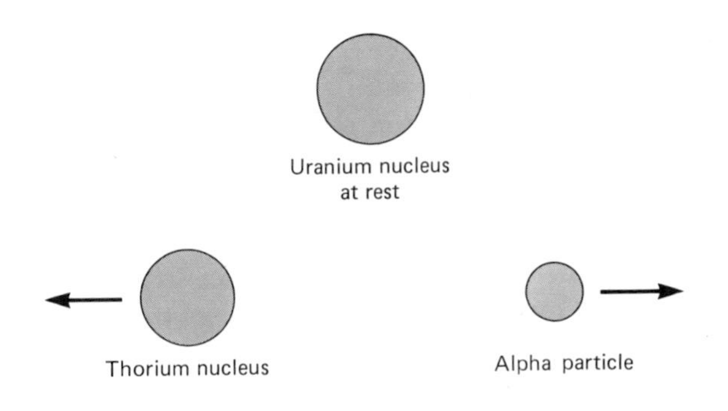

the reaction forces all balance, but a shock wave would propagate inward through the core, and the core would absorb the focused energy of the disturbance. Sufficient energy produces significant heating of the core. Inertial confinement of nuclear fusion reactions exploits this principle. A fusible spherical pellet of deuterium and tritium, for example, is irradiated uniformly with a pulse of some appropriate radiation or particles. The impact of the radiation causes a surface layer to escape outward (ablate). An inward-rushing shock wave heats the pellet sufficiently to instigate nuclear fusion. Because the pulse duration is a fraction of a billionth of a second, the inertia of the pellet is sufficient to confine the fusion reactions long enough for a net gain in energy. Light beams from lasers and beams of energetic electrons and ions are being used for the radiation impinging on the pellet.

Laser is a word derived from the first letter of the words Light Amplfication (by) Stimulated Emission (of) Radiation. The lasers of interest for the fusion process produce an extremely energetic beam of electromagnetic radiation, which is often visible. The radiation is very different from a beam of light from the sun or from an incandescent bulb in two important aspects. First, it is composed almost entirely of waves having the same wavelength; if visible, it appears as a pure color. Second, the peaks and valleys of the individual waves all line up. Radiation of this type is labeled coherent. Radiation from a neon sign is incoherent because there is a random relationship between the peaks and valleys of the individual waves. The mechanism for emission for both the laser and neon sign evolves from

* "The Prospects for Fusion Power," William C. Gough and Bernard J. Eastlund, *Scientific American* **224,** 2 (February 1971): 50.

atomic transitions from higher energy states to lower energy states.

Gas lasers composed of a mixture of neon and helium and with continuous power outputs of one or two milliwatts are commonly used in light experiments in an elementary physics laboratory. The lasers used in fusion research produce extremely energetic flashes of radiation. The energy in a pulse may be 1000 joules and the flash may last for only a tenth of a billionth (10^{-10}) of a second, corresponding to an instantaneous power of ten trillion watts. Lasers 10 to 100 times more powerful are planned. Thus the power is enormous. The lasers may be of either a gas or a solid type. Carbon dioxide is often used in gas lasers. A glass with a distribution of neodymium atoms is routinely used in solid lasers. The huge power outputs of lasers are useful for igniting fusion reactions. The laser would be focused on a suitable sample of a fusible material that absorbs the energy and, hopefully, instigates the fusion process.

A proposed scheme (Fig. 12.13) would focus a pulse of laser radiation on a tiny solid deuterium-tritium pellet about one millimeter in diameter. (A dime is about one millimeter thick.) The intense energy absorbed by the pellet creates a pressure disturbance that propagates inward rapidly compressing and heating the pellet. The pellet is

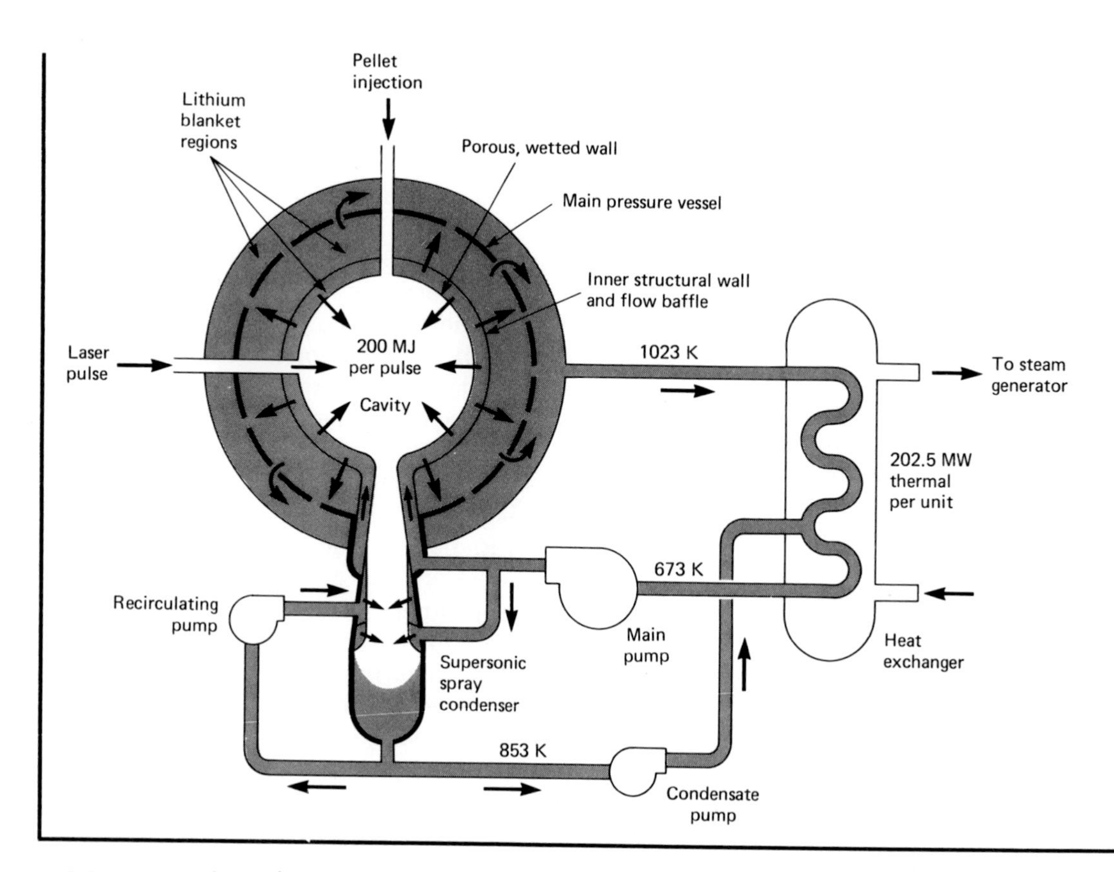

FIGURE 12.13 Laser-fusion power cycle concept. The laser beam strikes a fusible pellet placed near the center of the spherical structure. Energy from the fusion process is transferred to the lithium blanket and then to a conventional steam generator. (Drawing courtesy of the Los Alamos Scientific Laboratory.)

brought to ignition temperature, and fusion reactions develop. Pressures generated by the fusion reactions cause the pellet to explode. Neutrons from the fusion reactions possess the bulk of the energy produced. This energy is extracted by nuclear reactions in a lithium blanket which surrounds the cavity. A new pellet would be inserted and the process continued. Although laser fusion is far from a practicality, it is a promising scheme and its development is being pursued.

Electron bombardment performs much the same function as irradiation with intense laser light. Electrons accelerated through potential differences of around one million volts and producing currents of a half a million amperes can be focused to a spot a few millimeters in diameter. Absorption of the electron energy instigates a shock wave in much the same fashion as in laser-induced fusion.

REFERENCES

NUCLEAR BREEDER REACTORS

"Fast Breeder Reactors," Glenn T. Seaborg and Justin L. Bloom, *Scientific American* **223,** 5, 13 (November 1970).

"Plutonium: Reactor Proliferation Threatens a Nuclear Black Market," Deborah Shapeley, *Science* **172,** 143 (1971).

"Energy from Breeder Reactors," Floyd L. Culler, Jr., and William O. Harris, *Physics Today* **25,** 5, 28 (May 1972).

"The Breeder Reactor and Prudent Energy Planning," *NPA Report 186,* National Planning Association, Washington, D.C., 1980.

NUCLEAR FUSION

Fusion: Science, Politics, and the Invention of a New Energy Source, Joan Lisa Bromberg, M.I.T. Press, Cambridge, Mass., 1982.

"The Tokamak Approach in Fusion Research," Bruno Coppi and Jan Rem, *Scientific American* **227,** 1, 65 (July 1972).

"Fusion Power by Laser Implosion," John L. Emmett, John Nuckolls, and Lowell Wood, *Scientific American* **230,** 6, 24 (June 1974).

"Fusion Energy in Context: Its Fitness for the Long Term," J. P. Holdren, *Science* **200,** 168 (14 April 1978).

"The Promise of Fusion: What and When?" Stephen O. Dean, *Power,* **126,** 27 (May 1982).

REVIEW

1. Describe how plutonium is produced in a nuclear breeder reactor.
2. How does a nuclear breeder reactor differ in principle from a light-water reactor?
3. What is the attraction of the nuclear breeder reactor?
4. What does LMFBR stand for?
5. Name some desirable and undesirable features of liquid sodium when used as a heat transfer medium.
6. What is meant by "secondary criticality"?
7. Discuss the efforts to develop a commercial nuclear breeder reactor in the United States.
8. Describe the nuclear fusion of two protons, two deuterons, and a deuteron and a triton.
9. In the jargon of nuclear fusion, what is meant by ignition temperature, confinement time, plasma, and plasma density?
10. How is plasma confined in a nuclear fusion reactor?
11. Describe the probable features of the first practical nuclear fusion reactor.
12. Name some advantages and disadvantages of using nuclear fusion for a practical energy source.
13. What is a laser?
14. Describe inertial confinement of nuclear fusion reactions.
15. How is the laser to be used to induce nuclear fusion?
16. Plutonium-239 can be formed from the neutron bombardment of uranium-238. This bombardment process is called ____ and uranium is termed the ____.
    a) radioactivity; daughter
    b) breeding; fertile nucleus
    c) fusion; fusee
    d) fission; fissee
    e) radiolysis; fissile nucleus
17. In a nuclear breeder reactor ^{239}Pu is produced from the radioactive decay of ^{239}Np. Knowing the motivation for a breeder reactor, pick the half-life of ^{239}Np from the list below.
    a) 24,390 years
    b) 2.35 days
    c) 10 centuries
    d) 30.2 years
    e) Actually, ^{239}Np is stable.

18. The word "fast" in the expression fast-breeder reactor means that the
    a) reactor utilizes a fast-acting safety system.
    b) cooling must circulate unusually fast.
    c) fissionable nuclei must be moving quickly.
    d) reactor utilizes fast neutrons.
    e) neutrons are fast initially but are slowed by the moderator.

19. The nuclear breeder reactor will generate a usable fuel during its routine operation. The principle employed in this feature is
    a) capture of a neutron by an appropriate nucleus followed by beta ($_{-1}e$) decay to form the usable fuel.
    b) capture of a neutron by an appropriate nucleus followed by fission to form the usable fuel.
    c) the use of thermal energy to form a usable fuel via chemical reactions.
    d) the formation of a usable fuel using fusion reactions.
    e) capture of a neutron by an appropriate nucleus followed by alpha particle decay to form the usable fuel.

20. Because the fast breeder reactor must operate with fast neutrons, the ^{235}U content in the fuel elements
    a) must be reduced to about 0.1%.
    b) must be enriched to about 15–30%.
    c) must be carefully mixed with ^{239}Pu.
    d) must be enriched to about 99%.
    e) none of the above are correct; the fuel in a nuclear breeder reactor is not ^{235}U.

21. The basic energy-producing mechanism in a nuclear breeder reactor is
    a) nuclear fusion.
    b) neutron activation.
    c) nuclear fission.
    d) nuclear breeding.
    e) none of these.

22. The fast breeder reactor requires fast neutrons for its operation in order to
    a) optimize the likelihood of producing nuclear fission reactions.
    b) minimize the chance of a core meltdown.
    c) optimize the production of ^{239}Pu.
    d) promote nuclear fusion reactions.
    e) both (c) and (d) are correct choices.

23. Many technologists are attracted to the nuclear breeder reactor because
    a) it is able to use ^{239}Pu as a fuel while a conventional reactor cannot.
    b) it is a much more efficient way of utilizing our uranium supplies.
    c) it takes advantage of the nuclear fusion process.
    d) it can operate on naturally occurring ^{239}Pu.
    e) all of the above are correct.

24. The basic energy-producing mechanism in the sun is the fusion of two protons. To effect the fusion process energy must be supplied to the fusing protons because
    a) there is a tendency of the protons to repel each other.
    b) the origin of the energy from the sun is the energy required to bring the protons together.
    c) the nuclear force is repulsive.
    d) particles of the same mass always repel.
    e) of Einstein's equation $E = mc^2$.

25. Pick the reaction below that is termed nuclear fusion.
    a) $^1H + {}^1H \rightarrow {}^2H + {}_{+1}e$
    b) $n + {}^{238}U \rightarrow {}^{239}U$.
    c) $n + {}^{235}U \rightarrow Ba + Kr + 3n$
    d) $^{239}Np \rightarrow {}^{239}Pu + {}_{-1}e$
    e) $^3H \rightarrow {}^3He + {}_{-1}e$

26. We have learned that an atom consists of a positively charged nucleus surrounded by electrons. In a container of hydrogen, for example, if the electrons have all been stripped from the nuclei, we would call the resulting system a (an)
    a) plasma.
    b) Tokamak.
    c) un-ionized gas.
    d) hydrogenated gas.
    e) electrolytic gas.

27. The energy released from the fusion of two nuclei comes from
    a) the conversion of mass.
    b) intrinsic chemical energy.
    c) stored electromagnetic energy.
    d) stored thermal energy.
    e) unknown reactions.

28. If four hydrogen atoms are fused into a helium atom by a fusion generator, it's certain that
    a) if there is energy released, the mass of four hydrogen atoms is less than a helium atom.

b) if there is energy released, the mass of a hydrogen atom is greater than one-fourth of a helium atom.
c) all the mass of the hydrogen is converted into energy as is done by the sun.
d) no mass is converted into energy.
e) none of the statements above are correct.

29. If you could see an object such as a satellite revolving around the earth or a charge revolving in a circle in a nuclear fusion reactor, then you know that there must be
a) a force on the object that points toward the center of the circle.
b) no force on the object.
c) a force on the object that is always tangent to the circle.
d) no acceleration of the object.
e) a force created by the motion of the object.

30. Confinement of the plasma in a torus-shaped fusion reactor is based on the principle that
a) a moving charge in the presence of an externally applied magnetic field will feel a force.
b) moving charges produce their own electric fields.
c) moving charges are confined by the reactor's gravitational field.
d) the higher the temperature, the more likely fusion is to occur.
e) moving charges in the presence of a magnetic field produce their own magnetic fields.

31. Nuclei being fused in a nuclear fusion reactor must be prevented from touching the walls of any containment structure. This is to
a) prevent high-temperature plasma from melting the containment structure.
b) ensure that the plasma does not lose energy, thereby decreasing its temperature.
c) prevent corrosion of the containment pipes.
d) prevent nuclear fusion reactions with the containment pipes.
e) prevent spontaneous ignition of the containment pipes.

32. The nuclear fusion reaction in the first generation of nuclear fusion reactions will probably involve deuterium and tritium. While deuterium can be obtained from ______, tritium must be obtained from ______.
a) water; air
b) lithium; water
c) nuclear transmutations; iron ores
d) water; nuclear reactions involving lithium
e) iron ores; water

33. While a nuclear fusion reactor will not produce radioactive fission products, there will be radioactivity to be contended with because
a) the fusion reactor will use ^{235}U which is naturally radioactive.
b) of the production of ^{239}Pu.
c) the deuterium fuel is radioactive.
d) components of the reactor become radioactive.
e) of the 100,000,000° C temperature.

34. Laser radiation of the type used in laser-fusion is different from solar radiation or that of an incandescent light bulb, in that the laser radiation
a) has a wider spectrum of wavelengths.
b) has nearly a single wavelength or frequency.
c) has lower power output.
d) has more infrared radiation.
e) has more ultraviolet radiation.

Answers

16. (b)	17. (b)	18. (d)	19. (a)
20. (b)	21. (c)	22. (c)	23. (b)
24. (a)	25. (a)	26. (a)	27. (a)
28. (b)	29. (a)	30. (a)	31. (b)
32. (d)	33. (d)	34. (b)	

QUESTIONS AND PROBLEMS

12.2 THE NUCLEAR BREEDER REACTOR

1. In a water-cooled nuclear reactor, what is the minimum average number of neutrons that can be released in fission reactions in order to maintain a self-sustaining chain reaction?

   In a breeder reactor, what is the minimum average number of neutrons released in fission reactions in order to maintain a self-sustaining chain reaction as well as replacement of the spent fuels?
2. How does the basic energy-conversion method in a breeder reactor differ from that in a water-cooled reactor?
3. What practical roles do the half-lives of ^{239}U and ^{239}Np play in the breeding process shown in Fig. 12.1?
4. According to the theory of nuclear fission, $^{231}_{92}U_{139}$ is a suitable fuel for a nuclear chain reaction. It could

be bred from ^{230}Th much like ^{233}U is bred from ^{232}Th.

$^{230}_{90}Th_{140} + ^{1}_{0}n_{1} \rightarrow ^{231}_{90}Th_{141},$

$^{231}_{90}Th_{141} \rightarrow ^{231}_{91}Pa_{140} + {}_{-1}e + \bar{\upsilon}$

$^{231}_{91}Pa_{140} \rightarrow ^{231}_{92}U_{139} + {}_{-1}e + \bar{\upsilon}.$

What time considerations might rule out this scheme as a possible source of nuclear fuel?

5. It is more than just coincidence that the useful fissionable nuclei—^{233}U, ^{235}U, ^{239}Pu—have an odd number of nucleons. Another nucleus in this category is ^{239}Np formed from the decay of ^{239}U (see Fig. 12.1). What time consideration precludes using ^{239}Np atoms in nuclear reactor fuel?
6. This problem emphasizes the differences in concentration of energy in an ordinary energy source and a nuclear breeder reactor. A coil on a 1000-watt heating coil has a volume of about 500 cubic centimeters. The energy-producing core of a 1000-megawatt nuclear breeder reactor has a volume of about three cubic meters. Compare the watts per cubic centimeter for these two systems.
7. Using Fig. 12.1 as a guide, fill in the steps for generating fissionable ^{233}U starting from ^{232}Th.

$^{232}_{90}Th_{142} + n \rightarrow$ ______

______ $\rightarrow$ ______ $+ {}_{-1}e + \bar{\upsilon},$

______ $\rightarrow ^{233}_{92}U_{141} + {}_{-1}e + \bar{\upsilon},$

8. Sodium-23 ($^{23}_{11}Na_{12}$) is the only stable form of natural sodium. Sodium-22 ($^{22}_{11}Na_{11}$) and sodium-24 ($^{24}_{11}Na_{13}$) are radioactive and are produced in the liquid sodium circulating around the core of a breeder reactor. Construct nuclear reaction equations showing how these isotopes can be made by bombarding stable $^{23}_{11}Na_{12}$ with neutrons.
9. The Clinch River nuclear breeder electric power plant is expected to produce 1250 megawatts of thermal power and generate 350 megawatts of electric power. Compute the efficiency and compare with the efficiency of a water-cooled nuclear power plant.

12.6 THE NUCLEAR FUSION REACTOR

10. Why could solar energy be also categorized as nuclear energy?
11. Why would the gaseous diffusion method (Section 9.7) used to separate ^{235}U from ^{238}U also be a useful technique for separating deuterium from hydrogen?
12. Storing nuclear fission fuels requires some care because of the possibility of creating a critical arrangement. Why do you not have similar concern when storing nuclear fusion fuels?
13. A person traveling in a circle on a merry-go-round is analogous to a charged particle traveling in a circle in a magnetic field. What force keeps the person on the merry-go-round?
14. It is extremely difficult to reduce the intensity of a beam of neutrons with an absorber such as concrete. What special problem does this pose for capturing the energy released in deuterium–tritium nuclear fusion reactions?
15. All materials become radioactive if exposed to a "rain" of neutrons of sufficient intensity. Why might we expect considerable radioactivity to be generated in the metal containment structures of nuclear fusion reactors?
16. Thermal energy is produced in a continuous fashion in a nuclear fission reactor. But in the design concepts of nuclear fusion reactors using the Tokamak concept, energy is produced in bursts or pulses. What problems might this cause in a commercial nuclear fusion electric power plant?
17. Using the data below and the principle of energy conservation, show that 17.6 MeV of energy are released when a deuteron and a triton are fused to form an alpha particle and a neutron.

triton	=	3.016050 amu
deuteron	=	2.014102 amu
alpha particle	=	4.002603 amu
neutron	=	1.008665 amu
1 amu	=	931 MeV

18. Only one out of every 7000 hydrogen isotopes is of the deuterium variety, but there are still over a million-trillion-trillion-trillion (10^{42}) deuterium atoms in the oceans. If these deuterium atoms were used in deuterium–deuterium fusion reactions and each reaction liberated a millionth of a billionth (10^{-15}) of a Btu of energy, show that the total energy released is over a 100 million (10^8) Q (Q = billion-billion (10^{18}) Btu). How does this amount of energy compare with that available from fossil fuels?
19. A mass in grams equal to the atomic weight of an elemental substance always contains the same num-

ber of atoms. This number is equal to 6.02×10^{23} and is called Avogadro's number. Thus 235 grams of ^{235}U contain Avogadro's number of atoms. Why could you liberate more fusion energy from one gram of deuterium than fission energy from one gram of ^{235}U?

20. Early versions of Tokamak nuclear fusion reactors produced plasmas of 10 trillion (10^{13}) particles per cm^3, temperatures up to 10,000,000 K, and confinement times of about 0.03 seconds. How close are these machines to satisfying the Lawson criterion for a useful energy source?

21. One stable isotope of lithium is labeled $^{6}_{3}Li_3$. Show how a triton (a tritium nucleus) and an alpha particle (a helium nucleus) can be produced when ^{6}Li interacts with a neutron.

22. The energy in a pulse of electromagnetic radiation from a certain laser is 1000 joules. If the pulse lasts for one tenth of a billionth (10^{-10}) seconds, show that the instantaneous power is 10 trillion (10×10^{12}) watts.

13

Energy Conservation

13.1 MOTIVATION

Several messages have emerged as we look back on this study of energy for a technological society. Not the least of these is the simple fact that our country is consuming more energy than it produces. Dependency on external sources can be reduced, perhaps eliminated, by developing alternatives. But even given the incentive and financial backing, development takes several years. Energy conservation is an important measure for reducing the gap between energy production and energy consumption.

Generally, conserving connotes "doing without." It is not easy to persuade people to do without. Perhaps a more realistic approach is to view energy conservation as "doing better." If energy is literally wasted,

Heat lost through crevices and other openings in a house often account for 15% of a homeowner's fuel bill. Sealing cracks as this person is doing is a wise conservation effort. (Photograph courtesy of the Department of Energy.)

then doing better might mean doing without. But more often energy conservation means taking advantage of the best method for a given task. Elementary algebraic concepts help clarify these ideas.

A father may be planning to use the family car on a vacation trip. The number of gallons of fuel used by the car during the trip equals the gallons used per mile of travel multiplied by the length (miles) of the trip:

$$\text{gallons} = \left(\frac{\text{gallons}}{\text{mile}}\right) \times (\text{miles}). \tag{13.1}$$

The object of energy conservation is to reduce the quantity on the left side of the equation, that is, gallons. One way is to reduce the number of miles traveled. But then the father may not achieve his goal of traveling a certain distance. This is the fallacy in saving energy by being frugal but not achieving a task. The goal may still be attained and a fuel savings effected if a more efficient mode, using fewer gallons per mile, were opted for. That may mean selecting an unfamiliar vehicle and adopting a different life-style. For example, commuting to work might be more energy efficient if one were able to use either an electric car or mass transportation rather than a private automobile with an internal combustion engine. The best energy-conservation innovations are for naught unless they can be implemented. Ultimately this may mean legislation. Short of legislation, no scheme can be sold to the public unless it makes economic sense.

Relatively simple conservation measures can produce significant energy savings. Most energy conservation schemes are based on physical principles that we have discussed already. The application of these principles motivates this final chapter of our study of "Energy."

Energy consumption can be divided into four categories—residential and commercial, transportation, electric utilities, and industry. The utilization of energy in these categories was discussed in Chapter 1. Only about half of the energy fed into the system is converted to useful forms.* Some energy rejected is due to the intrinsic inefficiency of some of the energy conversion processes. But a substantial fraction of it is literally wasted. Therefore, opportunities exist for significant energy saving. Let us now examine each of the energy-consuming categories for ways of conserving energy through application of simple physical principles. The treatment is illustrative and not exhaustive. You may want to extend your knowledge of this subject by examining the references listed at the end of the chapter.

* There are differences of opinion on how much of the energy is rejected. However, it is always about 50% or more.

13.2 ENERGY CONSERVATION IN THE RESIDENTIAL AND COMMERCIAL SECTOR

The residential and commercial establishments used 25.7 million-billion (25.7×10^{15}) Btu of energy in 1981. About 14 million-billion Btu was used for space heating and cooling in structures most of which have temperature-controlling systems. These figures will be roughly the same for the coming five years or so. By simply lowering the thermostat 2° F in winter and raising it 2° F in summer it is estimated that 12% of the energy projected for space heating and cooling could be saved. Although this may seem unreasonable it is understandable.

A student's financial worth does not change if her income equals her spending. Similarly, the thermal energy content of a structure does not change if the heat gain equals the heat loss. If the thermal energy does not change, then the temperature does not change. So the key to maintaining constant temperature in a building is to balance heat gains with heat losses. In winter, heat gains are provided primarily by a heating unit—such as a furnace, solar collector, electric heaters—but people, stoves, and appliances also contribute to heat gains. Heat losses occur mostly by conduction but air movement (convection) through openings is significant.* In summer, heat gains are mostly by conduction and heat losses are provided by air conditioners, if used.

Heat losses from a building depend mainly on the indoor–outdoor temperature difference. As the

* We discuss these losses in more detail in the next section.

temperature difference increases, the heat losses and the energy used in the heating unit increase accordingly. No one can choose the outdoor temperature, but the indoor temperature can be changed to decrease the temperature difference. Historically, the accepted temperature for a dwelling was taken to be 70° F. So if the average outside temperature is 55° F, the average temperature difference is 15° F. If the thermostat setting were lowered 2° F so that the temperature inside is 68° F, then the average temperature difference decreases to 13° F. Because heat furnished by the heating system is proportional to the temperature difference, then with the lower thermostat setting the heating system would have to supply only $\frac{13}{15} = 0.87$ or 87% as much heat. Thus there would be a 13% saving in energy. If the average outdoor temperature were 40° F, lowering the thermostat from 70° F to 68° F produces a 7% energy saving. Thus, with only the effort required to change a thermostat setting a trivial amount, substantial energy can be saved.

Regardless of the quality of construction of a building, lowering the thermostat setting saves energy and money. A more important question is, "Are the heat losses and corresponding fuel bills inordinate for the structure even with a reduced thermostat setting?" An answer requires a closer look at conduction and convection.

Whenever an open window or door exposes the indoors of a building to the exterior, air and thermal energy are exchanged by convection (Section 7.3). The exchange may be a consequence of a natural movement created by the indoor–outdoor temperature difference, or it may be forced by winds blowing cold air into the building. It is impossible to eliminate all cracks and crevices for air exchange. Energy losses by convection are commonly termed infiltration energy losses, or just infiltration losses for short. Thirty percent of the heat loss in an average dwelling may occur through infiltration. Calculating infiltration losses is difficult but an energy-conscious person is wise to methodically seal cracks and crevices around doors, windows, foundations, etc., and to make sure chimney openings are blocked when not in use. If you have a $\frac{1}{8}$-inch crack under a 32-inch wide door it is like having a 2-inch by 2-inch square hole in the center of the door. If the opportunity exists for constructing a double-door entry to a dwelling, a homeowner in a colder climate is wise to opt for it.

Heat losses by conduction are the most important and are relatively easy to estimate (Fig. 13.1). A window, for example, limits heat conduction from the interior to the cooler exterior. Large windows conduct more heat than small windows, and thick windows conduct less heat than thin windows. Additionally, the heat flow depends on the quality of the windows as reflected in the thermal conductivity (Section 7.3). Summarizing in equation form we can express the heat flow in units of Btu per hour as

$$\frac{E_{\text{lost}}}{t} = \frac{kA(T_{\text{warm}} - T_{\text{cool}})}{d}. \tag{13.2}$$

In this equation E_{lost} records the heat flow, measured in Btu, in a time period t measured in hours. Symbol A reflects the size (area) measured in square feet, d is the thickness in feet, $(T_{\text{warm}} - T_{\text{cool}})$ is the indoor–outdoor temperature difference in °F, and k is the thermal conductivity in units of Btu ÷ (hours • feet • °F). Let's illustrate the use of Eq. (13.1) for

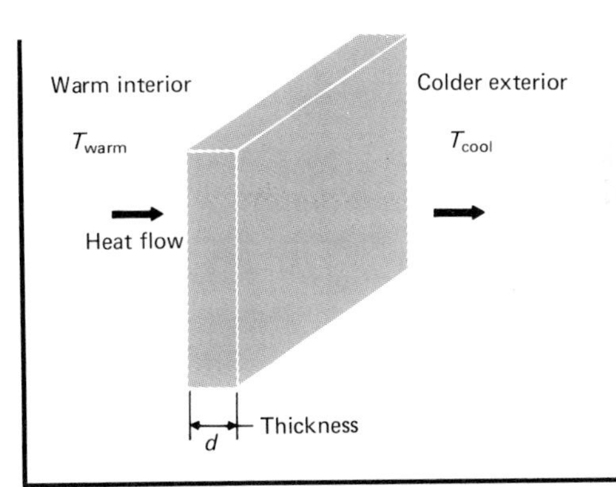

FIGURE 13.1 A slab of some material intercepting the flow of heat from a warm to a cool region. The slab might be a window, a wall, a ceiling, or a floor of a house. The heat flow (for example, Btu per hour) depends on the thickness of the slab, the area of a face of the slab, the temperature difference, and the type of material composing the slab.

a glass window 2 feet wide, 4 feet long and $\frac{1}{4}$-inch thick, and thermal conductivity 0.46 Btu/(hour • feet • °F). Let us assume the temperature difference is 10° F. Then

$$A = 2\,\text{ft} \times 4\,\text{ft} = 8\,\text{square ft}$$
$$d = \tfrac{1}{4}\text{-in} \times \tfrac{1}{12}\,\text{ft/in} = \tfrac{1}{48}\,\text{ft},$$
$$\frac{E_{\text{lost}}}{t} = \frac{0.46 \times 8 \times 10}{\frac{1}{48}} = \frac{1800\,\text{Btu}}{\text{hour}}.$$

Every hour 1800 Btu of energy flows through this window for the given conditions. To maintain a steady temperature, a heating unit must furnish 1800 Btu of heat each hour.

Equation (13.2) is very useful for many heat conduction calculations. For example, knowing the thermal conductivity of a variety of materials, you can compare heat losses for differing sizes and conditions. A homeowner is more likely to be considering adding additional windows (storm windows) or more insulation, and wants to know how much energy can be saved over the length of the heating season. He or she can then figure the cash savings and decide if the purchase is in order. For these considerations we manipulate Eq. (13.1) as

$$E_{\text{lost}} = \frac{kA}{d}(T_{\text{warm}} - T_{\text{cool}}) \cdot t \tag{13.3}$$

and take the time t to be the length of the heating season. The time, t, in the equation is expressed in hours but if we have the heating season in days, we simply multiply days by 24 to get hours. The product of the temperature difference $(T_{\text{warm}} - T_{\text{cool}})$ and time measured in days is termed "heating degree-days" (Section 10.2). Normally there are enough heat sources in a dwelling other than the heating system to maintain a comfortable indoor temperature if the outdoor temperature stays above 65° F. Therefore, for calculations of heat provided by the heating system, the indoor temperature is taken to be 65° F. Heating degree-days are recorded continuously throughout the United States and we find there are about 4000 heating degree-days in the central part of the country. There is about twice this number in the northern United States. The heating factor, H.F., is the number that 4000 is multiplied by to get an estimate of the heating degree-days for a particular area. A map of heating factors

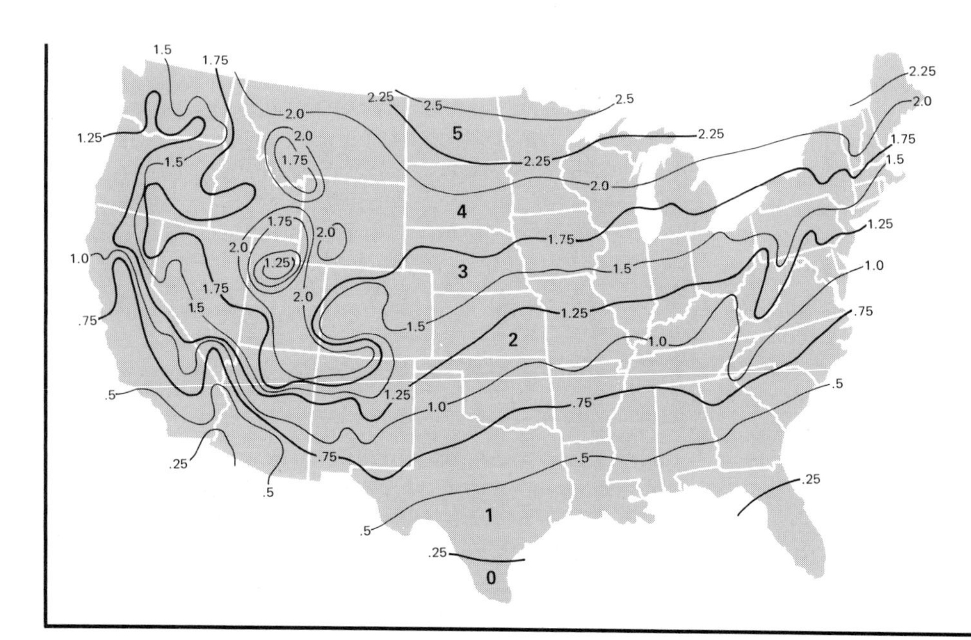

FIGURE 13.2 District heating factors in the United States. The heating factor is the same along any continuous line. A heating zone is an area bounded by two heavier black lines. Six heating zones numbered 0 through 5 have been labeled. The heating conditions are about the same for all communities located in a given heating zone.

is shown in Fig. 13.2. In New York City, for example, the heating factor is 1.5, meaning the approximate seasonal heating degree-days is 1.5(4000) = 6000. Familiarize yourself with the use of the map by picking the heating factor where you live.

Materials and thicknesses for construction of windows, walls, ceilings, and floors are fairly standard but the size (area) varies from house to house. If we know the flow of heat through one square foot of a particular type of construction material, say glass, we can compute the heat flow through a different size by simply multiplying by the area. The ability of one square foot of a material to resist the flow of heat is termed the R-value. R-values for common structural materials are presented in Table 13.1. As the R-value of materials increases, the heat flow through the materials decreases. Incorporating the size aspect, the R-value, and the heating degree-days (converted to hours in accordance with our heat conduction formula) into Eq. (13.3), we have for the total heat loss in a heating season

$$E_{\text{lost}} = \frac{A \times \text{H.F.} \times 4000 \times 24}{\text{R}} \text{ Btu per season.} \tag{13.4}$$

Because this equation is only approximate, let us round 4000 × 24 to 100,000 so that we have

$$E_{\text{lost}} = \frac{A \times \text{H.F.} \times 100{,}000}{\text{R}} \text{ Btu per season.} \tag{13.5}$$

This final equation is very practical and very easy to manipulate. To illustrate let's assume

single pane window for which R = 1,

size of 3 feet by 4.5 feet for an area of 13.5 square feet,

and a heating factor of 1.

Then

$$E_{\text{lost}} = \frac{13.5 \times 1 \times 100{,}000}{1}$$
$$= 1{,}350{,}000 \text{ Btu per season.}$$

A 42-gallon barrel of heating oil provides 5,800,000 Btu of energy. Hence about ten gallons of oil would need to be burned to provide the energy which was conducted through the window. If oil costs $1/gallon, then the cost would be ten dollars for the entire heating season. This cost could be cut in half by installing a double pane window having an R-value of 2. It would take a few years to pay for the window at five dollars a year but energy costs will escalate and the pay-off period will decline. Additionally, if the energy source is electricity, the savings are about twice as much.

For every doubling of a given thickness of insulation, the heat flow through it will be halved. But regardless of how much insulation is installed, there is no amount that will completely stop all heat flow through it. Therefore, the insulation used is a pragmatic decision. One has to decide if an additional amount of insulation will pay for itself through reduced fuel costs. In order to assist a homeowner or builder in this decision, guidelines for suggested amounts of insulation have been determined. These guidelines are presented as a suggested R-value for a type of construction in a particular heating zone (Fig. 13.2). Table 13.2 lists the recommended ceiling and floor R-values for the six heating zones specified in the United States. If in your area the suggested R-value for a floor is R-19, then the sum of the R-values in the construction materials of the floor should add up to 19.

An insulated wall in a house may have an R-value of 14 (Problem 14). Compared to a single-paned window, the energy loss per square foot of area is 14 times greater through glass than through the insulated wall. This is why in building an energy-efficient house careful consideration must be given to the number and the sizes of windows. Windows facing the sun receive energy during the day but they lose energy at a very large rate at night. It is important to cover windows at night to help reduce heat losses. Curtains and drapes having R-values of 3–4 are available for this insulating function. Windows not facing the sun lose energy day and night. Therefore, they should be as small as possible.

You need not be a homeowner to be conscious of heat losses. Look around where you live. If windows are drafty, seal them. Cover single-paned windows with clear plastic sheeting. The time invested is small and the returns are large.

TABLE 13.1 Insulation R-values for the common structural materials. A more complete compilation can be found in *Handbook of Fundamentals,* published by The American Society of Heating, Refrigerating and Air-Conditioning Engineers (ASHRAE).

	Material	Thickness (inches)	R-value $\left(\frac{ft^2 \cdot hr \cdot °F}{Btu}\right)$
Air film and spaces:	Air space, bounded by ordinary materials	$\frac{3}{4}$ or more	.91
	Air space, bounded by aluminum foil	$\frac{3}{4}$ or more	2.17
	Exterior surface resistance	—	.17
	Interior surface resistance	—	.68
Masonry:	Sand and gravel concrete block	8	1.11
		12	1.28
	Lightweight concrete block	8	2.00
		12	2.13
	Face brick	4	.44
	Concrete cast in place	8	.64
Building materials—General:	Wood sheathing or subfloor	$\frac{3}{4}$	1.00
	Fiber board insulating sheathing	$\frac{3}{4}$	2.10
	Plywood	$\frac{5}{8}$	.79
		$\frac{1}{2}$	.63
		$\frac{3}{8}$	.47
	Bevel-lapped siding	$\frac{1}{2} \times 8$	.81
		$\frac{3}{4} \times 10$	1.05
	Vertical tongue and groove board	$\frac{3}{4}$	1.00
	Drop siding	$\frac{3}{4}$	.94
	$\frac{3}{8}''$ gypsum lath and $\frac{3}{8}''$ plaster	$\frac{3}{4}$	.42
	Gypsum board	$\frac{3}{8}$	.32
	Interior plywood panel	$\frac{1}{4}$	.31
	Building paper	—	.06
	Vapor barrier	—	.00
	Wood shingles	—	.87
	Asphalt shingles	—	.44
	Linoleum	—	.08
	Carpet with fiber pad	—	2.08
	Hardwood floor	—	.71
Insulation materials (mineral wool, glass wool, wood wool):	Blanket or batts	1	3.70
		$3\frac{1}{2}$	11.00
		6	19.00
	Loose fill	1	3.33
	Rigid insulation board (sheathing)	$\frac{3}{4}$	2.10
Windows and doors:	Single window	—	approx. 1.00
	Double window	—	approx. 2.00
	Exterior door	—	approx. 2.00

Source: Data are from "Project Retrotech," a home weatherization guide published by the Department of Energy, Washington, D.C. 20461.

TABLE 13.2 Recommended ceiling and floor R-values for the six heating zones specified in the United States. To find the heating zone where you live, see Fig. 13.2.

Heating Zone	Ceiling	Floor
0, 1	R-26	R-11
2	R-26	R-13
3	R-30	R-19
4	R-33	R-22
5	R-38	R-22

Limiting conductive and convective heat losses will minimize the energy that needs to be supplied by the heating system. Making a house energy efficient always involves an outlay of capital. Hopefully, the investment is returned in the long run. A homeowner who cannot obtain the initial capital is forced to tolerate an energy-inefficient dwelling and to pay a price premium in monthly installments. In a mobile society where a house is occupied by a given owner for short periods, the owner will usually opt for energy inefficiency because a comparatively short occupancy precludes the recovery of the capital investment. An owner who can continually feed energy into the structure can get by. But if the energy source is limited or prohibitively expensive, then there is no recourse but to shore-up the structure.

Heat Pumps

We usually think of the energy content of gasoline in terms of Btu per gallon. Because each molecule in a gallon of gasoline contributes to the energy liberated when burned, we could just as easily say the energy in a gallon of gasoline equals the number of molecules multiplied by the energy liberated per molecule. Similarly each molecule in a gas or liquid contributes to the total thermal energy of a volume of the material. So we could express the total energy as

$$\text{energy} = \frac{\text{energy}}{\text{molecule}} \times \text{number of molecules}. \tag{13.6}$$

Clearly, there are two factors contributing to the energy. If one factor, say the number of molecules, is very large, the total energy could still be significant even if the other factor, the energy per molecule, is fairly small.

Because the energy of a molecule is directly related to the kelvin temperature (See Section 7.2), we are accustomed to thinking that a sizable quantity of energy necessarily implies a high temperature. For example, a heating coil on an electric stove is "red hot" and the flame of the burner on a gas stove will quickly sear your hand if you place it in the flame. Both the coil and burner provide substantial energy and are at a very high temperature. But it follows that there can still be significant energy in a low-temperature system if the number of molecules is large. On a freezing day the ground and the surrounding air contain massive amounts of thermal energy because the volume of either is extremely large. Experience tells us that this thermal energy will not flow spontaneously into a home which is at a higher temperature. The second law of thermodynamics guarantees this. On the other hand, there is nothing in the laws of physics that says we cannot do work (expend energy) and extract the thermal energy from the earth and deposit it in a dwelling. The basic question is, can we put more energy into the dwelling than was used to extract the energy from the ground? If the answer is yes, then there is a net energy gain and we would be wise to do it *if it can be justified economically.* If the answer is no, then one would be better off to use the input energy for the system as energy for the dwelling.

A motor-driven air conditioner maintains an indoor temperature lower than the outdoor temperature by removing thermal energy from the indoors and depositing it outdoors. The motor does work to effect the heat movement from a lower temperature to a higher temperature, and the homeowner pays the electric power company for the electric energy used by the motor. A heat pump for warming a house exploits the same principle (Fig. 13.3). In fact, a heat pump can be used to air condition a home in summer and heat the home in winter. In winter, thermal energy is removed from the cold outdoors to warm the indoors. In summer, the system is reversed and thermal energy is removed from the indoors in order to lower the indoor temperature.

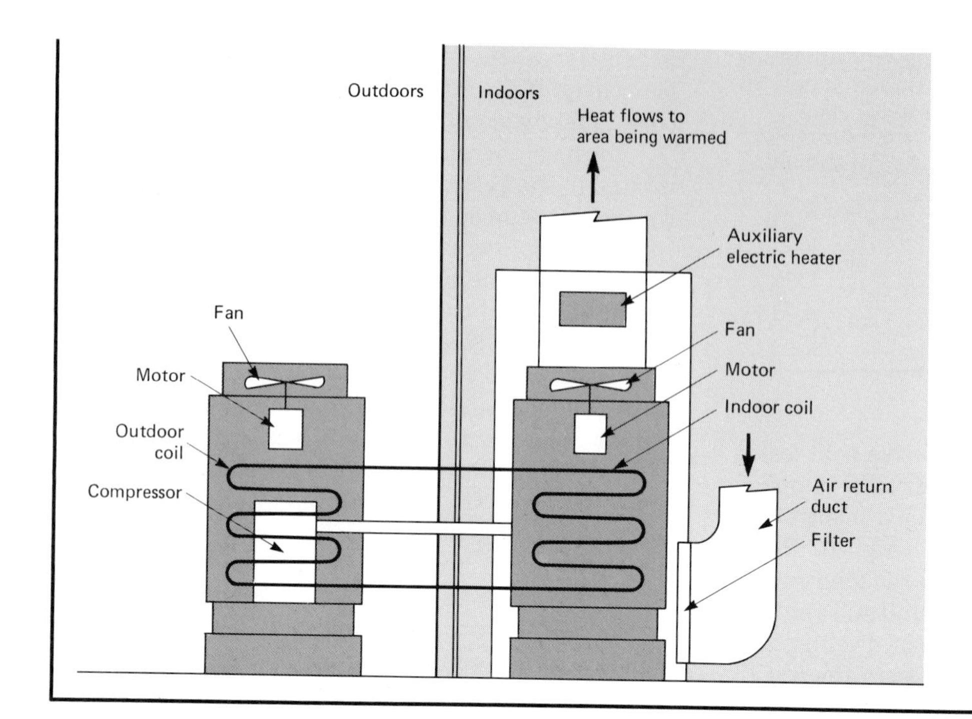

FIGURE 13.3 Layout of a heat pump installation. The system functions like a household refrigerator. An easily vaporized liquid absorbs heat from the exterior. The vapor is forced into the interior where it is condensed into a liquid. The thermal energy released in condensation remains to be circulated through the building. Energy is literally pumped from the cooler outdoors to the interior of a building being heated. An auxiliary electric heater is provided as a supplemental heat source.

Energy is invested in the heat pump in order to deposit thermal energy in a building. The performance is measured by dividing the thermal energy deposited by the energy invested. We choose to call this measure the performance.*

$$\text{COP} = \frac{\text{thermal energy deposited in the warm interior}}{\text{energy invested}}. \tag{13.7}$$

If a heat pump deposited two Btu of thermal energy in the interior of the house and one Btu of energy was used to operate the heat pump, we would say its performance is two. Just as for a heat engine, the laws of thermodynamics limit the performance of a heat pump. For a heat engine the maximum efficiency is (Section 7.3)

$$\epsilon\,(\text{maximum}) = \frac{T_{\text{warm}} - T_{\text{cold}}}{T_{\text{warm}}} \tag{13.8}$$

For a heat pump the maximum performance is

$$\text{COP}\,(\text{maximum}) = \frac{T_{\text{interior}}}{T_{\text{interior}} - T_{\text{exterior}}} \tag{13.9}$$

where T_{interior} and T_{exterior} are the interior and exterior temperatures of the building.

Now for some numbers. If the outside temperature is 41° F (278 K) and the inside temperature is 68° F (293 K), then

$$\text{COP}\,(\text{maximum}) = \frac{293}{293 - 278} = \frac{293}{15} = 20.$$

This means that one unit of input energy would cause 20 units of energy to be deposited in the building. A practical heat pump cannot perform like the ideal heat pump. Still it is often possible to achieve performances of between two and four, and this is significant. For example, electric heating is the most expensive form of household space heating because the efficiency of converting the energy of coal to electricity is about 33%. If the electric energy were used to run a heat pump rather than to produce heat directly with a heating coil, then, in effect, the energy

* Earlier editions of this book used the word effectiveness rather than performance. Performance was chosen in this edition to be more in line with the coefficient of performance, abbreviated COP, meaning the same thing in engineering discussions.

lost at the power plant can be recovered. Installing a heat pump requires a capital outlay and its worth must be judged on an economic basis. It is important to note that the performance of a heat pump decreases as the indoor–outdoor temperature difference increases. Accordingly, heat pumps are not practical in very cold climates. Technical improvement in heat pumps has made them increasingly attractive in mild climates. Most are reversible so that they provide space heating in winter and air conditioning in summer.

In principle, a heat pump is a heat engine operating backwards. A heat engine operating in a cycle delivers energy by drawing heat from a source, converting some of the energy to mechanical energy, and rejecting heat at a lower temperature. A heat pump reverses these steps. Heat is drawn from a source, energy is fed to the heat pump, and heat is deposited at a higher temperature. No heat engine performs better than the Carnot engine. And no heat pump performs better than the ideal heat pump. These two thermodynamic principles have very significant implications. The task of any heating system is to provide thermal energy to the inside of a building at the indoor temperature (T_{interior}) when the outdoor temperature (T_{exterior}) is lower. Let us call Q the thermal energy that has to be provided to maintain the required inside temperature. The thermal energy Q could be provided by a variety of systems—gas furnace, electric heaters, etc. If Q is furnished by the ideal heat pump, then using Eqs. (13.7) and (13.9) the energy needed to run the ideal heat pump is

$$W = Q\left(1 - \frac{T_{\text{exterior}}}{T_{\text{interior}}}\right). \tag{13.10}$$

No heating system can perform this task using less energy. For example, if the heat requirement of a house is 5000 Btu/hour and the exterior and interior temperatures are 41° F (278 K) and 68° F (293 K), then the minimum energy requirement is

$$\text{W} = 5000\left(1 - \frac{278}{293}\right) = 5000\left(\frac{15}{293}\right) = \frac{256\ \text{Btu}}{\text{hour}}.$$

Any real system—gas furnace, electric heater, real heat pump—uses more energy to do this task. Let us call W_{actual} the actual energy expended for the task. W_{actual} will always be larger than W determined from Eq. (13.6). The ratio of the minimum amount to the actual amount is a measure of the performance of the system. We call this ratio the second law efficiency.

$$\begin{matrix}\text{Second law}\\ \text{efficiency}\end{matrix} = \frac{\begin{matrix}\text{minimum amount}\\ \text{of energy required}\end{matrix}}{\begin{matrix}\text{actual amount}\\ \text{of energy used}\end{matrix}} = \frac{W}{W_{\text{actual}}} \tag{13.11}$$

We replace W using Eq. (13.10) and arrive at

$$\begin{matrix}\text{Second law}\\ \text{efficiency}\end{matrix} = \frac{Q}{W_{\text{actual}}}\left(1 - \frac{T_{\text{exterior}}}{T_{\text{interior}}}\right). \tag{13.12}$$

Let us focus on the term Q/W_{actual}. Remember, Q is the heat required for the task (heating the building), and W_{actual} is the amount of energy actually expended to do the task. Thus ratio Q/W_{actual} is the energy conversion efficiency concept we discussed in Chapter 2. Thus Eq. (13.12) can be interpreted as

$$\begin{matrix}\text{Second law}\\ \text{efficiency}\end{matrix} = \begin{pmatrix}\text{energy}\\ \text{conversion}\\ \text{efficiency}\end{pmatrix} \times \left(1 - \frac{T_{\text{exterior}}}{T_{\text{interior}}}\right). \tag{13.13}$$

Let us examine the implications of Eq. (13.13). A typical household gas furnace system has an energy conversion efficiency of 70%; that is, 70% of the energy derived from the burning gas appears as useful heat in the house. The second law efficiency for this system for the temperatures used earlier is

$$\begin{matrix}\text{Second law}\\ \text{efficiency}\end{matrix} = 70 \times \left(1 - \frac{278}{293}\right) = 3.6\%.$$

What may we conclude from the enormous difference between the energy conversion efficiency and the second law efficiency? It means simply that the gas furnace is very inappropriate for the designated task of supplying energy between two areas having a temperature difference of 15 K. Viewed another way, the flame temperature of 2000 K for the burning gas is simply not required for the assigned task.

Certainly, the gas furnace does the job but at the expense of using a much higher temperature than is required. There are other more productive uses of the high temperature than for providing household heat. For example, the high-temperature gas could be used to run a gas turbine coupled to an electric generator. The much lower temperature gas leaving the turbine can then be used for space heating. Such a scheme is called cogeneration of energy and has important implications for energy conservation (Section 13.4). Although heat pumps and cogeneration are technologically sensible, they are not easy to implement; but it is innovations like these that offer the greatest prospects for real contributions to energy conservation.

13.3 ENERGY CONSERVATION IN THE TRANSPORTATION SECTOR

At the outset, energy usage was divided into four categories, labeled electric utilities, residential and commercial, industry, and transportation. Of these four categories, transportation offers the greatest opportunity for energy conservation at the personal level. Transportation accounts for about one-fourth of total energy use in the United States. About 97% of this energy comes from burning gasoline and lower forms of petroleum. Passenger transportation accounts for about 60% of the total energy used for transportation purposes and nearly 90% of this energy is used by automobiles. Nearly half the energy used by automobiles involves trips of less than ten miles. Some 87% of the workforce choose automobiles or trucks for commuting to work. Nearly 70% of the workforce choose to travel alone.*

It is no accident that the private automobile commands great respect among the populace. The automobile is a symbol of freedom at the most personal level. At their convenience, a large fraction of the United States public has both the option and the resources to drive a multitude of short mileage trips or to travel thousands of miles on the interstate highway system. While reasons abound for fostering the person–automobile relationship, some simple facts force us to reassess this love affair. Pollution is one aspect already discussed. Energy efficiency, to be examined now, is another.

Seeking an energy efficient form of transportation is akin to shopping for food where we try to get the most amount of sustenance for the least amount of money. When energy is of concern, we shop for a transportation vehicle that carries the greatest number of passengers for as large a distance as possible with the least amount of fuel. A way of incorporating this idea in a passenger efficiency is

$$\text{passenger efficiency} = \frac{(\text{passengers transported}) \times (\text{miles traveled})}{\text{gallons of fuel used}}. \tag{13.14}$$

Note that this definition involves miles traveled divided by gallons of fuel used. This ratio is the customary figure of merit for gasoline use by an automobile. Thus if an automobile obtains 12 miles for each gallon of gasoline consumed while carrying two passengers, we record its passenger efficiency as

$$\text{passenger efficiency} = (2)\cdot(12) = 24\,\frac{\text{passenger miles}}{\text{gallon}}.$$

Now this number does not mean a lot until it is compared with other forms of transportation. This comparison (see Fig. 13.4) shows that the automobile, as currently used, ranks near the very bottom of the list.

Equation (13.14) reveals two things contributing to the low passenger efficiency for the private automobile. One is the conscious decision of an owner to generally include only himself or herself in the vehicle (the average number of persons per vehicle is between one and two in urban driving). The other is a tendency for using vehicles getting few miles per gallon. Whether we decide to include more passengers remains to be seen. But a reversal of the trend toward using less efficient automobiles is a reality. The federal government has mandated a

---

* Because of the changing nature of these data, the numbers are only approximate. The data were taken from *Transportation Energy Conversion Data Book:* Edition 2, D. B. Shonka, A. S. Loeble, and P. D. Patterson, Oak Ridge National Laboratory, Oak Ridge, Tennessee (October 1977).

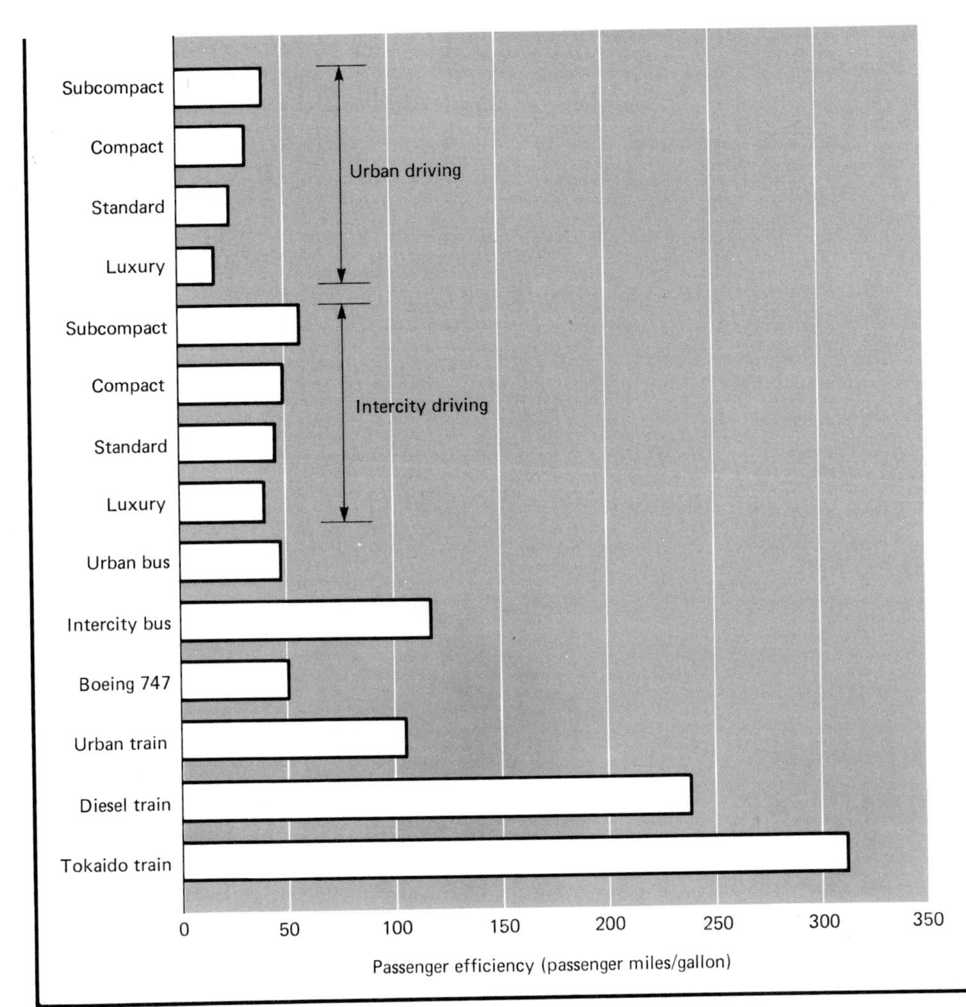

FIGURE 13.4 Passenger efficiencies for some transportation modes of interest. Note the significant differences in efficiency for subcompact and luxury cars in urban driving. This difference is not as pronounced for intercity driving. This is because the luxury car carries more passengers and performs better in this type of driving. The energy advantages for mass transit are evidenced by the high passenger efficiencies for bus and train travel. The data were taken from *Transportation Energy Conservation Data Book: Edition 2*, D. S. Shonka, A. S. Loeble, and P. D. Patterson, Oak Ridge National Laboratory, Oak Ridge, Tennessee.

steady yearly increase in the fuel economy of vehicles used on United States highways. As it stands now, the fuel economy of 1985 model vehicles must be 129% better than that of the 1974 models. These standards will not be met by simply making the internal combustion engine more efficient. That limit is determined by the laws of physics and there is not a lot of room for improvement. Some gain is possible through using a different type of engine, such as a diesel engine, but the biggest change will involve the type of cars and the way we use them. This is because massive cars and high speeds have fostered excessive gasoline consumption. Application of basic physical principles illustrates why.

Pushing a box on a flat surface is analogous to a car being "pushed" along a flat highway. Everyday experience reveals that the energy put forth depends on the weight of the box or the car, its speed, the distance moved, and the nature of the surface of the box and floor. The energy required increases as the weight of the box or car and distance traveled increase. In a quantitative analysis, for a given speed the energy is directly proportional to weight and distance.* Thus for a given automobile moving with

* *Problems of Our Physical Environment,* Joseph Priest, Addison-Wesley, Reading, Mass., 1973.

constant speed, the energy (or, equivalently, the gasoline) used per mile of travel depends directly on the weight. If the weight is reduced, the gasoline used per mile of travel will decrease. Figure 13.5 shows data compiled by the EPA for 1977 model cars. The direct relation between fuel economy and automobile weight is clear. Under normal driving the number of miles obtained per gallon of gasoline doubles if the automobile weight is cut in half. In an era of high gasoline prices and federally mandated fuel economy, this has instigated a trend toward lighter vehicles. Cars are made smaller to reduce the weight and many body components are made of light-weight aluminum or plastic. There are even tests at using nonmetals for building engines.

A moving automobile is held back by a force exerted on it by the air through which it moves. You can feel this force when you hold your hand out of the window of a moving car and you can perceive that the force increases as the speed increases. As a result the energy used also increases as the speed increases. The exact dependence on speed is fairly complex but a car traveling between 75 and 80 miles per hour gets about half the number of miles per gallon as one traveling 50 miles per hour. This is why a 55-mile-per-hour speed limit, if enforced, can be an extremely effective energy conservation measure. Additionally, there is the consolation of knowing that reduced speeds significantly lower the number of accidents and fatalities.

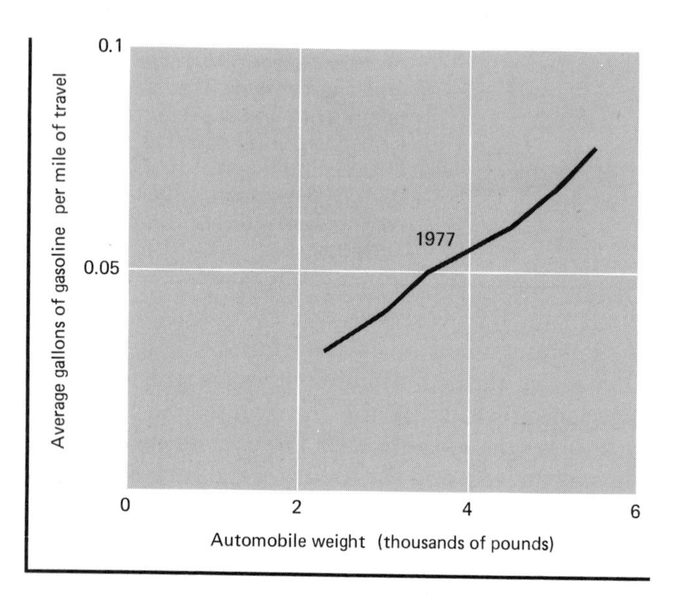

FIGURE 13.5 Relationship between fuel consumption and automobile weight for combined city/highway driving. Fuel consumption is expressed in gallons of gasoline per mile of travel. A value of 0.05 gallons per mile means the vehicle obtained 1/0.05 = 20 miles per gallon of gasoline used.

That mass transportation can be significantly more energy efficient than the private car is graphically illustrated in Fig. 13.4. Some progress has been made toward bringing about a shift to more efficient energy systems. Examples are the Bay Area Rapid Transit (BART) in the Oakland–San Francisco area and the METRO system in Washington, D.C. But mass transportation systems are extremely expensive and widespread utilization will be slow in coming. At present, using buses is probably the best way to solve mass transit problems. If buses are forced to compete for the same traffic lanes as automobiles, they are bound to be slow and uncomfortable to ride. To alleviate this, many cities provide lanes for the exclusive use of buses (Fig. 13.6). Apart from these more drastic solutions, substantial savings could be made by simply getting more passengers in automobiles (car pooling) and walking or bicycling for many short trips.

A similar efficiency analysis can be performed for bulk freight transportation. The idea is to transport as much bulk weight as possible over the greatest distance for the least amount of energy. Expressed as an equation,

$$\text{freight efficiency} = \frac{(\text{weight in tons}) \times (\text{distance traveled})}{\text{gallons of fuel expended}}. \tag{13.15}$$

A truck carrying a ten-ton cargo and getting six miles for each gallon of gasoline burned has a freight efficiency of 60. Figure 13.7 shows the freight efficiencies for several common modes of transport. As for passenger efficiency, there is a dramatic difference in efficiencies. Clearly there are reasons such

FIGURE 13.6 This photograph shows how the median strip of a freeway can be utilized by buses or trains to facilitate mass transit of people. (Photograph courtesy of United States Department of Transportation.)

as expediency and reliability for choosing a given form of transportation. However, for those situations where time is not a major factor and all other factors are equal, considerable energy can be saved by choosing the most efficient energy method.

The idea of feedback has been used on several occasions. A change in the output of any system can conceivably affect its input in either a positive or negative fashion. Automobiles constitute the output of a very influential, complex, industrial system. Alterations in the character of automobiles can produce consequential changes in the system. For example, if the public decides for energy conservation reasons that it no longer wants the massive cars that the industry is tooled to produce, then the industry falters until it can be retooled to make smaller cars. Consequently unemployment rises in the automobile industry and in the multitude of industries tied to it. This results in a general decline in the economy.

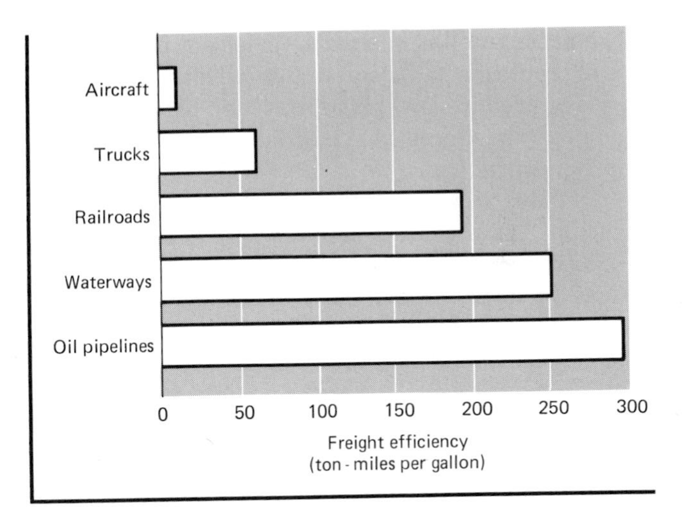

FIGURE 13.7 Freight efficiencies for some selected modes of transportation. From "System Energy and Future Transportation," Richard A. Rice, *Technology Review* (January 1972).

13.4 ENERGY CONSERVATION IN THE INDUSTRIAL SECTOR

Industry consumed 39% of the total energy used in 1981. Nearly 60% was used for heating processes and production of steam. Because industry is a major energy consumer and much of the energy is used for relatively simple processes, the potential for significant energy savings is large. Practical implementation of conservation schemes is difficult, but reductions continue. This is accomplished mostly by replacing old inefficient equipment, better design of new equipment, and more conscientious maintenance of existing equipment. In the longer term, the heat rejected from electric power plants could be used or, possibly, a self-contained total energy system that would produce electric energy on site with the rejected heat going directly into heating processes and steam production.

Cogeneration

Steel-making requires heat for melting iron ore. The auto industry needs heat for forming many components. Many industries, including the steel and auto industries, use steam for cleaning and heating. These are examples of a wide variety of industrial uses of process heat. About 40% of industrial energy is used to produce low-pressure steam. Process heat is generally produced on site by burning fossil fuels, largely natural gas.

All industries use electricity, usually generated off-site by large central power plants. Turbines at the central generating station reject about two units of heat for every unit of electric energy delivered. But the temperature of the waste heat is generally too low to be useful for process heat. Even if industrial electricity were generated on-site in smaller units operating at maximum efficiency, the heat rejected would still be of low value because the temperature would be too low. But if the efficiency for generating electricity were sacrified somewhat, then the rejected heat would be of sufficiently high temperature to be valuable for process heat. This on-site sequential production of electricity and process heat is called cogeneration. If a steam turbine is used, low-pressure steam exiting the turbine can be shuttled off and used as needed. With a gas turbine, combustion gases after passing through the turbine can be used to produce low-pressure steam. The energy for producing electricity and low-pressure process steam by cogeneration is about half that required to produce the two things separately. And both thermal and air pollution are lessened.

Cogeneration supplied about 30% of industrial electricity in the 1920s. But the favorable economics and political climate for generating electricity in large central generating facilities discouraged cogeneration in the interim. Today, only about 8% of industrial electricity is produced by cogeneration.

Incentives for replacing existing technology for generating electricity and producing process heat with cogeneration are not overwhelming. Central generating power plants are reliable sources of electricity. Low-pressure steam boilers are simply constructed and essentially maintenance-free. Cogeneration systems involve turbines and boilers that are expensive and complex. They are not maintenance free and trained technicians are required to operate them. Thus the capital outlay and maintenance personnel costs tend to cancel out savings in energy costs. Cogeneration makes greatest economic sense where local electricity costs are high. This generally means areas in which oil and natural gas are used extensively for generating electricity.

It is also possible to sacrifice efficiency in a central generating power plant to provide steam for widespread space heating of buildings. But unless the heating districts are close to the power plants, it becomes expensive to transport the steam heat.

13.5 ENERGY CONSERVATION IN THE ELECTRIC UTILITIES SECTOR

The overall efficiency for electric power production from existing plants is about 33%. A modern coal-fired electric power plant is nearly 40% efficient. Thus considerable savings will be made as existing plants are upgraded or replaced. Nuclear power plants will relieve the load on fossil fuel resources, but will increase the burden on the environment for absorbing waste heat because light-water reactors are only about 32% efficient. Nuclear breeder reactors and high-temperature gas-cooled reactors could increase the efficiency to around 40%. This generation of reactors, though, will not make an impact for at least a decade. Even with these advanced reactors, significant energy will always be wasted because of the use of steam turbines. And the steam turbine is bound to be around for several decades. Thus, in the long run, it is imperative to devise schemes for making use of the rejected heat from electric power plants.

Electric power transmitted at 345,000 volts for 200 miles is 98% efficient. Even though the transmission efficiency is high, the energy loss is not trivial because the total energy transmitted is large. The transmission of electric energy is the most expensive of all energy-transmission systems because the transmission lines, their supporting structures, and the land that they remove from other uses are extremely expensive. Some seven million acres of land are used for overhead transmission lines. Land problems worsen as demands for electric energy increase. Although the land may be available, environmental opposition to the structures is mounting. Installing the transmission lines underground would remove most of these objections but then costs would soar. Furthermore, because considerable heat is generated in conventional underground transmission lines, their capacity is lower because the earth cannot remove the heat as efficiently as the air around overhead transmission lines. Three schemes have been proposed for increasing the capacity of underground transmission. One method provides a compressed-gas insulation that more effectively removes the heat produced. The other two reduce the electrical resistance by cooling the conductors. Energy savings result because power loss is directly related to electrical resistance. These latter two technologies are extremely interesting because the temperatures considered are about −320° F and −453° F. These are the temperatures of liquid nitrogen (−320° F) and liquid helium (−453° F). The resistance of a common electrical conductor decreases uniformly as the temperature decreases. Typically the resistance decreases by a factor of ten in going from room temperature to −320° F. So the power loss would decrease by a factor of ten. But realize that energy is required to cool the cable. In fact as much energy is used to cool the cable as is gained through reduction in heating loss. However, the safe capacity of the line for carrying electric current increases significantly and this is the attractive feature. There are some metals, called superconductors, whose electrical resistance completely vanishes at extremely low temperatures. A niobium-tin metal alloy is a superconductor often mentioned for use in a power transmission line. The resistance of this alloy vanishes at −427° F. Again, the electricity required to cool the wires offsets the gain from heat energy losses but the safe current-carrying capacity increases dramatically (Fig. 13.8). Cooled transmission lines of the type described here are not practical at this time. However, they are being seriously considered and research and development is progressing on them.

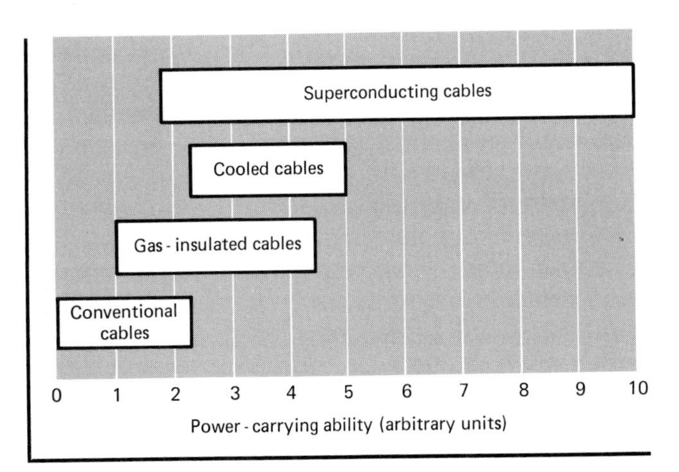

FIGURE 13.8 Comparison of the power-carrying ability of several types of power transmission cables. Note that a range of ability is given for each type. (Adapted from Figure 31, *Energy and the Future,* Allen L. Hammond, William D. Metz, and Thomas H. Maugh II, American Association for the Advancement of Science, Washington, D.C. (1973).

13.6 CONCLUDING REMARKS

These few illustations of energy-saving measures clearly show the potential and importance of conservation. Most of the schemes involve action by the energy industry, but there are many opportunities for personal energy conservation. When the urge arises to travel 80 miles per hour or to take an extended shower or to excessively cool or light a house, you should always bear in mind that the energy source very likely took millions of years to produce and that the supply is rapidly depleting. When you have the option to purchase a product in a disposable bottle or aluminum can, you should realize that energy was required to make the container and that energy is required to dispose of it. It is incumbent on us to recycle these products as energy and resource conservation measures. You should think seriously about walking or bicycling rather than motoring for the many short trips now taken with automobiles. Although our technological society has given us many problems resulting from energy conversion, it has also given us many benefits. It is important that we control this technology rather than letting it control us. Energy conservation and more effective utilization of energy is a meaningful way to begin achieving this goal.

REFERENCES

Energy: The Conservation Revolution, J. H. Gibbons and W. U. Chandler, Plenum, New York, 1981.

Our Energy: Regaining Control, M. H. Ross and H. Williams, McGraw-Hill, New York, 1981.

"Residential Energy Use Alternatives: 1976 to 2000," Eric Hirst, *Science* **194,** 1247 (17 December 1976).

"Transportation Energy Conservation Policies," Eric Hirst, *Science* **192,** 15 (2 April 1976).

A New Prosperity: Building a Sustainable Energy Future (The SERI Solar Conservation Study), H. Kelly and K. Gawell, Brick House, Andover, Mass., 1981.

Industrial and Commercial Cogeneration, Office of Technology Assessment, Washington, D.C. 20510 (February 1983).

"System Energy and Future Transportation," Richard A. Rice, *Technology Review* **74,** 31 (January 1972).

"The Fuel Economy of Light Vehicles," Charles L. Gray, Jr., and Frank von Hippel, *Scientific American* **244,** 48 (May 1981).

"Superconductors in Electric Power Technology," T. H. Geballe and J. K. Hulm, *Scientific American* **243,** 138 (November 1980).

The most detailed physics treatment of the second law efficiency is presented in *Efficient Use of Energy: The APS Studies on the Technical Aspects of the More Efficient Use of Energy,* American Institute of Physics, New York, 1975.

Less mathematical discussions are presented in

"Energy Efficiency: Our Most Underrated Energy Resource," Marc H. Ross and Robert H. Williams, *Bulletin of the Atomic Scientists* **32,** 30 (November 1976).

"Energy I," Barry Commoner, *The New Yorker,* February 2, 1976.

"Industry Can Save Energy without Stunting Its Growth," Tom Alexander, *Fortune,* May 1977.

REVIEW

1. How was energy consumption categorized for purposes of evaluating potential areas for conservation?
2. Roughly, how much of the total energy converted goes into useful forms? What happens to the energy that goes unused?
3. What is the ultimate disposition of virtually all the energy converted?
4. Why does lowering the thermostat during heating periods tend to conserve energy?
5. What is insulation and how is it used in a building?
6. What is a heat pump and what are its attractive features?
7. Explain the concepts of passenger efficiency and freight efficiency.
8. How do the weight and speed of an automobile affect its gasoline consumption?
9. What economic effects can occur if a major change is made in the character of automobiles?
10. Name some ways that industry might be able to conserve energy.
11. What is the meaning of cogeneration? Why will there always be considerable waste of energy in the generation of electricity as long as a steam turbine is used?
12. What are some problems associated with the transmission of electric power?
13. What is a superconductor? How might a superconductor be utilized in the electric power industry?
14. The cost in dollars of operating a car can be expressed as cost = miles $\cdot A \cdot B$ where

a) $A = \frac{\text{gallons}}{\text{mile}}$ and $B = \frac{\text{gallons}}{\text{dollar}}$.

b) $A = \frac{\text{miles}}{\text{gallon}}$ and $B = \frac{\text{dollars}}{\text{gallon}}$.

c) $A = \frac{\text{gallons}}{\text{mile}}$ and $B = \text{dollars} \cdot \text{gallons}$.

d) $A = \frac{\text{gallons}}{\text{mile}}$ and $B = \frac{\text{dollars}}{\text{gallons}}$.

e) $A = \text{gallons} \cdot \text{mile}$ and $B = \text{dollars} \cdot \text{gallons}$.

15. Using principles of physics, one knows that when the temperature in a room is maintained at 70° F on a cold day,
    a) heat losses to the outside are balanced by heat input from the heating system.
    b) the heating system must necessarily run continuously.
    c) the convective heat losses are balanced by the conductive heat losses.
    d) the rate at which energy enters the house equals the rate at which energy is supplied by the heating system.
    e) the heating system must operate at a temperature of 70° F.

16. Raising the thermostat setting during the summer saves energy because
    a) the air conditioner will not run as long.
    b) the air conditioner must remove less energy from the building.
    c) the air conditioner must lower the room temperature fewer degrees.
    d) the electric utilities company gives lower rates if you agree to raise the thermostat setting.
    e) (a), (b), and (c) are all correct.

17. The two factors to be considered in conserving energy used by lights in a dormitory room are ____ and ____, and these two factors should be ____.
    a) watts per hour; hours; multiplied.
    b) watts; hours; divided
    c) joule • hours; hours; multiplied
    d) watts; hours; squared
    e) kilowatts; hours; multiplied

18. In a certain house, a 1000-watt light bulb is normally on for 30 hours each week. If electricity costs 8¢/kWh and, for conservation reasons, it is decided to keep the bulb lit only half as much, then the savings per week would be
    a) 24¢. b) 12¢. c) $120. d) $240.
    e) $1.20.

19. The R-value of $\frac{5}{8}$-inch thick plywood is 0.8. Knowing this, it follows that the R-value of $\frac{3}{8}$-inch thick plywood is
    a) 0.3. b) 0.8. c) 1.33. d) 0.375.
    e) 0.48.

20. With all other factors being equal, the heat losses through a material will ____ if the R-value doubles because the heat losses ____ the R-value.
    a) halve; are directly related to
    b) double; are directly related to
    c) quadruple; are inversely related to the square of
    d) be unchanged; are unrelated to
    e) halve; are inversely related to

21. For heat loss by conduction, builders use a symbol "R" value.
    a) R is the "resistance to heat flow."
    b) R is the rigidity of or stiffness of the insulation used.
    c) R is a relative thickness compared to rock wool.
    d) R is a regional insulation requirement established by the federal government.
    e) R is a sum of all "U" values for the wall.

22. Infiltration energy losses in a home are caused by
    a) cold air entering in through air filters.
    b) keeping refrigerator doors open for excessive lengths of time.
    c) conduction of heat through ceilings.
    d) convection currents emanating from the furnace to the outside.
    e) energy losses through openings between the inside and outside of the house.

Questions 23, 24, and 25 are related and make use of the formula

$$\text{Annual energy lost} = \frac{\text{area} \cdot \text{heating factor}}{\text{R-value}} \cdot 100{,}000 \text{ Btu.}$$

The ceiling in the room of a homeowner's house at present has an R-value of 10. The area of the ceiling is 300 square feet and the house is in an area having a heating factor of 1.5. Fuel for the house is provided by fuel oil (energy content 100,000 Btu per gallon) at a cost of $1 per gallon.

23. At present, the amount of energy that escapes

annually through the ceiling is

a) 45 Btu.
b) 45 million Btu.
c) 45 thousand Btu.
d) 4.5 million Btu.
e) none of the above.

24. The cost of the fuel that provided the energy that escaped through the ceiling is

a) \$450
b) \$45
c) \$4.50
d) \$0.45
e) none of the above.

25. If the R-value of the ceiling is increased to 20, then at the same fuel cost per gallon the homeowner can expect an annual saving of

a) \$22.50
b) \$225.
c) \$2.25.
d) \$0.225.
e) none of the above.

26. Riding a bicycle at 13 miles/hour requires an energy expenditure of about 660 food calories per hour. There are four Btu in a food calorie. Therefore the number of Btu used per mile of travel is

a) 2145.
b) 13.
c) 34,320.
d) 203.
e) 0.079.

27. In Problem 26, the passenger efficiency is ____ passenger-miles per gallon. (Assume a gallon of gasoline produces 125,000 Btu.)

a) 616
b) 12.7
c) 0.005
d) 0.00047
e) 0.08

28. Heat pumps deliver more energy (in the form of heat) to the house than is supplied or produced by the electric company. Thus you get more heat than you would pay for in resistance heating, and this extra heat is removed from the cold outdoors and added to the hotter interior.

a) true
b) false

29. A heat pump is a device that literally pumps heat from the cold exterior of a house to the inside of a house. Generally, electric energy is used to run the pump. For a heat pump to be practical then, it would have to

a) transfer more energy to the house than was used to run the pump.
b) supply less energy to the house than was used to run the pump.
c) supply the same energy to the house that was used to run the pump.
d) run continuously.
e) violate the second law of thermodynamics.

30. A car traveling 60 miles per hour requires one hour to travel 60 miles. If the speed were reduced to 50 miles per hour, the extra time required for the 60-mile trip would be

a) 1.2 hours.
b) 0.2 hours.
c) zero.
d) $1\frac{1}{3}$ hour.
e) $\frac{2}{3}$ hour.

31. The automobile fuel used per mile of travel depends on the weight of the vehicle. Pick the graph below that best describes the relation between gallons per mile and weight.

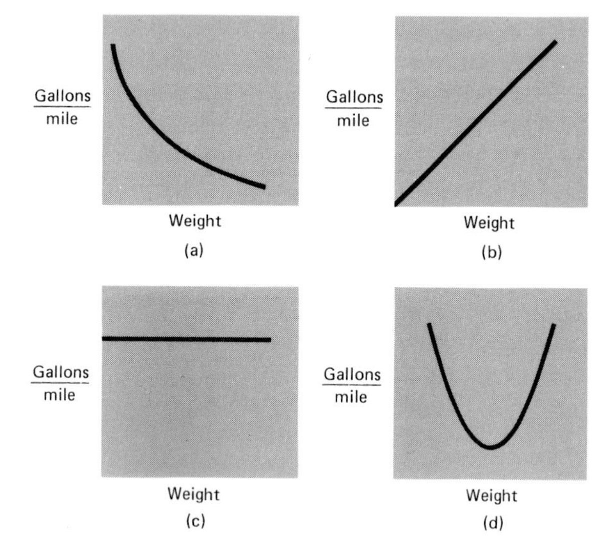

32. A 55-mph speed limit saves energy because

a) road friction decreases with increasing speed.
b) air entering the carburetor has more oxygen when the speed is lower.
c) wind resistance increases as vehicle speed increases.
d) the braking distance of a vehicle decreases as speed decreases.
e) gasoline burns better when the vehicle maintains a speed of 55 mph.

33. The number of gallons of gasoline consumed per mile of travel by an automobile will

a) quadruple for each doubling of vehicle weight.
b) be independent of vehicle weight.
c) double if the vehicle weight is doubled.
d) halve if the vehicle weight is doubled.
e) not change regardless of weight changes.

Answers

14. (d)	15. (a)	16. (e)	17. (e)
18. (e)	19. (e)	20. (e)	21. (a)
22. (e)	23. (d)	24. (b)	25. (a)
26. (d)	27. (a)	28. (a)	29. (a)
30. (b)	31. (b)	32. (c)	33. (c)

QUESTIONS AND PROBLEMS

13.2 ENERGY CONSERVATION IN THE RESIDENTIAL AND COMMERCIAL SECTOR

1. Clothing limits heat loss from a person's body in a cold climate. How does the type and thickness of wearing apparel affect heat loss?
2. Why do you intuitively expect that heat lost by conduction through a wall depends on the difference in temperature between the interior and exterior of the building?
3. From your personal reactions, explain why you can or cannot tell if the temperature in a room is 70° F or 68° F.
4. Consider buying one of two houses that are nearly equivalent except one costs $3000 more because of improved insulation and a more efficient heating system. You can save on the cheaper one but fuel bills will be higher. What considerations enter into your choice?
5. Explain qualitatively why heat pumps are more effective when there is a small temperature difference between the inside and the outside of a building.
6. Lowering a thermostat setting during heating periods conserves energy. How does raising the thermostat setting during cooling periods conserve energy?
7. How is the R-value for a heat conductor analogous to the resistance value for an electrical conductor?
8. Ohm's law, discussed in Section 4.5, relates electric current to potential difference and electrical resistance ($I = V/R$). In what ways is the heat flow equation (Eq. 13.2) similar to the equation for Ohm's law?
9. A newspaper article suggests that every 1° F lowering of a thermostat produces a 5% increase in energy savings. Assuming that the outdoor and indoor temperatures are 50° F and 70° F, respectively, show that this is indeed the case as long as temperatures do not fall too far below 70° F.
10. What fraction of the total energy used for air conditioning a home could be saved if the thermostat setting were raised 3° F from its nominal 70° F setting? Assume that the outdoor temperature is 85° F.
11. If the thickness of insulation in the walls of a house is doubled, how much change would you expect in the heat loss through the walls?
12. The R-value for $\frac{5}{8}$-inch plywood is 0.79 (Table 13.1). Would you expect the R-value for $\frac{3}{8}$-inch plywood to be greater than or less than 0.79? Figure out the R-value for $\frac{3}{8}$-inch plywood and check your answer with that quoted in Table 13.1.
13. The ceiling of a room in a house measures 12 feet by 20 feet. The ceiling is composed of $\frac{3}{8}$-inch gypsum board covered with three inches of loose fill.
    a) Calculate the heat loss through the ceiling in an area having a heating factor of 1.25.
    b) If the loose fill is increased to six inches, how much will the heat loss be reduced?
    c) If the house is heated by burning fuel oil, how many gallons of oil will be saved after the additional three inches of loose fill are installed?
14. The construction of a well-insulated house wall is shown below. Using Table 13.1, determine the overall R-value for the wall. What sort of an energy penalty would a homeowner pay if the $3\frac{1}{2}$ inches of blanket insulation were not used?

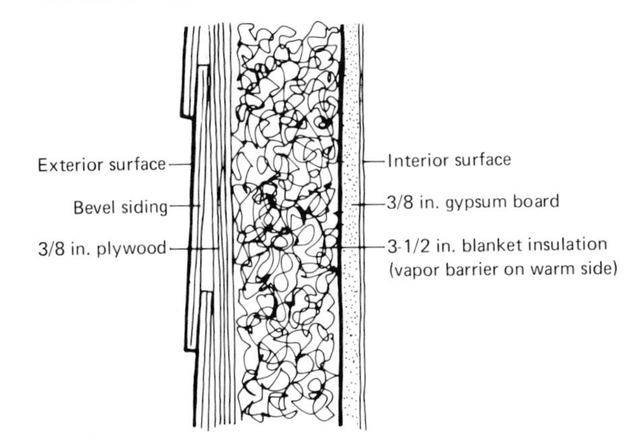

15. A central heating system has the task of providing energy for 100 identical apartments. How is the total energy required related to the energy used per apartment? Without worrying about performing the assigned task, how could the total energy use be decreased? Insisting that the task be accomplished,

what is the alternative if a decrease is desired for the total energy use?

16. An average house has about 1500 square feet of floor space. The walls of most contemporary houses are eight feet high. In the central United States, this house will need about 100,000,000 Btu of heat during the heating season. Figure the total area of the walls, ceiling, and floors of this house assuming that it is 50 feet long, 30 feet wide, and 8 feet high. Using Eq. (13.3), determine the average R-value for the house. Does the R-value obtained seem reasonable?
17. A certain heat pump for a house has a performance of three. How many units of heat are fed to the house for every unit of energy used to operate the heat pump?

13.3 ENERGY CONSERVATION IN THE TRANSPORTATION SECTOR

18. What features of the automobile entice 86% of the working community to use it for commuting to work?
19. Using Eq. (13.15) as a guide, define an energy efficiency for transporting coal.
20. What social and economic forces hinder construction of mass transit systems?
21. When you hold your hand outside the window of a moving car, does the force felt depend on the orientation of your hand? On the basis of this experience, would you expect all automobiles to experience the same air resistance when traveling at the same speed?
22. What considerations other than energy conservation enter into choosing a form of transportation?
23. What is the passenger efficiency of a car carrying three passengers and obtaining 20 miles for each gallon of gasoline consumed?
24. A person traveling 12 miles per hour on a bicycle uses about 200 Btu of energy per mile of travel.
    a) Assuming that the energy equivalent of one gallon of gasoline is 100,000 Btu, determine the gasoline equivalent of 200 Btu of energy.
    b) How many "miles per gallon" do you get by biking at 12 miles per hour?
    c) Determine the passenger efficiency for bike riding and compare it with those shown in Fig. 13.4.
25. a) How much time does it take to travel 75 miles at an average speed of 55 miles per hour?
    b) How much time is saved if the 75-mile trip is traversed at an average speed of 70 miles per hour?

13.5 ENERGY CONSERVATION IN THE ELECTRIC UTILITIES SECTOR

26. Electric power usually peaks around 5:00 P.M. Why does this peaking contribute to the inefficient use of energy in the electric utilities industry?

Appendixes

Appendix A

TABLE A.1 Historical dependence of the United States on various sources of energy. Energy is expressed in quads. One quad = 10^{15} British thermal units.

Total energy consumption		Fuelwood		Coal		Petroleum and natural gas		Hydropower		Nuclear	
Year	*Quads*	*Quads*	*Percent*	*Quads*	*Percent*	*Quads*	*Percent*	*Quads*	*Percent*	*Quads*	*Percent*
1860	3.1	2.6	83.5	0.5	16.4	—	0.1	—	—	—	—
1880	5.0	2.9	57.0	2.0	41.1	0.1	1.9	—	—	—	—
1900	9.6	2.0	21.1	6.8	71.3	0.5	5.0	0.3	2.6	—	—
1920	21.3	1.6	7.5	15.5	72.8	3.4	16.1	0.8	3.6	—	—
1940	25.0	1.4	5.3	12.5	50.1	10.2	40.9	0.9	3.7	—	—
1950	34.0	1.2	3.3	12.9	36.7	19.7	55.9	1.4	4.1	—	—
1960	44.6	—	—	10.1	22.8	32.8	73.5	1.7	3.7	—	—
1970	67.1	—	—	12.7	18.9	51.5	76.8	2.7	4.0	0.2	0.3
1975	70.6	—	—	12.8	18.2	52.6	74.6	3.2	4.6	1.8	2.6
1976	74.2	—	—	13.7	18.6	55.2	74.5	3.1	4.2	2.0	2.7
1980	75.91	—	—	15.46	20.4	54.59	71.9	3.11	4.1	2.67	3.5

Sources: Energy in Focus: Basic Data, Federal Energy Administration (1977), and *1981 Annual Report to Congress,* Vol. 2, Energy Statistics, Energy Information Administration, U.S. Department of Energy, Washington, D.C. (May 1981).

TABLE A.1 (cont.)

Year	Total energy consumption	Coal		Natural gas		Petroleum		Hydropower		Nuclear power		Geothermal		Wood and waste	
		Quads	*Million short tons*	*Quads*	*Trillion cubic feet*	*Quads*	*Million barrels*	*Quads*	*Billion kWh*	*Quads*	*Billion kWh*	*Quads*	*Billion kWh*	*Quads*	*Billion kWh*
1951	36.11	13.20	505.9	7.05	6.81	14.43	2,561	1.45	106.6	0	0	0	0	0	0
1952	35.83	11.84	454.1	7.55	7.29	14.96	2,661	1.50	112.0	0	0	0	0	0	0
1953	36.76	11.87	454.8	7.91	7.64	15.56	2,774	1.44	111.6	0	0	0	0	0	0
1954	35.73	10.17	389.9	8.33	8.05	15.84	2,831	1.39	114.0	0	0	0	0	0	0
1955	39.17	11.52	447.0	9.00	8.69	17.25	3,086	1.41	120.3	0	0	0	0	0	0
1956	40.75	11.72	456.9	9.61	9.29	17.94	3,212	1.49	129.8	0	0	0	0	0	0
1957	40.80	11.14	434.5	10.19	9.85	17.93	3,215	1.56	137.0	0	0	0	0	0	0
1958	40.65	9.83	385.7	10.66	10.30	18.53	3,328	1.63	146.9		0.2	0	0	0	0
1959	42.41	9.79	385.1	11.72	11.32	19.32	3,477	1.59	144.7		0.2	0	0	0	0
1960	44.08	10.12	398.0	12.39	11.97	19.92	3,586	1.65	153.7	0.01	0.5		0		0.1
1961	44.72	9.89	390.3	12.93	12.49	20.22	3,641	1.68	157.5	0.02	1.7		0.1		0.1
1962	46.80	10.17	402.2	13.73	13.27	21.05	3,796	1.82	172.2	0.03	2.3		0.1		0.1
1963	48.61	10.69	423.5	14.40	13.97	21.70	3,921	1.77	169.1	0.04	3.2		0.2		0.1
1964	50.78	11.25	445.7	15.29	14.81	22.30	4,034	1.91	182.3	0.04	3.3		0.2		0.1
1965	52.99	11.89	472.0	15.77	15.28	23.25	4,202	2.06	196.8	0.04	3.7		0.2		0.3
1966	55.99	12.48	497.7	17.00	16.45	24.40	4,411	2.07	199.0	0.06	5.5		0.2		0.3
1967	57.89	12.24	491.4	17.94	17.39	25.28	4,585	2.34	224.6	0.09	7.7	0.01	0.3		0.3
1968	61.32	12.66	509.8	19.21	18.63	26.98	4,902	2.34	225.2	0.14	12.5	0.01	0.4		0.4
1969	64.53	12.72	516.4	20.68	20.06	28.34	5,160	2.66	254.5	0.15	13.9	0.01	0.6		0.3
1970	66.83	12.66	523.2	21.79	21.14	29.52	5,364	2.65	252.9	0.24	21.8	0.01	0.5		0.4
1971	68.30	12.01	501.6	22.47	21.79	30.56	5,553	2.86	273.1	0.41	38.1	0.01	0.5		0.3
1972	71.63	12.45	524.3	22.70	22.10	32.95	5,990	2.94	283.6	0.58	54.1	0.03	1.5		0.3
1973	74.61	13.30	562.6	22.51	22.05	34.84	6,317	3.01	289.7	0.91	83.5	0.04	2.0		0.3
1974	72.76	12.88	558.4	21.73	21.22	33.45	6,078	3.31	316.9	1.27	114.0	0.05	2.5		0.3
1975	70.71	12.82	562.6	19.95	19.54	32.73	5,958	3.22	309.3	1.90	172.5	0.07	3.2		0.2
1976	74.51	13.73	603.8	20.35	19.95	35.17	6,391	3.07	295.5	2.11	191.1	0.08	3.6		0.3
1977	76.33	13.96	625.3	19.93	19.52	37.12	6,727	2.51	241.0	2.70	250.9	0.08	3.6	0.01	0.5
1978	78.18	13.85	625.2	20.00	19.63	37.97	6,879	3.14	303.2	3.02	276.4	0.06	3.0		0.3
1979	78.91	15.11	680.5	20.67	20.24	37.12	6,757	3.14	303.4	2.71	255.2	0.08	3.9	0.01	0.5
1980	75.91	15.46	702.7	20.39	19.83	34.20	6,242	3.11	300.1	2.67	251.1	0.11	5.1		0.4
1981	73.91	16.01	727.7	19.93	19.42	32.00	5,840	2.97	287.0	2.90	272.3	0.12	5.7		0.4

Source: 1981 Annual Report to Congress, Energy Information Administration, United States Department of Energy (May 1982).
A blank entry means the consumption was less than 0.005 quad.

TABLE A.2 Contribution of the major sources of energy to the transportation, industrial, residential and commercial, and electric utilities sectors for the year 1981. Energy is expressed in quads. One quad $=10^{15}$ Btu.

Sector	Coal	Gas	Petroleum
Residential and commercial	0.167	7.404	3.125
Industrial	3.142	7.963	8.123
Transportation	—	0.630	18.548
Electric utilities	12.707	3.764	2.202
Nuclear electric (2.901)			
Hydroelectric (2.937)			

Source: Monthly Energy Report—1982, United States Department of Energy.

TABLE A.3 Breakdown of the end-use sectors according to major user type. Energy is expressed in quads. One quad $= 10^{15}$ Btu.

	Total quads		Total quads		Total quads
Residential and commercial	*25.687*	*Industrial*	*28.853*	*Transportation*	*19.0*
Space heat	44.7%	Process steam	35.5%	Automobiles	51.6%
Water heat	10.7	Direct heat	24.0	Trucks	20.5
Air conditioning	10.0	Electric drive	22.0	Aircraft	6.8
Refrigeration	8.0	Electrolysis	3.0	Rail	3.2
Lighting	14.6	Feedstocks	13.5	Pipelines	9.5
Other	11.9	Other	2.0	Ships	1.6
				Buses	1.1
				Other	5.8

Percentages were extracted from *Energy in America's Future,* Johns Hopkins Universty Press, Baltimore (1979). Energy data, which include transmission losses, are from *Monthly Energy Report—1982,* United States Department of Energy.

TABLE A.4 Comparison of energy use and gross national product (GNP) per capita for selected nations, 1979. The conversion of coal to barrels of oil assumes 27.8 $\times 10^3$ Btu per kg of coal, 5.8 $\times 10^6$ Btu per barrel of oil.

Nation	E (kg/person) (1980)	E (bbl/person)	GNP/person
United States	10,410	49.9	10,775
Canada	10,241	49.0	9,586
West Germany	5,727	27.4	12,419
Australia	6,032	28.9	8,126
France	4,351	20.8	10,720
Finland	5,135	24.6	8,701
Japan	3,690	17.7	8,627
Israel	2,368	11.3	4,969
United Kingdom	4,835	23.2	7,192
Saudi Arabia	1,677	8.03	7,375
Venezuela	3,375	16.2	3,622
Greece	2,137	10.2	4,093
Spain	2,530	12.1	5,300
Mexico	1,770	8.48	1,749
Peru	619	2.96	716
Jordan	632	3.03	931
Malawi	48	0.230	212
Brazil	761	3.64	761
Egypt	473	2.27	473
India	191	0.915	191
Iraq	606	2.90	606

Source: Energy data are from *1980 Yearbook of World Energy Statistics,* United Nations, New York, 1982. GNP data are from *World Statistics in Brief,* United Nations Statistical Pocketbook, 6th edition.

TABLE A.5 Yearly production and consumption of energy of all forms in the United States. Energy is expressed in quads. One quad = 10^{15} Btu. The conversion to barrels of oil assumes 10^{12} Btu per 0.172 million barrels of oil.

Consumption			Production			
Year	*Quads*	*Billions of barrels*	*Quads*	*Billions of barrels*	*Resident population (millions)*	*Per capita (bbls/person)*
1950	34.0	5.85	34.4	5.92	152.3	38.4
1955	39.7	6.83	39.1	6.73	165.9	41.2
1960	44.6	7.67	41.6	7.16	180.7	42.4
1961	45.3	7.79	42.3	7.28	183.0	42.6
1962	47.4	8.15	43.9	7.55	185.8	43.9
1963	49.3	8.48	46.0	7.91	188.5	45.0
1964	51.2	8.81	47.6	8.19	191.1	46.1
1965	53.3	9.17	49.1	8.45	193.5	47.4
1966	56.4	9.70	51.9	8.93	195.6	49.6
1967	58.3	10.03	54.8	9.43	197.5	50.8
1968	61.8	10.63	56.6	9.74	199.4	53.3
1969	65.0	11.18	58.7	10.10	201.4	55.5
1970	67.1	11.54	62.0	10.66	204.0	56.6
1971	68.7	11.82	61.0	10.49	206.8	57.2
1972	71.9	12.37	62.1	10.68	209.3	59.1
1973	74.609	12.83	62.433	10.74	211.4	60.7
1974	72.759	12.51	61.229	10.53	213.3	58.6
1975	70.707	12.16	60.059	10.33	215.5	56.4
1976	74.510	12.82	60.091	10.34	217.6	58.9
1977	76.332	13.13	60.293	10.37	219.8	59.7
1978	78.175	13.45	61.231	10.53	222.1	60.4
1979	78.910	13.57	63.851	10.98	224.6	60.6
1980	75.913	13.06	65.499	11.27	227.2	57.5
1981	73.779	12.69	65.069	11.19	229.3	55.3

Source: The data are from *Annual Report to Congress, 1982,* United States Department of Energy.

TABLE A.6 United States crude oil and natural gas reserves.

Year	Crude oil *Billion barrels*	Natural gas *Trillion cubic feet**
American Petroleum Institute and American Gas Association Data		
1951	27.5	192.8
1952	28.0	198.6
1953	28.9	210.3
1954	29.6	210.6
1955	30.0	222.5
1956	30.4	236.5
1957	30.3	245.2
1958	30.5	252.8
1959	31.7	261.2
1960	31.6	262.3
1961	31.8	266.3
1962	31.4	272.3
1963	31.0	276.2
1964	31.0	281.3
1965	31.4	286.5
1966	31.5	289.3
1967	31.4	292.9
1968	30.7	287.3
1969	29.6	275.1
1970	39.0	290.7
1971	38.1	278.8
1972	36.3	266.1
1973	35.3	250.0
1974	34.2	237.1
1975	32.7	228.2
1976	30.9	216.0
1977	29.5	208.9
1978	27.8	200.3
1979	27.1	194.9
Energy Information Administration Data		
1977	31.8	207.4
1978	31.4	208.0
1979	29.8	201.0
1980	29.8	199.0

* Does not include natural gas held in underground storage.
Source: The data are from *1981 Annual Report to Congress,* Energy Information Administration, United States Department of Energy.

TABLE A.7 United States annual production of electric energy.

Year	Production (billion kWh)
1951	371
1952	399
1953	443
1954	472
1955	547
1956	601
1957	632
1958	645
1959	710
1960	753
1961	792
1962	852
1963	917
1964	984
1965	1,055
1966	1,144
1967	1,214
1968	1,329
1969	1,442
1970	1,532
1971	1,613
1972	1,750
1973	1,861
1974	1,867
1975	1,918
1976	2,038
1977	2,124
1978	2,206
1979	2,247
1980	2,286
1981	2,293

Source: The data are from *1981 Annual Report to Congress,* Energy Information Administration, United States Department of Energy.

Appendix B

TABLE B.1 Some useful energy and power units. Units in column 1 are convertible to equivalent units by multiplying by the numbers given in columns 4 and 5. Thus, in joules, 2000 Btu are equivalent to

$$2000\ \text{Btu} \cdot 1055 \frac{\text{joules}}{\text{Btu}} = 2{,}110{,}000\ \text{joules}.$$

Special energy unit	Study area of main use	Symbol	Equivalent in joules	Other useful equivalents
kilowatt-hour	electricity	kWh	3,600,000	3413 Btu, 860 kcal
calorie	heat	cal	4.186	
kilocalorie (food calorie)	heat	kcal	4186	1000 cal
British thermal unit	heat	Btu	1055	252 cal
electron volt	atoms, molecules	eV	1.602×10^{-19}	
kiloelectron volts	X-rays	keV	1.602×10^{-16}	1000 eV
million electron volts	nuclei, nuclear radiation	MeV	1.602×10^{-13}	1,000,000 eV
quintillion	energy reserves	Q	1.055×10^{21}	10^{18} Btu or billion billion Btu
quadrillion	energy reserves	quad	1.055×10^{18}	10^{15} Btu or a million billion Btu

Special power unit	Study area of main use	Symbol	Equivalent in watts
British thermal unit per hour	heat	Btu/hr	0.293
kilowatt	household electricity	kW	1000
megawatt	electric power plants	MW	1,000,000
gigawatt	total U.S. or world power production	GW	1,000,000,000
horsepower	automobile engines, motors	hp	746

ENERGY EQUIVALENTS

One gallon of gasoline = 126,000 Btu.
One pound of bituminous coal = 13,100 Btu.
One cubic foot of natural gas = 1,030 Btu.
One 42-gallon barrel of oil = 5,800,000 Btu.
One therm = 100,000 Btu.
One quad = 10^{15} Btu.

Variations of these energy equivalent values will appear in the literature.
The values quoted here are typical.

Appendix C

Powers of Ten

Hand-held calculators are enormously useful for routine numerical calculations. However, there is a limit to the number of digits incorporated in a number processed by the calculator. Generally, this number is ten. For example, the number 10,000,000,000 cannot be handled as such by most hand-held calculators because it has 11 digits. But don't despair. Most hand-held calculators will process large and small numbers if they are entered in a notation utilizing "powers of ten." The purpose of this appendix is to illustrate the importance of the "powers of ten" notation and to show how they are used to great advantage in modern hand-held calculators.

The range of numbers encountered in assessing energy concepts is staggering. For example, the radius of the nucleus of the aluminum atom is 0.0000000000036 cm, and the United States electric energy production in 1981 was 2,293,000,000,000 kWh. Multiplying, dividing, and just keeping track of numbers like these is tedious without good bookkeeping.

A number like 2,293,000,000,000 represents a measurement or an estimate and there is, therefore, some uncertainty in the number. If the number were known to an accuracy of 1%, which normally would be quite good, it would mean that the true value would, perhaps, be between 2,270,000,000,000 and 2,316,000,000,000. At best, only the first four digits, that is, 2, 2, 9, and 3, are significant. The remaining zeros denote the relative size or magnitude. For example, the number might be written as 2293 billion kWh. Billion denotes the magnitude and is used as a bookkeeping procedure, albeit not a particularly practical one in the sense that it does not lend itself to easy multiplication and division. To illustrate a better method, consider the numbers 0.0015 and 1500. These numbers could also be recorded as

$$1.5 \div 1000 \qquad \text{and} \qquad 1.5 \times 1000.$$

The 1000 figure denotes the magnitude. Now 1000 can be generated by multiplying $10 \times 10 \times 10$. A shorthand way of symbolizing this operation is 10^3. Using this notation, we can write

$$0.0015 = 1.5 \div 10^3 = \frac{1.5}{10^3}$$

$$1500 = 1.5 \times 10^3.$$

It is conventional and meaningful to write $\frac{1}{10^3}$ as 10^{-3} so that $\frac{1.5}{10^3}$ becomes 1.5×10^{-3}. The 3 in 10^3 indicates the third power of 10. The beauty of this method is that the appropriate power of ten can be obtained by counting the positions the decimal point is moved.

$1.5 \times 10^{-3} = 0.0015$ — Decimal point of 1.5 is moved 3 places to left.

$1.5 \times 10^3 = 1500.$ — Decimal point of 1.5 is moved 3 places to right.

To convince yourself of the utility of the method, express the following numbers as a number times ten to a power.

0.00000001 cm (approximate diameter of an atom)

30,000,000,000 m/sec (speed of light)

Express the following numbers in decimal form

9.11×10^{-28} g (mass of the electron)

3.31×10^4 cm/sec (speed of sound in air at one atmosphere pressure and 0° C)

You should be convinced of the usefulness of the method if for no other reason than compactness. The real utility is in multiplication and division because it reduces much of the effort to that of addition and subtraction. To illustrate, consider

$(1000) \times (1000) = 1{,}000{,}000$ (long method)

$10^3 \times 10^3 = 10^{3+3} = 10^6$ (short method)

The general rule is $10^a \cdot 10^b = 10^{a+b}$, where a and b are the powers of 10. There are no restrictions on a and b; they can be positive, negative, and even fractional. The reason for writing $\frac{1}{10^3}$ as 10^{-3} can now be seen.

Example:

$$\frac{1{,}000{,}000}{1000} = 1000 \text{ (long method)}$$

$$\frac{10^6}{10^3} = 10^6 \cdot 10^{-3} = 10^{6-3} = 10^3 \text{ (short method)}$$

Example:

$$\begin{aligned}(5.1 \times 10^4) \cdot (3.2 \times 10^{-1}) &= 5.1\,(3.2) \times 10^{4-1} \\ &= 16.32 \times 10^3 \\ &= 1.632 \times 10^4\end{aligned}$$

Example:

$$\begin{aligned}(5.1 \times 10^4) \div (3.2 \times 10^{-1}) &= \frac{5.1 \times 10^4}{3.2 \times 10^{-1}} \\ &= \left(\frac{5.1}{3.2}\right) 10^{4+1} \\ &= 1.59 \times 10^5\end{aligned}$$

To use numbers expressed as powers of ten in a hand-held calculator, one merely has to feed the calculator the significant figures and then the power of ten. To signal that a power of ten is being entered a key generally labeled EE is pressed. For example, 5.1×10^4 would be entered as follows:

Enter 5.1 on the keyboard.

Press EE key.

Enter 4 on the keyboard.

The display would then read

5.1 04

which is interpreted as 5.1×10^4. If this number is to be multiplied by 3.2×10^{-1}, then one would signal the appropriate multiplication key and

Enter 3.2.

Press EE.

Enter -1.

Execute multiplication.

The answer would be revealed as 1.632 04 which is interpreted as

1.632×10^4 or 16,320.

Numerical calculations are simplified enormously with hand-held calculators. Take advantage of them. To gain confidence, perform the following operations using the power of ten notation. Check your results using both a hand-held calculator and the conventional method for multiplication and division.

$(0.000303) \cdot (43200) =$

$(250{,}000) \div 0.0025 =$

$(0.0101) \cdot (0.0005) \div (0.0002) =$

Devise some exercises of your own and practice the method.

Some useful numerical prefixes

Numerical unit	*Prefix*	*Symbol*	*Power of ten designation*
trillion	tera	T	10^{12}
billion	giga	G	10^9
million	mega	M	10^6
thousand	kilo	k	10^3
hundredth	centi	c	10^{-2}
thousandth	milli	m	10^{-3}
millionth	micro	μ	10^{-6}

Some useful physical constants

Constant	*Symbol*	*Value*
Boltzmann's constant	k	1.38×10^{-23} joules per kelvin
Speed of light	c	3×10^8 meters per second
Planck's constant	h	6.63×10^{-34} joule • seconds

Glossary

Glossary

Acceleration: Time rate of change of velocity.

Acid rain: Abnormally acidic rainfall or snowfall. The acids of interest are usually dilute sulfuric and/or nitric.

Actinides: Atoms, usually radioactive, produced from the bombardment of $^{235}_{92}U$ and $^{238}_{92}U$ by neutrons in a nuclear reactor.

Active solar system: A method or device that utilizes solar energy for heating and cooling with the aid of pumps and fans for circulating heat.

Aerosol: Literally, a gaseous suspension of solid or liquid particles. Environmentally, a maximum size such as 0.0001 m is sometimes included.

Albedo: The fraction (or percentage) of incident electromagnetic radiation reflected by a surface.

Alpha particle: Originally designated as a radioactive particle emitted naturally by several elements such as radium and uranium. It is indistinguishable from the nucleus of the helium atom having two protons and two neutrons.

Alternating current (ac): An electric current that alternates direction at regular intervals.

Ampere (A): The ampere is the unit of electric current. One ampere is a rate of flow of charge equal to one coulomb per second ($I = Q/t$).

Anthracite: The highest rank coal. Anthracite has a heat content greater than 15,000 Btu/pound.

Arithmetic mean: For N values of something of interest, the arithmetic mean is the sum of all N values divided by the number of values.

Atom: An atom consists of a dense, positively charged nucleus surrounded by a system of electrons equal in number to the number of nuclear protons. The atom is bound together by electric forces between the electrons and the nucleus.

Atomic Energy Commission (AEC): An advisory board formed in the United States in 1946 for the domestic control of nuclear energy. Discontinued in 1974 and replaced by the Nuclear Regulatory Commission (NRC) and the Energy Research and Development Agency (ERDA). ERDA's functions were transferred to the Department of Energy (DOE) in August 1977.

Atomic mass unit (amu): A unit of mass equal to $\frac{1}{12}$ the mass of the carbon isotope with six protons and six neutrons. 1 amu $= 1.6604 \times 10^{-24}$ g.

Atomic number: The number of electrons (or protons) in a neutral atom.

Available energy: For given operating temperatures for a heat engine, available energy is the theoretical maximum amount of mechanical energy that can be derived.

Background radiation: Nuclear radiation in the environment that comes primarily from cosmic rays, building materials, and the earth.

Barrel: Unit of volume measurement. The barrel referred to in oil measurements contains 42 gallons.

Base load generation of electricity: Electricity generated by an electric power plant with generators operating at constant output with little hourly or daily fluctuation.

Beta decay: The process in which an electron is created from the transformation of a neutron, or a positron is created from the transformation of a proton bound in a nucleus.

Beta particles: Charged particles emanating from the nuclei of atoms. They are created at the time of emission. The negative beta particle is identical to an electron which orbits the nucleus. The positive beta particle (positron) is identical to an electron except that it is positively charged.

Binding energy: In nuclear physics, binding energy is the energy required to separate a nucleus into its constituent neutrons and protons.

Bioconversion: The conversion of organic wastes into methane through the action of microorganisms.

Bituminous coal: Coal having heat content ranging from about 13,000 Btu/pound to about 15,000 Btu/pound.

Blanket: Fertile material such as $^{238}_{92}U$ or $^{232}_{90}Th$ that is placed around the core of a nuclear reactor. Neutron-induced reactions in the blanket lead to the formation of $^{239}_{94}Pu$ and $^{233}_{92}U$.

Boiling Water Reactor (BWR): A reactor in which water in thermal contact with the reactor core is brought to a boil. The vapor from the boiling water is then used to drive a steam turbine.

Bound system: A system, such as a nucleus, atom, or molecule bound (or held) together by some force(s).

Breeder reactor: A nuclear fission reactor that generates fissionable fuel by bombarding appropriate nuclei with neutrons formed in the energy producing operation of the reactor.

British thermal unit (Btu): The engineering unit of heat. The amount of heat required to raise the temperature of one pound of water by one degree Fahrenheit.

Buoyant force: An upward force exerted by a fluid on an object immersed in the fluid.

Calorie (cal): The amount of heat required to raise the temperature of one gram of water by one degree Celsius. The food calorie is equivalent to one thousand calories defined in this manner.

CANDU: An acronym for CANadian Deuterium Uranium reactor that uses natural uranium for fuel and heavy water as moderator and coolant.

Carbon dioxide (CO_2): A molecule containing one carbon and two oxygen atoms. It is the product of combustion of carbon in fossil fuels and is of concern because of the "greenhouse effect."

Carbon monoxide (CO): A molecule containing one carbon and one oxygen atom. It is a product of incomplete combustion of carbon in fossil fuels and is of concern because of its effect on human body functions.

Catalyst: A substance which modifies the rate of a chemical reaction without being consumed in the process.

Catalytic converter: A device using catalytic chemical reactions to convert toxic gases to harmless gases.

Centrifuge isotope separation: The separation of $^{235}_{92}U$ from $^{238}_{92}U$ by ultra high-speed centrifuges.

Chain reaction: As applied to a nuclear reactor, a self-sustaining, multistage nuclear reaction instigated by neutrons and sustained by other neutrons generated in the reactions.

Chlorofluorocarbon: A type of molecule containing chlorine, fluorine, and carbon as constituents. Freon™ is a chlorofluorocarbon used as a refrigerant.

Clean Air Act: (Public Law 88–206) Enacted in 1963, the act granted permanent authority for federal air pollution control activities. The act has been amended several times.

Clinch River Breeder Reactor: A proposed commercial demonstration of the liquid metal fast breeder reactor.

Coal: A natural dark brown to black solid formed from fossilized plants. It is primarily carbon.

Coal gasification: The conversion of coal into a gaseous fuel that is a mixture of carbon monoxide, hydrogen, and methane.

Coal liquefaction: The conversion of coal into liquid hydrocarbons by the addition of hydrogen.

Cogeneration: The generation of two forms of useful energy in a single energy conversion process. For example, a turbine may produce both mechanical energy for an electric generator and heat for a building.

Compton effect: Scattering of an X-ray or gamma ray photon by an atomic electron. The photon energy is less after the scattering.

Condensation: The change of state of a vapor to a liquid. Energy is liberated in condensation.

Condenser: A device at the exit of a steam turbine used to extract heat from water vapor condensing to a liquid.

Conduction: The transmission of something through a passage or medium without motion of the medium. Heat is conducted between adjacent parts of a medium if their temperatures are different. Moving electrons in a metal constitute electric conduction.

Conservation of energy: An important physical principle that requires that there be no change in the total energy in any energy transformation.

Control rod: A rod often made of boron or cadmium that can be inserted into the core of a nuclear reactor to control the fission reactions by selective absorption of neutrons.

Convection: Transfer of heat from one place to another by actual motion of the heated material.

Converter: A device such as an automobile engine or nuclear reactor that converts the intrinsic potential energy of an appropriate fuel into another form of energy.

Cooling tower: A tower used to collect the heated water effluent from the condenser of a steam turbine and to transfer heat to the atmosphere.

Coulomb (C): The metric unit of electric charge. In terms of the charge of the electron, the smallest unit in nature, the sum of the charges of 6 $\frac{1}{4}$ billion-billion electrons equals one coulomb.

Criticality: A condition in a nuclear reactor in which the neutron generation and loss rates are equal.

Critical mass: The minimum mass of fissionable material that will sustain a nuclear chain reaction.

Crankcase blowby: The leakage of gaseous or liquid hydrocarbons between a piston and its cylinder during operation of an internal combustion engine.

Crude oil: The liquid part of petroleum.

Curie (Ci): A unit of radioactivity amounting to 37 billion (37×10^9) disintegrations per second.

Cyclone collector: A particulate-collecting device in which a whirlwind of air throws particles out of an air stream to the walls of a container on whose surface they collect or adhere.

Density: The amount of something divided by the volume it occupies. Mass density is the mass of a substance divided by the volume it occupies.

Department of Energy (DOE): The 12th Cabinet agency. Established on October 1, 1977, it absorbed all functions of the Federal Energy Administration, Federal Power Commission, and Energy Research and Development Administration and selected energy activities of the Department of Interior, Department of Defense, Interstate Commerce Commission, Department of Commerce, and Department of Housing and Urban Development.

Deuterium: An isotope of hydrogen having one proton and one neutron in its nucleus.

Deuteron: The nucleus of the deuterium atom.

Diesel engine: An internal combustion engine named for its inventor Rudolf Diesel, German mechanical engineer and inventor, 1858–1913. Requiring no spark plugs, the diesel engine achieves ignition of a fuel–air mixture by high compression.

Direct current (dc): An electric current for which charges flow in one direction.

Disintegration: A transformation involving the nucleus of an atom resulting in a less massive configuration by emission of radiation that may be either particle or electromagnetic.

Doubling time: The time it takes for something that is changing exponentially to double in amount.

Dry-bulb temperature: The temperature measured by a dry thermometer.

Efficiency: The useful output of any system divided by the total input.

Electric current: Electric charges in motion. The ampere is the measuring unit.

Electric field: Qualitatively, a region of space where an electric charge feels a force.

Electric force: A force between two objects each having the physical property of charge.

Electricity: In the popular sense, it is electric current used as a source of power.

Electric potential difference: The work done on a charge to move it between two points divided by the strength of the charge. The volt is the measuring unit.

Electrical resistance: The resistance provided by the structure of a conductor to the flow of electric charges.

Electrolysis: The decomposition of a liquid into its atomic constituents as a result of an electric current in the liquid. The electrolysis of water produces hydrogen and oxygen.

Electron volt (eV): A unit of energy. It is the energy acquired by an electron accelerated through a potential difference of one volt. $1 \text{ eV} = 1.602 \times 10^{-19}$ joules.

Electrostatic precipitator: An apparatus for the removal of suspended particles from a gas by charging the particles and precipitating them through application of a strong electric field.

Emergency core cooling system (ECCS): A system designed to safely dissipate heat from the core of a nuclear reactor in the event of a sudden loss of the normal cooling facilities.

Energy: The capacity for doing work. The joule is the metric measuring unit.

Energy conservation: Distinguished from the principle of conservation of energy, it incorporates the idea of saving, or not wasting, energy.

Enriched uranium: Uranium in which the percentage of ^{235}U has been increased above the 0.7% level found in natural uranium.

Environmental Protection Agency (EPA): A federal agency officially established December 2, 1970 under a presidential reorganizational plan. Placed in EPA were the Interior Department's Federal Water Quality Administration; the HEW Department's National Air Pollution Control Administration, Bureau of Solid Waste, and Bureau of Water Hygiene; the pesticide registration, research, and regulation functions of the Agriculture, Interior, and HEW Departments, and certain radiation functions of the Nuclear Regulatory Commission; the Federal Radiation Council; and HEW's Bureau of Radiological Health. The 1970 Clean Air amendments also created an Office of Noise Abatement within EPA.

Epilimnion: The upper stratum (layer) of a lake or sea.

Equilibrium: A condition involving no change in the translational and rotational motion of a system.

Ethanol (C_2H_6O): A component of fermented and distilled liquors and can be made from petroleum hydrocarbons and grains.

Evaporation: The change of state of a liquid as it passes to vapor. The liquid cools in the process.

Exponential growth: The growth in some quantity characterized by doubling of its value at regular intervals of time. For example, if some quantity doubles its value every ten years, it is growing at an exponential rate.

Fast neutrons: A loosely defined classification of neutrons according to speed (or energy). A neutron with energy greater than 10 keV is called a fast neutron.

Feedback: The return of a portion of the output of any process or system to its input.

Fertile nucleus: A nucleus that serves as the target for generating fissionable fuel in a nuclear breeder reactor.

First law efficiency: Recognizing explicitly the first law of thermodynamics, first law efficiency is identical to the popular meaning of efficiency, i.e., the useful output of an energy converter divided by the input energy.

First law of thermodynamics: An extension of the principle of conservation of energy to include heat and internal energy.

Fission: The splitting of an atomic nucleus into fragments, two of which are usually much more massive than the remaining fragments. Energy is released in the process.

Fission fragment: An atom, usually radioactive, that is formed from the fission (or splitting) of a heavy nucleus.

Flue gas desulfurization: The use of chemical reactions to remove sulfur oxides from the gases produced in the burning of coal.

Fluidized bed combustion: The burning of pulverized coal in a sulfur-capturing bed of limestone-based particles.

Fly ash: Fine particles that are generated from noncombustible products in the burning of coal.

Fossil fuel: Fuel such as coal or petroleum derived from remnants of plants and organisms of a past geological age.

Freight efficiency: A measure of the amount of freight that can be moved some distance by a given mode of transportation for an expenditure of a certain amount of fuel (energy). It is defined as the number of tons of freight moved multiplied by the number of miles obtained per gallon of gasoline used.

Fusion: In nuclear terms, a nuclear reaction in which nuclei are combined (fused) to form other nuclei. Energy is released in the process.

Gamma ray: A quantum (or photon) of electromagnetic radiation. It is distinguished from other electromagnetic radiation by its energy. The energy limits are loosely defined, but typically photons with energies greater than 100 keV are called gamma rays.

Gas cooled fast breeder reactor (GCFBR): A fast breeder reactor which uses a gas to transfer the heat from the reactor core.

Gaseous diffusion: The diffusion or migration of a gas through a semiporous membrane. Because the rates of diffusion of different types of molecules depend on the masses of the molecules, the process is used to separate different molecular species such as $^{235}UF_6$ and $^{238}UF_6$.

Gasohol: An automotive fuel that is usually a mixture of 10% ethanol and 90% gasoline.

Genetic effect: A damaging effect to a person's genes which could produce an abnormality in the offspring. Genetic damage can be caused by nuclear radiation.

Geometric mean: For N values of something of interest, the geometric mean is the Nth root of the product of all N values.

Geothermal energy: Thermal energy in the earth's crust.

Gravitational collector: A device for collecting noncombustible products released in the burning of coal. It utilizes the gravitational attraction between the earth and the particles to effect the separation.

Gravitational field: Qualitatively, a region of space where a mass feels a force due to an interaction with another mass.

Gravitational force: A force between two objects each having the physical property of mass.

Gravitational potential energy: Potential energy of a mass due to an elevated position above the earth's surface. Numerically, gravitational potential energy is the product of weight and height ($E_p = mgh$). The metric units are joules.

Greenhouse effect: An effect that produces a warming of a system by trapping electromagnetic radiation within the system. A common greenhouse warms primarily by trapping infrared radiation within the structure.

Gross National Product (GNP): The total market value of all the goods and services produced by a nation during a specific period.

Half-life: The time required for one-half of a given number of radioactive atoms to disintegrate.

Heat: A form of energy transfer associated with the motion of atoms and molecules.

Heating degree days: Used as a factor in determining the heating requirements for a building, it is the product of the average inside-outside temperature difference and length of a heating period. Temperature difference is measured relative to a base temperature of 65° F.

Heating factor: A numerical factor used to relate heating requirements in a given location to those in the central United States.

Heating zone: A geographical zone where heating requirements are roughly the same for a given type of structure.

Heat plume: The hot water discharge of a steam turbine into a body of water is confined to a "feather-shaped" region called a heat plume.

Heat pump: A device used to supply heat to a building by pumping energy from the cooler exterior. In principle, it operates like a household refrigerator.

Heat sink: A system to which heat can flow.

Heavy water reactor (HWR): In a heavy water reactor, the hydrogen component of the water molecules in the moderator is of the second heaviest (in mass) isotopic form ($^2_1H_1$, i.e., deuterium).

Hertz (Hz): The hertz is a unit of frequency equal to one cycle per second. Named for Heinrich Hertz, German physicist, 1857–1894.

High-level radioactive waste: Intensely radioactive materials, primarily fission fragments and actinides, produced by nuclear reactions in a nuclear reactor.

High-Temperature Gas-cooled Reactor (HTGR): A graphite moderated helium-cooled reactor using highly enriched fuel.

Horsepower (hp): A unit of power equivalent to 746 watts.

Humidity: A qualitative term pertaining to the water vapor in air.

Hydrocarbon: A chemical compound containing only hydrogen and carbon.

Hydroelectric power: Electric power that is produced from the conversion of the kinetic energy of water.

Hydrogen economy: An economy that is based on energy produced when hydrogen is burned in an oxygen atmosphere.

Hypolimnion: The lower stratum (layer) of a lake or sea.

Income energy: Energy provided to supplement or replace the energy locked up within the earth. The sun is the earth's source of income energy.

Inertial confinement: Taking advantage of the inertial property of mass to confine nuclear fusion reactions in a pellet of frozen deuterium or tritium, for example.

Insolation: The solar radiation crossing a horizontal area, measured in units such as Btu per day per square foot or watts per square meter.

Insulation: Material used to restrict heat flow between two regions, for example, between the interior and exterior of a house.

Internal combustion engine: An engine, such as an automotive gasoline piston or rotary engine, in which fuel is burned within the engine proper rather than in an external furnace as in a steam engine.

Internal energy: Thermodynamically, internal energy is the kinetic and potential energy of the atomic and molecular constituents of a system.

Isotopes: Atoms with nuclei having the same number of protons, but a different number of neutrons.

Joule (J): The joule is a metric unit of work or energy. The work done by a force of one newton moving an object one meter in the direction of the force. Named for James Prescott Joule, British physicist, 1818–1889.

Kelvin (K): Unit of temperature based on a freezing point of water of 273 and a boiling point of water of 373. Named for Lord Kelvin, William Thomson, British physicist, 1824–1907.

Kerogen: A solid organic material occurring in oil shale that yields a crude oil when heated in the absence of oxygen.

Kilowatt-hour (kWh): A unit of energy equal to a power of one kilowatt (1000 W) acting for one hour.

Kinetic energy: Energy due to motion. Numerically, it is one-half the product of the mass of a body times the square of its velocity ($E_k = \frac{1}{2} mv^2$).

Lapse rate: The variation of temperature with respect to height above the earth.

Laser: A word coined from the first letters of the main words of "*l*ight *a*mplification by *s*timulated *e*mission of *r*adiation." It is a device that produces a beam of electromagnetic radiation having very uniform wavelength. In some lasers, the beam is very intense.

Laser isotope separation: The separation of $^{235}_{92}U$ from $^{238}_{92}U$ by selective excitation of the molecules containing uranium atoms.

Light water reactor (LWR): In a light water reactor, the hydrogen component of the water molecules in the moderator is of the lightest (in mass) isotopic form ($^1_1H_0$).

Lignite: The lowest rank coal. Lignite has heat content of about 8000 Btu/pound.

Liquefied natural gas (LNG): Natural gas converted to a liquid by cooling to $-259°$ F.

Liquid metal fast breeder reactor (LMFBR): A nuclear breeder reactor using fast neutrons to produce the fissionable fuel and a liquid metal, usually sodium, as the heat transfer medium.

London smog: A heavily polluted atmosphere consisting mainly of particulates and sulfur dioxide emitted from the combustion of coal and petroleum products.

Loss of cooling accident (LOCA): An accident in a nuclear reactor that evolves from an accidental loss of coolant.

Low-level radioactive waste: Radioactive material produced in the routine operation of a nuclear reactor. It must be handled carefully but is substantially less problematical than the high-level radioactive wastes.

Low-sulfur coal: Coal with sulfur content generally less than 1.0%.

Magnetic field: Qualitatively, a region of space where the pole of a magnet or a moving charge feels a force.

Magnetic field line: An imaginary line in a magnetic field along which a tiny compass needle aligns.

Magnetohydrodynamics (MHD): The science concerned with the motion of electrically conducting fluids through electric and magnetic fields.

Mass: The measure of an object's resistance to acceleration.

Meltdown: A nuclear reactor accident in which the radioactive fuel overheats, melting the fuel cladding.

Methane (CH_4): The principal constituent of natural gas.

Methanol (CH_3OH): A fuel derived from wood or coal. It can be used as an additive to gasoline to make a suitable fuel for conventional cars and trucks.

Microgram per cubic meter ($\mu g/m^3$): One microgram (10^{-6} g) of a given substance in a one cubic meter volume of air.

Micrometer (μm): One millionth (10^{-6}) of a meter. Sometimes called a micron.

Moderator: A mechanism used to reduce the speed of neutrons in a nuclear reactor.

Molecule: A bound system of two or more atoms.

Natural gas: The gaseous component of petroleum. It is primarily methane (CH_4) and is commonly used as a household and industrial fuel.

Neutrino: A massless, electrically neutral particle emitted in beta decay.

Neutron: An electrically neutral subatomic particle. Protons and neutrons are the constituents of the atomic nucleus.

Neutron capture: The capture of a neutron by a nucleus of an atom.

Newton (N): A newton is a metric unit of force. The force required to accelerate one kilogram, at one meter per second, each second. Named for Sir Isaac Newton, English scientist, 1642–1727.

Nitrogen oxides (NO_x): In energy considerations, molecules produced by the oxidation of atmospheric nitrogen in the combustion of fossil fuels.

Nuclear fuel cycle: The series of steps leading to fuel for nuclear power reactors and disposal of the fuel following power generation.

Nuclear parks: A concentration of several nuclear power plants into a localized region.

Nuclear reactor: A device for converting nuclear energy released from nuclear fission reactions to heat.

Nuclear Regulatory Commission (NRC): Established in 1975 to perform the licensing and regulation of nuclear power plants and the safeguarding of nuclear materials and facilities.

Nuclear waste: The radioactive products formed by fission and other processes in a nuclear power reactor.

Nucleon: A proton or a neutron.

Nucleus (of an atom): The positively charged central region of an atom. It is composed of neutrons and protons and contains nearly all the mass of the atom.

Ocean Thermal Energy Converter (OTEC): A scheme for extracting thermal energy from the surface of the ocean and converting the thermal energy into electric energy.

Octane rating: A numerical rating of the ability of a gasoline to reduce an audible knock (noise) in an internal combustion engine.

Off-peak generation of electricity: The generation of electricity in periods of low demands by consumers.

Ohm: The ohm is the unit of electrical resistance. If a potential difference of one volt across some electrical element causes a current of one ampere, the electrical resistance is one ohm ($R = V/I$).

Ohm's law: If the electrical resistance of a device, a toaster element, for example, does not depend on the current in the device, then it is said to obey Ohm's law.

Oil shale: A brownish rock containing a solid hydrocarbon material called kerogen that can be converted to a crude oil product.

Once-through fuel cycle: Terminates the nuclear fuel cycle prior to reprocessing the spent fuel elements.

Ozone (O_3): A highly reactive molecule containing three atoms of oxygen.

Ozone layer: An atmospheric layer of ozone located at altitudes between 20,000 and 30,000 meters.

Pair production: The production of an electron and a positron from the interaction of a gamma ray photon with an atom. To produce an electron-positron pair, the photon must have a minimum energy of 1.022 MeV.

Particulate: Existing in the form of minute separate particles.

Parts-per-million (ppm): A measure of the concentration of a substance. On a molecular basis, a one ppm concentration of a given molecule means that one of every one million molecules is a molecule of the specified type.

Passenger efficiency: A measure of the number of people that can be moved some distance by a given mode of transportation for an expenditure of a certain amount of fuel (energy). It is defined as the number of passengers moved multiplied by the miles obtained per gallon of gasoline used.

Passive solar system: Utilization of solar energy for heating and cooling without the use of mechanical pumps and fans for circulating heat.

Peaking unit: An auxiliary electric power system that is used to supplement the prime power system during peak periods of electricity demand.

Performance: For a heat pump, the performance is the total energy deposited in an area of interest, a room, for example, divided by the energy expended to cause the energy transfer. For a refrigerator or air conditioner, the performance is the energy removed from a cool region, a room or the inside of a refrigerator, for example, divided by the energy expended to cause the energy transfer.

Petroleum: A natural flammable mixture of solid, liquid, and gaseous hydrocarbon products found principally beneath the earth's surface. Gasoline, fuel oil, lubricating oil, and paraffin wax, as examples, are refined from petroleum.

Photochemical oxidants: Primarily ozone, but also including other chemical compounds created by smog conditions.

Photochemical (Los Angeles) smog: A heavily polluted atmosphere consisting mainly of eye-irritating chemicals produced from the interaction of automobile exhaust gases with sunlight.

Photoelectric effect: In nuclear physics, the absorption of a photon by an atom and a subsequent ejection of an atomic electron.

Photon: A quantum of electromagnetic energy.

Photosynthesis: The process by which plants convert electromagnetic (solar) energy, carbon dioxide, and water to carbohydrates and oxygen.

Photovoltaic cell: Commonly called a solar cell, it is a device that converts electromagnetic (solar) energy directly to electric energy.

Pollutant: A constituent of some substance (air, water, and land, for example) that alters the value of the substance.

Potential difference: The work required to move an amount of charge between two positions divided by the size of the charge ($V = W/Q$).

Potential energy: Energy that is potentially convertible to another form of energy, usually kinetic.

Power: Rate of doing work. The watt is the metric measuring unit.

Pressure: The force on a surface divided by the area over which the force acts ($P = F/A$). Typical units are newtons/m^2 and pounds/in^2.

Pressurized water reactor (PWR): A nuclear reactor in which the water in contact with the reactor core is allowed to become pressurized to prevent boiling and thereby allows a higher operating temperature.

Price–Anderson Act: Enacted in 1957, the act provides insurance for the owners (at the owner's expense) of a nuclear power plant against liability claims made in the event of a nuclear accident.

Primary air quality standards: Levels of air quality that are judged necessary, with an adequate margin of safety to protect the public health.

Proton: A subatomic particle containing one unit of positive charge. Protons and neutrons are the constituents of the atomic nucleus.

Proven energy reserve: A source of energy known to be available at current prices and extractable with current technology.

Pumped-storage system: An elevated reservoir of water in which the water has been pumped into the reservoir. Ultimately, the water is channeled through water turbines that drive electric generators.

Quad: A million billion (10^{15}) British thermal units.

Radiation: The emission and propagation of waves (such as light and sound) or particles (such as beta and alpha).

Radiation absorbed dose (rad): A unit of energy absorbed from ionizing radiation. One rad is equal to 0.00001 joule per gram of irradiated material.

Radioactivity: The emission of particles or electromagnetic radiation from unstable atomic nuclei.

Rate: The change in some quantity divided by the time required to produce the change.

Reactor core: The central portion of a nuclear reactor containing the fuel elements and the control rods.

Relative biological effectiveness (RBE): A measure of the capacity of a specific ionizing radiation to produce a specific biological effect.

Relative humidity: The amount of water vapor in the air at a specific temperature divided by the maximum amount of water vapor that the air can contain at that temperature.

Reprocessing: The mechanical and chemical processes by which $^{235}_{92}U$ and $^{239}_{94}Pu$ are recovered from spent reactor fuel.

Roentgen equivalent man (rem): A biological unit of absorbed ionizing radiation that accounts for biological damage. It is the radiation dose measured in rads multiplied by the relative biological effectiveness (RBE).

R-value: A measure of the resistance to the flow of heat through a medium. Generally expressed in $ft^2 \cdot hr \cdot {}^\circ F$ per Btu.

Salt dome: A geologic formation frequently associated with the occurrence of oil and/or natural gas.

Scrubbing: A wet chemical process for removing gaseous pollutants from the emissions produced by burning fossil fuels.

Second law efficiency: Recognizing explicitly the second law of thermodynamics, second law efficiency compares the actual efficiency of an energy converter with the maximum efficiency guaranteed by the second law of thermodynamics.

Second law of thermodynamics: The law can be formulated in several ways. Two practical formulations are (1) heat does not flow spontaneously from a cooler to a hotter object and (2) in a cyclical device, heat cannot be transformed wholly to work.

Secondary air quality standards: Levels of air quality that are judged necessary to protect public welfare. For example, by protecting property, materials, and economic values.

Settling velocity: The velocity (or speed) with which a particle in the atmosphere settles to the ground.

Smog: A word derived from smoke and fog describing a polluted atmosphere that has resulted from the combustion of fossil fuels.

Solar cell: A popular name for a photovoltaic cell.

Somatic effect: A biologically damaging effect to the human body. Somatic damage can be caused by nuclear radiation.

Specific heat: The amount of heat required to raise the temperature of a unit mass (or weight) of a substance by one degree. Typical units are calories per gram per degree Celsius and Btu per pound per degree Fahrenheit.

Speed: Distance traveled divided by the time required to achieve that distance. It is distinguished from velocity by disregarding any reference to the direction traveled.

Spent fuel: Fuel elements removed from a nuclear reactor after the useful lifetime.

Stratified charge engine: Basically a conventional, reciprocating, internal combustion engine with two vertically connected combustion chambers above each cylinder. Ignition of a fuel mixture rich in gasoline vapor is instigated in the upper chamber. The flame progresses to the lower chamber and ignites a somewhat leaner fuel mixture.

Subbituminous coal: Coal having a heat content ranging from about 8,300 Btu/pound to about 13,000 Btu/pound.

Subsidence: The sinking or subsiding of an air mass.

Sulfur dioxide (SO_2): A stable molecule consisting of one atom of sulfur and two atoms of oxygen. It is emitted as an unwanted by-product in the combustion of coal.

Super-Phenix: The name given to a commercial nuclear breeder reactor scheduled for completion in France in 1984.

Superconductor: A material that presents no resistance to the flow of electrons if the temperature of the material is maintained at a sufficiently low temperature, typically about the temperature of liquid helium (4.2 K).

Surface tension: A molecular force at the surface of a liquid that restricts the escape of molecules from the liquid.

Synergism: A situation in which the combined action of two or more agents acting together is greater than the sum of the agents acting separately.

Synthetic fuel: Fuels derived from conversion process rather than mining or drilling. Includes burnable liquids and gases produced from coal, shale, tar sands, municipal wastes, and biomass.

Tar sands: Viscous, oily, liquid hydrocarbon products intermixed in the pore spaces of sandstone.

Temperature: A measure of the sensation of hotness or coldness referred to a standard scale such as the height of a column of mercury in a tube.

Temperature inversion: An atmospheric condition in which the temperature increases with increasing altitude.

Tetraethyl lead: A lead compound, $(C_2H_5)_4Pb$, added to gasoline to increase the octane rating.

Therm: A natural gas energy unit equivalent to 100,000 British thermal units.

Thermal conductivity: A property of a medium—solid, liquid, or gas—measuring the ability to conduct heat. Typical units are (calories/second) per cm • °C or (Btu/second) per ft • °F.

Thermal energy: The random kinetic energy of an atom or molecule that is numerically equal to $\frac{3}{2}kT$ where k is Boltzmann's constant and T is the kelvin temperature.

Thermal pollution: The addition of heat to the environment, usually the water environment, to the extent that the value of the environment is altered.

Thermal radiation: The electromagnetic radiation emitted by any object maintained at a temperature above zero kelvins.

Thermal stratification: The layering of a body of water into sections of nearly constant temperature.

Thermodynamics: The science of the relationship between heat and other forms of energy.

Thermostat: A temperature-controlling device that signals a heating or cooling supply to add or remove heat depending on whether or not the temperature is to be raised or lowered.

Tidal power plant: An electric power plant that uses the incoming and outgoing kinetic energy of tides to drive water turbines that drive electric generators.

Tokamak: Derived from the Russian words for toroidal magnetic chamber, it designates a type of nuclear fusion reactor.

Transformer: A device used to change the size of an alternating current.

Transuranic elements: All of the elements having more nuclear protons than uranium has.

Tritium: An isotope of hydrogen having one proton and two neutrons in its nucleus.

Triton: The nucleus of the tritium atom.

Troposphere: The lower region of the earth's atmosphere extending up to an average altitude of about 40,000 ft (eight miles).

Turbidity: In meteorology, any condition in the atmosphere that reduces its transparency to radiation, especially visible radiation.

Turbine: A device in which the kinetic energy of a fluid is converted to rotational kinetic energy of a shaft by impulses exerted on vanes attached to the shaft.

Unproven energy reserve: A source of energy thought to be available because of geologic information and oil-exploration experience.

Uranium hexafluoride (UF_6): A gaseous compound of uranium and fluorine used in fuel enrichment processes.

Velocity: The displacement of an object divided by the time required to achieve the displacement. It is distinguished from speed by accounting for the direction of the displacement.

Viscous force: A frictional force exerted on an object moving in a fluid.

Visual range: The distance, under daylight conditions, at which the apparent contrast between an object chosen as the target of observation and its background becomes equal to the threshold of contrast of the observer.

Volt (V): The volt is the unit of potential difference. If one joule of work is required to move one coulomb of charge between two positions, the potential difference between the positions is one volt ($V = W/Q$).

Voltage: A popular expression for potential difference.

Watt (W): The watt is the metric unit of power. A rate of doing work of one joule per second is a watt ($P = W/t$).

Wave: A disturbance or oscillation propagated with a definite velocity from point to point in a medium.

Weight: The force of gravity on a mass. Numerically, weight is equal to the product of mass and acceleration due to gravity ($W = mg$). Typical units are newtons (metric) and pounds (British).

Wet-bulb temperature: The temperature measured by a thermometer covered with a cloth moistened with water.

Work: The product of the distance an object is moved times the force operating in the direction that the object moves.

X-ray: A quantum (or photon) of electromagnetic radiation. It is distinguished from other electromagnetic radiation by its energy. The energy limits are loosely defined but typically photons with energies between about 1 and 100 keV are called X-rays.

Yellowcake: A yellowish solid containing 80–90% U_3O_8 that results from the milling of uranium ore.

Index

Index